Design Automation of Cyber-Physical Systems

Mohammad Abdullah Al Faruque
Arquimedes Canedo

Editors

Design Automation of Cyber-Physical Systems

Editors
Mohammad Abdullah Al Faruque
University of California, Irvine
Irvine, CA, USA

Arquimedes Canedo
Siemens Corporate Technology
Princeton, NJ, USA

ISBN 978-3-030-13052-7   ISBN 978-3-030-13050-3   (eBook)
https://doi.org/10.1007/978-3-030-13050-3

This Springer imprint is published by the registered company Springer Nature Switzerland AG.
The registered company address is: Gewerbestrasse 11, 6330 Cham, Switzerland

Introduction: Research Challenges in the Design Automation of Cyber-Physical Systems

Mohammad Abdullah Al Faruque and Arquimedes Canedo

Cyber-physical systems (CPS) are all around us – from smart watches and home automation devices to traffic infrastructure, power grid, and transportation systems. The penetration of CPS into the world is accelerating thanks to better Internet connectivity, power-efficient computation, higher-capacity memory, and software functions. In the past, CPS development was mainly driven by companies. Today, all kinds of people from different backgrounds are proactively creating new CPS thanks to the Internet, crowdfunding, crowdsourcing, availability of inexpensive electronics, software tools, and access to additive manufacturing and other forms of flexible manufacturing. As the complexity of products increases and the time-to-market cycles shrink, CPS design automation tools and methodologies become a necessity. In this book, design automation tools refer to software tools for designing cyber-physical systems. Design automation methodologies refer to workflows where these tools are used by engineers to analyze the CPS along one or more phases of their life cycle.

CPS is a well-established discipline. There are several journals and conferences specialized in specific aspects including the theory of CPS (ACM TCPS, ICCPS), hybrid computation and control (HSCC), sensing (Sensys, IPSN), real-time computation and communication (RTAS), and embedded systems (ESWEEK). Unfortunately, the design automation aspect is not the primary focus of any of these venues. On the other hand, design automation conferences (DAC, DATE, ICCAD) have been mainly focused on electronic design automation. In recent years, these conferences have expanded their reach, and currently, they have dedicated tracks and special sessions on cyber-physical systems.

M. A. Al Faruque
University of California, Irvine, Irvine, CA, USA

A. Canedo
Siemens Corporate Technology, Princeton, NJ, USA

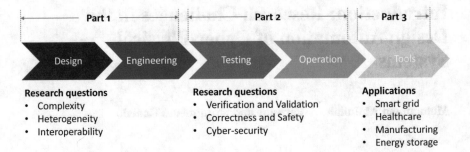

Fig. 1 Organization of this book in terms of the CPS design automation lifecycle phases

This book aims at covering the gap between cyber-physical systems and design automation communities and focuses on the most important research questions in this intersection. This book is organized in three parts corresponding to the CPS lifecycle phases as shown in Fig. 1. Part I consists of Design and Engineering, Part II consists of Testing and Operation, and Part III consists of Application-Specific Design Automation Methodologies and Tools. Each chapter is written by leading researchers in the field and provides a focused discussion on the latest design automation tools and methodologies. All the contributing authors of this book have provided examples and use cases that illustrate how the presented design automation tools and methodologies are used in practice.

Part I: Design and Engineering

The first two phases of the CPS design automation life cycle are design and engineering. The design automation of the design phase consists of tools that allow system experts to specify the purpose, or functionality, of the system and its subcomponents. During design, one of the main research challenges is managing the complexity that arises when requirements are associated to functions, and these functions are decomposed into lower-level functions. Traditionally, the concept design of CPS was done "on paper," and unfortunately, there is very little computer support. A recent trend in CPS design automation is the use of computational concept design tools. These tools formalize the allocation of requirements to functions and provide traceability. Chapter 1 presents such an approach that relies on synthesis algorithms to automate the allocation of requirements to functions.

During the engineering phase, functions are allocated to specific implementations. In the engineering phase, the complexity increases further as functions can be allocated to more than one implementation. For example, a communication channel can be implemented through electrical, optical, or electromagnetic means. The heterogeneity of components, technologies, protocols, materials, algorithms, and communication presents a challenge to design automation tools. To manage this challenge, Chapter 2 presents a platform-based approach to deal with alternative system architectures. The third research challenge is the interoperability among CPS. Inspired by the hourglass-shaped architecture of the Internet, Chapter 3 presents a model-based approach for the engineering of networked CPS.

Part II: Testing and Operation

The testing phase is extremely challenging for design automation because the combination of cyber and physical components in a CPS makes the state space extremely large. While formal methods can be applied to cyber components, these are not suitable for physics-based components. Similarly, simulation is useful to explore the physics but does not address all the cyber concerns and the interactions between the two domains. Further complicating matters, the use of artificial intelligence and autonomy in CPS is pushing the limits of validation and verification of CPS. Addressing this problem, Chapter 4 presents an approach using formal methods to reason about the correctness of CPS applications using a combination of bounded time reachability analysis, simulation-guided reachability analysis, and deductive techniques.

During the operation phase, the CPS is deployed and interacting with its environment. A major challenge to design automation is to establish a baseline on the safe and correct behavior of the CPS. An important characteristic of this phase is the generation of large amounts of data that can be leveraged for correctness and safety. Chapter 5 addresses this research question with a data-driven safety verification approach of CPS. The third research challenge is cybersecurity. The interaction of the CPS with the environment creates exposures to known and unknown attack vectors both cyber and physical. This represents an important research question that Chapter 6 addresses with a model-based system assurance approach for the design of cyber-resilient CPS.

Part III: Application-Specific Design Automation Methodologies and Tools

Every CPS application domain such as manufacturing, healthcare, smart grid, and energy storage has unique requirements that design automation must take into account to be the most effective. For example, smart grids are very large CPS that are distributed in large geographical areas, and their main function is to control the energy demand response. Therefore, design automation tools for smart grids have specialized in control. Chapter 7 presents the latest developments in optimal design of distributed controllers for large-scale power grids. Similarly, healthcare systems due to their safety-critical nature have very strict requirements on their software and hardware. Chapter 8 presents model-based approaches to software design of rehabilitation systems. The design automation of manufacturing parts has been dominated by computer-aided design tools. Chapter 9 presents a deep learning approach to reason about manufacturing parts. Many CPS are battery-operated, and this presents unique challenges to the design automation tools. Chapter 10 presents the latest developments in design automation tools for energy storage systems.

Contents

Part I
Design and Engineering

Chapter 1
Concept Design: Modeling and Synthesis from Requirements to Functional Models and Simulation

Jiang Wan, Nafiul Rashid, Arquimedes Canedo, and Mohammad Abdullah Al Faruque

1.1 Introduction

Cyber-physical systems (CPS) are the new generation of automated systems that come with the tight coupling of the cyber and physical world. Examples include automotive systems, smart grids, healthcare monitoring, robotics, etc. Unlike embedded systems, which are generally standalone devices, a complete CPS is a combination of interacting physical components with physical inputs and outputs, forming a network using cyber components [27]. The main advantage of CPS is that it allows different physical components from disparate domains to interact. This flexibility brings new challenges for the design of CPS as it requires the collaboration of multiple domain experts/engineers from different fields. On the other hand, the competition among market peers and time-to-market of the products requires faster design, simulation, development, and deployment of CPS. The only way to meet these requirements is by developing integrated and automated engineering tools. The purpose of these tools is to bring the design requirements of CPS from different domains under one umbrella at the very early design stage.

J. Wan · N. Rashid (✉) · M. A. Al Faruque
University of California, Irvine, Irvine, CA, USA
e-mail: nafiulr@uci.edu

A. Canedo
Siemens Corporate Technology, Princeton, NJ, USA

© Springer Nature Switzerland AG 2019
M. A. Al Faruque, A. Canedo (eds.), *Design Automation of Cyber-Physical Systems*, https://doi.org/10.1007/978-3-030-13050-3_1

1.1.1 Motivation

One of the most technologically advanced and complex cyber-physical systems currently being produced is Automotive CPS. The traditional method of developing automobiles with completely mechanically driven systems is obsolete nowadays. Modern automobiles are now developed with the marriage of cyber and physically driven systems. The cyber components consist of the networked systems [5] and the software (electronic control units [ECUs]) that controls the physical components. For example, hundreds of cooperating cyber components interacting with the multi-physics physical processes in an automobile contribute to the rapid advances in various areas such as safety, fuel consumption, efficiency, etc. Multiple domain experts from various organizations collaborate to design a modern automobile, which can consist of hundreds of ECUs [1, 13]. It is a challenging task for companies to collaborate to improve the automotive design process. Therefore, it is important to create design automation tools to facilitate multi-disciplinary collaboration [15].

The objective of developing state-of-the-art automotive design tools is to *reduce the critical path* [45]. The detailed design phase [7, 22, 41] that includes precise engineering specifications created by the domain experts is called the critical path. Engineers from every domain have their preferred design tools to work with. For example, control engineering is done using LabView [25], LMS [28], Modelica [31], and Matlab/Simulink [39]; electrical engineers use Electronic Design Automation (EDA) design tools; mechanical engineering is supported by Computer-Aided Design (CAD) and Engineering (CAE) tools; and software engineers use UML [30] and in-house software development environments [9]. However, the incompatibility between different domains has made it difficult to combine these tools to perform system-level analysis and simulations [17]. Therefore, model-based systems engineering (MBSE) methodologies and tools [19] have become more popular as they allow high-fidelity multi-disciplinary system-level simulations.

1.1.2 Functional Modeling

A functional model decouples the design specifications (functions) of the systems from the behavior and/or architecture and reflects what each system does. Detailed domain-specific knowledge is discouraged in a functional model. Thus, it facilitates collaboration among different disciplines, bringing the domain experts' minds to the same level of abstraction [45]. The high abstraction level in the functional models makes them a suitable formalism for CPS design. Functional models abstract the details of the continuous and discrete dynamics of CPS and allow cross-domain collaboration [22, 41, 44]. Moreover, functional modeling is a systems engineering activity [18] that allows systems and their subsystems to be described in terms of their respective functionalities. Although functional models are very useful, they may have additional security issues which are not within the scope of this chapter.

Various researchers have addressed the inherent security issues in CPS design as a whole [2] and at the functional level [43].

To adopt the use of functional models as a system engineering practice, the formalization of a functional model [21, 35] is very important. To facilitate that, the Functional Basis language [21, 40] has been successfully implemented for function decomposition in the early design stage [10, 24, 36, 46]. The use of a constrained vocabulary as well as well-defined functions and flow semantics may help establish the functional modeling practice among different designers. Thus, it allows them to rely on the same language to design and analyze systems. As a result, Functional Basis language has become the de-facto standard for functional modeling [34]. Functional Basis defines three flow categories (*material*, *energy*, and *signal*) that expand into a total of 18 flow types, and 8 function categories with a total of 32 primitive functions as presented and discussed in [45]. Although Functional Basis language has been successfully used to provide a common platform of communication among different domain experts, the current major research challenge is making it executable for performing extensive design space exploration. Therefore, generating an executable system-level model from the functional model to perform design space exploration is essential for the adoption of this methodology in industry [45]. The functional model has already been used to synthesize architecture for automotive design [20]. However, this synthesis process is not automatic and does not utilize the functional model to help the design space exploration at the architecture level. Notably, an earlier work [12] has also demonstrated the possibility of using Functional Basis language to enable the early stage design automation for CPS.

1.1.3   Simulation in CPS

To design, validate, and test complex systems like CPS, simulation is widely used during the early design stage across industries. Simulations allow the engineers to analyze systems virtually in the cyber domain instead of implementing an actual prototype of these systems in the physical domain. Thus, simulations are quite cost-effective and efficient for the early design stage evaluation of the systems.

There are many state-of-the-art design automation tools [4, 28] that use various system simulation languages. For example, VHDL-AMS [14], Simscape [38], and Modelica [31] are some of the system simulation languages that allow modeling and simulation of multi-disciplinary systems. However, the lack of early stage design tools causes the simulation models to be generated manually by the domain experts. This manual process is very time consuming and less effective for multi-disciplinary systems as each of the disciplines has their own domain-specific tools and language. Although, SysML [33] is proposed as a domain-independent system-level modeling language [3, 23, 46], it has limitations in terms of expressing and executing those models [6].

This chapter discusses a novel approach mentioned in [45] for the creation of CPS design tools. Throughout the chapter, we will use automotive CPS and its related terms as an example of CPS. The rest of the chapter is organized as follows: Section 1.2 presents the detailed implementation of a functional model synthesis tool [45] that enables directly selecting architectures and generating high-fidelity multi-disciplinary simulation models from a functional model. Furthermore, Sect. 1.2.1 discusses the feedback function to facilitate the generation of closed loop system-level simulation models, an essential concept for control, software, sensors, and actuators of a CPS. Sections 1.2.2.4 and 1.2.3 present the detailed implementation of the synthesis tool in two separate steps. Section 1.2.2.4 discusses the contextualization-based mapping technique that translates functions to architecture components based on the contextualization of the components and (feedback) flows. Section 1.2.3 discusses the technique to generate the corresponding simulation models from the selected architectures. Section 1.2.4 presents the *refinement process* used to generate the high-fidelity simulation models. An example evaluation of a CPS use case is presented in Sect. 1.3 to demonstrate the usability of the discussed functional model synthesis tool. Finally, Sect. 1.4 concludes the chapter.

1.2   Functional Model Synthesis Tool

The main objective of the functional model synthesis tool is to support the automatic generation of CPS simulation models from the functional models. A **functional model** is a labeled directed multigraph, where each node is a function in the model and each edge represents the flow from source to target node. On the other hand, a **simulation model** is a strongly typed component with well-defined CPS ports and connectors that obey the energy conservation principles in various domains and allow components to exchange physical energy and data. The generated simulation models are comparable to the manual selection as done by the experts.

This chapter capitalizes on the method [45] that exploits the key insights of the two approaches mentioned in [11, 12]. In [11], researchers have developed a context-sensitive mapping technique to synthesize general purpose low-fidelity simulation models without leveraging any domain-specific knowledge. In [12], they have developed a high-level synthesis technique that utilizes the domain-specific knowledge in automotive architectures to generate medium fidelity simulation models. Combining the strength of these two approaches, researchers in [45] developed a novel functional model synthesis tool for synthesizing high-fidelity multi-domain CPS simulation models. The overall design flow of the tool is presented in Fig. 1.1. The feedback function and architecture templates used in the tool provide the basis for an architecture-driven mapping of functions to components. This mapping is further contextualized by the context-sensitive synthesis technique to generate simulations.

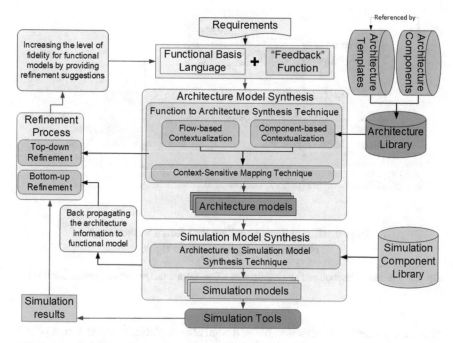

Fig. 1.1 The design flow of the functional model synthesis tool [45]

1.2.1   Feedback Function

In principle [35], a function is a process applied to an input flow to produce an output flow. That is why most functional models lack feedback information. With the approach mentioned in this chapter, when the functional model is transformed into a system-level simulation model, the synthesis technique applies design rules to create feedback of flows and components with the help of a feedback function as a complementary to the Functional Basis language. Figure 1.2 shows the syntax and semantics of the proposed feedback function. The feedback function has 2 properties: (1) The syntax feeds back a flow produced by a successor function to a predecessor function (a function executed earlier in time relative to a successor); (2) The semantics define the reuse or returning of material, energy, or signal flow to a predecessor function.

1.2.2   Synthesizing Architecture Models

As shown in Fig. 1.1, once the functional model is ready, the next step is to synthesize it into architecture models. The researchers from electronic design

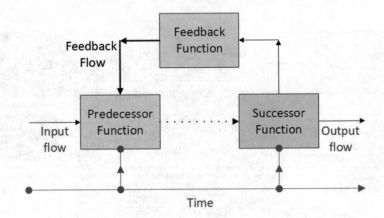

The Functional Basis specifies that functions execute
chronologically from left to right.

Fig. 1.2 Feedback function can be visualized as a flow from a successor to a predecessor function
[45]

automation refer the architecture-based design also as platform-based design [42]. It
allows the collaborative development of complex systems across different organizations [8]. The reusability of components offered by these architectures/platforms
saves billions of dollars for the companies annually [16]. The method uses the
existing knowledge of architectures to develop the architecture synthesis technique
of the functional model synthesis tool. The architecture synthesis method supports
the early design stage by synthesizing all the system-level potential architectural
solutions that satisfy the design intent depicted by the designer in the functional
model.

An **architecture component** is a pair of functional model, and a list of
constraints that specify relevant architectural parameters and properties such as the
number of cylinders in an engine and data path width in an ECU. An **architecture
template** is a multigraph, where each node is an architecture component, and each
edge is the connector connecting the source component and target component.
Each of the architecture components is associated with a list of constraints that
this architecture must meet. An **architecture library** consists of two sub-libraries,
where one sub-library is a collection of architecture components and another is
a collection of architecture templates. And **user given requirements** is a set of
requirements expressed in temporal logic that determine the expected system's
characteristics.

The architecture synthesis technique uses the architecture library, composed
by architecture components and templates, to allocate functions to candidate
architectures. The functional model created using the Functional Basis language
allows the functional model synthesis tool to validate the design contracts [37]
because the interfaces are strongly typed and can be mapped to CPS contexts (i.e.,

electrical, mechanical, signals, etc.) [11]. As a result, the contextualization-based mapping technique is used for mapping functional models to candidate architectures in the following subsections.

1.2.2.1   Contextualization Based on Input/Output Flows

In order to reliably generate high-quality simulation models, finding the correct function-to-component mapping for a given functional model is very important. Therefore, every function within the functional model of a CPS must be contextualized by its input and output flows. Two types of flows are defined for every function: **primary** and **secondary**. **Primary flows** are the flows that are inherent to a given function. The primary flows are fixed for every function, and they add no new information to the system. Therefore, for flow-contextualization, secondary flows are necessary. **Secondary flows** are the non-essential inputs/outputs of a function. Secondary flows reduce the many-to-many function-to-component relation down to a one-to-one mapping.

1.2.2.2   Contextualization Based on Components

The system-level components provided by academic and commercial libraries [26, 31, 32] are reusable but are not sufficient for automatic synthesis (which is explained in detail in Sect. 1.2.4.2). Therefore, to automatically generate the correct simulation models, it is very important to define the level of component granularity required by the synthesis techniques. For example, each simulation component (such as Modelica) defines both the structure (i.e., a capacitor) and its dynamic behavior using differential equations. Additionally, a component's connectors (or ports) specify the equations to honor energy conservation principles. Finally, components have an annotation field that can be used to store information about the component such as its documentation or icon.

The name-space of the component in a library is used to classify its domain (e.g., *Modelica.Electrical.Analog.Basic* and *Modelica.Mechanics.Components*) and to locate the component (e.g., resistor and damper). Since the technique works at the component level, the equations and behaviors associated with the component are never modified. The type of connectors in a component is used to determine the correct physical interface and generate compliant simulation models. Connector types are also useful to generate the energy conservation laws when a feedback function relates various components. A component's annotation field can be used as the means to associate and store the mappings of components-to-functions. Given the required level of component granularity, this technique imports the functions of a functional model and builds an abstract syntax tree to access a component's connectors, equations, techniques, and annotations during the mapping process.

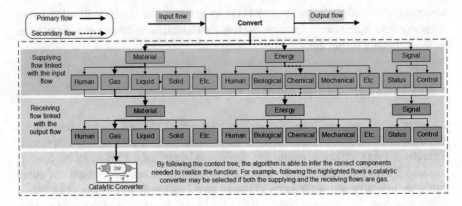

Fig. 1.3 A context tree for the "Convert" function of the automotive system

1.2.2.3 Context-Sensitive Mapping of Functions to Components

Context-sensitive mapping enables the mapping from functions to components. The mapping requires a specific function, a functional model, and a set of potential components to be mapped as input. First, it parses the input components into an abstract syntax tree (AST), where equations, techniques, connectors, and annotations are accessible for the technique. One context tree may be built for each function while mapping the input function to the component (an example of the context tree is shown in Fig. 1.3). Based on the context given by the function's signature (root node) and the secondary input/output flows (inner nodes), the realization mechanism may be deduced from basic engineering principles and added as the leaf nodes. The context tree may then be traversed (starting at the root) to create a path according to the existing secondary flows and the appropriate realization mechanism. This path may represent the flow-based contextualized function and all the functions and flows in the path may be mapped to the input components, thus may create a set of appropriate mapping. The details of the context-sensitive mapping may be found in [45] along with the pseudo code of the mapping technique.

1.2.2.4 Function to Architecture Synthesis

With the help of the context-sensitive mapping technique, the high-level synthesis technique may be developed for synthesizing functional model to architectural models. Given a functional model and the user-defined requirements as inputs, the objective of the function to architecture synthesis technique is to find the set of architecture templates in architecture library that fully or partially map to the functional model.

The function to architecture synthesis technique will generate both full and partial mappings from a functional model into a set of architecture templates based on the knowledge contained in the architecture library. The synthesis technique may also provide top-down and bottom-up refinements as design suggestions for functional model and architecture templates. The synthesis technique may perform a branch and bound synthesis [29] as follows: First, it can prune the selection of architecture templates whose constraints do not meet the requirements. Based on the selected architecture templates, it will aggregate the maximum possible mapping from functional model to architecture templates using the context-sensitive mapping technique.

After the architecture template design space has been expanded, the synthesis technique will perform a multi-step pruning process on each architecture template. Once the pruning is done, the synthesis technique will refine all partial mappings in a top-down manner. And the bottom-up refinement can be done on the functions that exist on the selected architecture templates but not in the functional model. The details of the refinement process are presented in Sect. 1.2.4. The refinement suggestions can help the designers to increase the fidelity of the original functional model. The details of the function to architecture synthesis technique can be found in [45] along with the pseudo code.

1.2.3   Architecture to Simulation Model Synthesis

Once the candidate architectures have been identified from the functional models using the function to architecture synthesis technique, the next step is to generate the corresponding simulation models.

The architecture to simulation model synthesis technique will generate simulation models for all the architecture-mapped functional models identified by the function to architecture synthesis technique. First, it will broaden the simulation design space for each of the architecture templates generated in Sect. 1.2.2.4. Whenever an architecture-mapped function cannot be mapped to any simulation component, a top-down refinement of simulation models may be constructed (see details of the refinement process in Sect. 1.2.4). Then, it will expand the simulation design space by creating individual simulation models for all possible combinations of simulation components that match an architecture-mapped functional model. Finally, it will add connections to every model according to the topology in their corresponding architecture component. Interested readers may refer to [45] for the details of the synthesis technique.

1.2.4   Process of Refinement

Designing CPS is a complex process that involves multiple iterations. For example, in automotive CPS one design is refined multiple times by multiple personnel from various organizations. The synthesis technique presented in this chapter supports the refinement of the low-fidelity functional models to achieve high fidelity.

For example, after the initial candidate architectures have been identified, the architecture models may contain a set of functions that are not modeled in the original functional model. Thus, a refined functional model can be created by back-propagating this architecture information to the functional model. This refinement is referred to as bottom-up refinement as information is being propagated from a lower level of abstraction (architecture) to a higher level of abstraction (functions) to achieve high fidelity. The **level of fidelity** is the amount of qualitative and quantitative information that can be obtained from a model. On the other hand, **incompleteness** of a model is the lack of fidelity necessary to answer a specific engineering question. And, **refinement** is the process by which the level of fidelity of a model is increased to make it less incomplete, and thus the ability to answer more detailed engineering questions.

1.2.4.1 Top-down Refinement Process

As the name implies, a top-down refinement increases the level of fidelity by propagating the information from a higher level of abstraction (functions) to a lower level of abstraction (architecture). For example, sometimes a functional model contains a set of functions that are not fulfilled by any of the components in the architecture library. To satisfy those functional requirements of that particular functional model, a new architecture component with the relative complement of the functions is created by the new architecture component in the architecture library.

1.2.4.2 Bottom-up Refinement Process

The bottom-up refinement analyzes the ports and interface of a simulation component to determine the functions automatically. In addition to that, the flow (energy, material, and signals) transformations occurring within the component's internal structure is also determined. Using the functions performed by the simulation components, the bottom-up step classifies the components in a library. Thus, it helps to achieve a correlation between functions, architectures, and simulation components that design tools may use to synthesize simulation models for candidate system architectures.

The type of energy, material, or signals that one component exchanges with another through its ports can be obtained using the components' interface analysis. The interface analysis infers the functions achieved by a given simulation component as shown in Fig. 1.4. For example, the *Electric Motor* component has three ports: thermal, electrical, and rotational mechanical energy. Typically, conjugate variables that represent effort/flow are used to exchange energy between the simulation components. Electrical energy, rotational mechanical energy, and thermal energy are represented by the conjugate variables' voltage/current, torque/angular velocity, and temperature/heat flow, respectively. As the relationship between the functional level and simulation component level energy is known, the technique

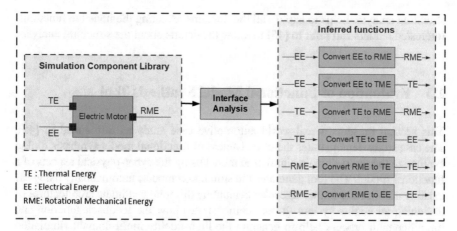

Fig. 1.4 Extraction of functions from simulation components [45]

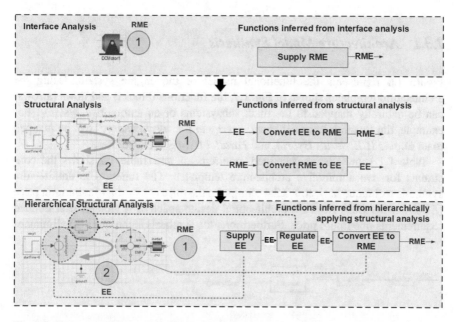

Fig. 1.5 Structural analysis on simulation components reveals additional domains and the energy, material, and signal transformations between them [45]

determines that the *Electric Motor* is able to perform the function of "*Convert*" energy from any port to the other ports. This leads to six inferred functions.

Sometimes, this technique also performs a structural analysis—a hierarchical traversal and flattening—of the internal simulation component structure to expose all the domains (e.g., electrical, mechanical, thermal, signals, etc.) that are not visible through the interface of the simulation components. Figure 1.5 shows how structural analysis helps to expose all the domains revealing the invisible functions. Interested readers can refer to [45] to know the details about the structural analysis.

1.3   Evaluation of Functional Model Synthesis Tool

This section presents a real-world automotive case study as introduced in [45]. The case study demonstrates the effectiveness of functional model synthesis tool to synthesize the multi-domain simulation model using the cyber-physical aspects of a functional model. The tool generates the simulation models automatically using the existing architecture knowledge after evaluating different candidate architectures of designing an engine system. It also demonstrates how the feedback function and the refinement process help to generate the high-fidelity multi-domain simulation models.

1.3.1   Architecture Model Synthesis

Figure 1.6 represents the functional model of the engine system. Using the architectural templates shown in Fig. 1.7, the functions (blocks) and flows (arrows) can be naturally mapped to the main subsystems of an automotive system . For example, the function "*convert chem. energy to rot. mech. energy*" can be mapped to an *Engine ICE*, *Series Hybrid*, and *Parallel Hybrid* architectures.

Table 1.1 shows the user-defined requirements and Table 1.2 shows the constraints for five automotive architecture templates. The function to architecture synthesis technique in Sect. 1.2.2.4 eliminates the third constraint of minimum price (*Electric Fuel Cell*) from the architectural design space as it violates user-defined requirements maximum price. Furthermore, the synthesis technique also eliminates

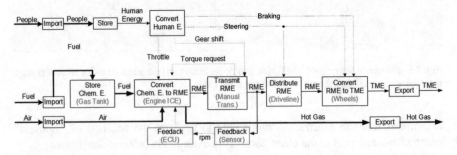

Fig. 1.6 An example functional model of an automotive power-train, associated requirements, and architectural constraints [45]

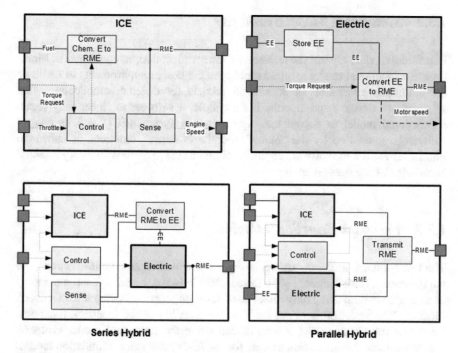

Fig. 1.7 Architecture templates for the engine block [45]

Table 1.1 User-defined requirements

Description	Value
Minimum energy efficiency	35%
Maximum weight	2000 kg
Maximum price	$40,000

Table 1.2 Constrain list of architecture templates

Description	ICE	Series hybrid	Parallel hybrid	Electric	EFC[a]
Max. energy efficiency	20%	37%	37%	60%	50%
Min. total weight	2000 kg	2000 kg	2000 kg	2000 kg	2000 kg
Min. price	$20,000	$30,000	$30,000	$30,000	$50,000
Input energy	Chem.	Chem., Elec.	Chem., Elec.	Elec.	Elec.
Output energy	Mech.	Mech.	Mech.	Mech.	Mech.
Energy price	$3/gallon	$0.23/kWh	$0.23/kWh	$0.23/kWh	$0.15/kWh

[a]*EFC* electric fuel cell

the *Electric* architecture template because neither "*Store EE*" nor "*Convert EE to RME*" functions exist in the functional model. It is to note that the *ICE* architecture template fully matches the functional model, whereas the *Series Hybrid* and *Parallel Hybrid* templates partially match the functional model.

1.3.2   Simulation Model Synthesis

Functional models support the technology-independent design requirements. Moreover, the functional model synthesis tool uses the design requirements to facilitate the automatic generation of simulation models for different architectures that satisfy those design requirements. For example, a software engineer may create a functional model to design the engine control system (ECU and its control software) represented by the *"Sense"* and the *"Control"* functions as shown in Fig. 1.6. However, the software engineer might be able to do so with a very minimal knowledge of mechanical engineering.

1.3.3   Feedback Function's Usability

The two feedback functions shown in Fig. 1.6 are used as an additional support to the Functional Basis language [45]. Researchers in [45] designed a functional model of an engine system using only Functional Basis language and without the feedback function. The intention is to demonstrate the usability of the feedback functions. Then that functional model is synthesized using the functional model synthesis tool to generate the simulation models for the *ICE* architecture. Simulation models are synthesized from both functional models (with feedback and without feedback function) and their performances are compared. The fuel consumption is much less with feedback functions compared to the one without feedback functions as shown in Fig. 1.8. Furthermore, Fig. 1.9 demonstrates that engine models synthesized with feedback functions generate much less emissions (CO, NOx, and HC) as compared to the ones without the feedback functions. The reason is, the feedback functions are mapped to additional control units which cause more efficient use of the engine.

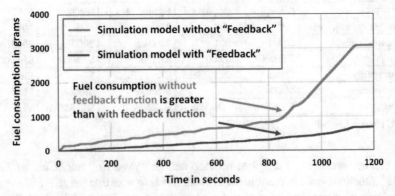

Fig. 1.8 Comparison of the fuel consumption between the synthesized simulation models with/without "Feedback" [45]

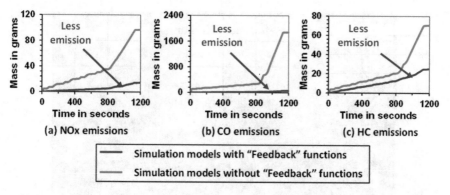

Fig. 1.9 Comparison of emissions between the simulation models with/without "Feedback" [45]

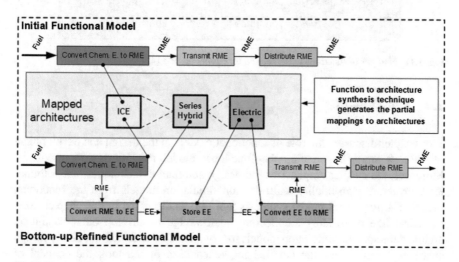

Fig. 1.10 Bottom-up refinement of a functional model [45]

1.3.4   Refinement Process Analysis

The usefulness of the bottom-up refinement of the discussed methodology is illustrated in Fig. 1.10. As shown in Fig. 1.6, original functional model does not have the functions "*Convert RME to EE,*" "*Convert EE to RME,*" and "*Store EE.*" However, the function to architecture synthesis technique in Sect. 1.2.2.4 generates the partial mappings to architectures (*ICE*, *Hybrid*, and *Electric*) and system engineers get the refined functional models suggestions. For example, new functions such as "*Convert RME to EE,*" "*Store EE,*" and "*Convert EE to RME*" are created using the *Hybrid* architecture mapping. The number of functions in the refined functional models generated from three different mappings is presented in Fig. 1.11.

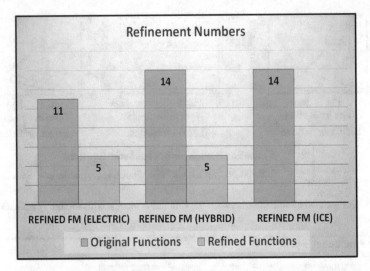

Fig. 1.11 Number of functions in the refined functional models [45]

1.4 Conclusion

This chapter discusses the role of functional models in the design and development of CPS. It starts by creating the functional model from design requirements. Moreover, the chapter emphasizes the design automation techniques that automatically generate high-fidelity multi-domain simulation models from the functional model. The functional model synthesis tool [45] as discussed in this chapter takes the advantage of existing architectures' knowledge to develop other candidate, architecture-specific simulation models for complex CPS and facilitates early design space exploration. For the detailed implementation of the tool, the concept of a "Feedback" function [45] is also discussed to capture feedback flows that are essential for complex systems using control, software, sensors, and actuators. A context-sensitive mapping technique is also discussed to construct an appropriate engineering context that facilitates the selection of function-to-component mappings by considering the surrounding flows of a function. Moreover, a refinement process is also discussed that backpropagates the results as design suggestions to the system-level designers. Finally, the usability of the discussed functional model synthesis tool is presented using an automotive engine system.

References

1. Abelein, U., Lochner, H., Hahn, D., & Straube, S. (2012, March). Complexity, quality and robustness-the challenges of tomorrow's automotive electronics. In *Design, Automation & Test in Europe Conference & Exhibition (DATE), 2012* (pp. 870–871). Piscataway: IEEE.

2. Al Faruque, M., Regazzoni, F., & Pajic, M. (2015, October). Design methodologies for securing cyber-physical systems. In *Proceedings of the 10th International Conference on Hardware/Software Codesign and System Synthesis* (pp. 30–36). New York: IEEE Press.
3. Alvarez Cabrera, A. A., Erden, M. S., & Tomiyama, T. (2009). On the potential of function-behavior-state (FBS) methodology for the integration of modeling tools. In *Proceedings of the 19th CIRP Design Conference Competitive Design*. Bedford: Cranfield University Press.
4. ANSYS Tool Kits for Automotive Solutions. http://www.ansys.com/Industries/Automotive/
5. AUTOSAR Automotive Open System Architecture. https://www.autosar.org/
6. Bassi, L., Secchi, C., Bonfe, M., & Fantuzzi, C. (2011). A SysML-based methodology for manufacturing machinery modeling and design. *IEEE/ASME Transactions on Mechatronics, 16*(6), 1049–1062.
7. Boucher, M., & Houlihan, D. (2008). *System design: New product development for mechatronics*. Boston: Aberdeen Group.
8. Broy, M., Gleirscher, M., Kluge, P., Krenzer, W., Merenda, S., & Wild, D. (2009). Automotive architecture framework: Towards a holistic and standardised system architecture description. White paper. IBM Corporation. Technical Report, Technische Universität München. TUM-I0915.
9. Broy, M., Kruger, I. H., Pretschner, A., & Salzmann, C. (2007). Engineering automotive software. *Proceedings of the IEEE, 95*(2), 356–373.
10. Bryant, C. R., Stone, R. B., McAdams, D. A., Kurtoglu, T., & Campbell, M. I. (2005). Concept generation from the functional basis of design. In *ICED 05: 15th International Conference on Engineering Design: Engineering Design and the Global Economy* (p. 1702). Barton: Engineers Australia.
11. Canedo, A., Schwarzenbach, E., & Al Faruque, M. A. (2013, April). Context-sensitive synthesis of executable functional models of cyber-physical systems. In *Proceedings of the ACM/IEEE 4th International Conference on Cyber-Physical Systems* (pp. 99–108). New York: ACM.
12. Canedo, A., Wan, J., & Al Faruque, M. A. (2014, November). Functional modeling compiler for system-level design of automotive cyber-physical systems. In *2014 IEEE/ACM International Conference on Computer-Aided Design (ICCAD)*, (pp. 39–46). Piscataway: IEEE.
13. Charette, R. N. (2009). This car runs on code. *IEEE Spectrum, 46*(3), 3.
14. Christen, E., & Bakalar, K. (1999). VHDL-AMS-a hardware description language for analog and mixed-signal applications. *IEEE Transactions on Circuits and Systems II: Analog and Digital Signal Processing, 46*(10), 1263–1272.
15. Cooprider, A. (2014). Automotive embedded systems tutorial-part I. In *Proceedings of the 51th Design Automation Conference, DAC* (Vol. 2014).
16. Dahmus, J. B., Gonzalez-Zugasti, J. P., & Otto, K. N. (2001). Modular product architecture. *Design Studies, 22*(5), 409–424.
17. Derler, P., Lee, E. A., & Vincentelli, A. S. (2012). Modeling cyberphysical systems. *Proceedings of the IEEE, 100*(1), 13–28.
18. Erden, M. S., Komoto, H., van Beek, T. J., D'Amelio, V., Echavarria, E., & Tomiyama, T. (2008). A review of function modeling: Approaches and applications. *Ai Edam, 22*(2), 147–169.
19. Fortney, G. (2014, August). Model based systems engineering using validated executable specifications as an enabler for cost and risk reduction. In *Proceedings of the 2014 Ground Vehicle Systems Engineering and Technology Symposium (GVSETS)*.
20. Helms, B., & Shea, K. (2012). Computational synthesis of product architectures based on object-oriented graph grammars. *Journal of Mechanical Design, 134*(2), 021008.
21. Hirtz, J., Stone, R. B., McAdams, D. A., Szykman, S., & Wood, K. L. (2002). A functional basis for engineering design: Reconciling and evolving previous efforts. *Research in Engineering Design, 13*(2), 65–82.

22. Komoto, H., & Tomiyama, T. (2012). A framework for computer-aided conceptual design and its application to system architecting of mechatronics products. *Computer-Aided Design, 44*(10), 931–946.
23. Kruse, B., Mnzer, C., Wlkl, S., Canedo, A., & Shea, K. (2012, August). A model-based functional modeling and library approach for mechatronic systems in SysML. In *ASME 2012 International Design Engineering Technical Conferences and Computers and Information in Engineering Conference* (pp. 1217–1227). New York: American Society of Mechanical Engineers.
24. Kurtoglu, T., & Campbell, M. I. (2009). Automated synthesis of electromechanical design configurations from empirical analysis of function to form mapping. *Journal of Engineering Design, 20*(1), 83–104.
25. LabVIEW System Design Software. http://www.ni.com/labview/
26. Lawrence Berkeley National Laboratory - Modelica Buildings Library. http://simulationresearch.lbl.gov/modelica
27. Lee, E. A. (2008, May). Cyber physical systems: Design challenges. In *11th IEEE Symposium on Object Oriented Real-Time Distributed Computing (ISORC)* (pp. 363–369). Piscataway: IEEE.
28. LMS Imagine.Lab AMESim. http://www.lmsintl.com/
29. McFarland, M. C., Parker, A. C., & Camposano, R. (1990). The high-level synthesis of digital systems. *Proceedings of the IEEE, 78*(2), 301–318.
30. Mellor, S. J., Balcer, M., & Foreword By-Jacoboson, I. (2002). *Executable UML: A foundation for model-driven architectures.* Boston: Addison-Wesley Longman Publishing Co.
31. Modelica Association, Modelica Standard Library. https://modelica.org/libraries/Modelica/
32. Modelon - Vehicle Dynamics Library. http://www.modelon.com/
33. OMG Systems Modeling Language (SysML). http://www.omgsysml.org/
34. Oregon State University, Design Engineering Lab, Design Repository. http://designengineeringlab.org/
35. Pahl, G., & Beitz, W. (2013). *Engineering design: A systematic approach.* New York: Springer Science & Business Media.
36. Rudov-Clark, S. D., & Stecki, J. (2009, March). The language of FMEA: On the effective use and reuse of FMEA data. In *Sixth DSTO International Conference on Health & Usage Monitoring* (pp. 9–12).
37. Sangiovanni-Vincentelli, A., Damm, W., & Passerone, R. (2012). Taming Dr. Frankenstein: Contract-based design for cyber-physical systems. *European Journal of Control, 18*(3), 217–238.
38. Simscape. http://www.mathworks.com/products/simscape/
39. Simulink. http://www.mathworks.com/products/simulink/
40. Stone, R. B., & Wood, K. L. (2000). Development of a functional basis for design. *Journal of Mechanical Design, 122*(4), 359–370.
41. Uckun, S. (2011). Meta II: Formal co-verification of correctness of large-scale cyber-physical systems during design. Palo Alto Research Center, Technical Report, 1–43.
42. Vincentelli, A. S. (2002). *Defining platform-based design.* EEDesign of EETimes.
43. Wan, J., Canedo, A., & Al Faruque, M. A. (2015, September). Security-aware functional modeling of cyber-physical systems. In *2015 IEEE 20th Conference on Emerging Technologies & Factory Automation (ETFA)* (pp. 1–4). Piscataway: IEEE.
44. Wan, J., Canedo, A., & Al Faruque, M. A. (2017). Cyberphysical codesign at the functional level for multidomain automotive systems. *IEEE Systems Journal, 11*(4), 2949–2959.
45. Wan, J., Canedo, A., & Al Faruque, M. A. (2017). Functional model-based design methodology for automotive cyber-physical systems. *IEEE Systems Journal, 11*(4), 2028–2039.
46. Wolkl, S., & Shea, K. (2009, January). A computational product model for conceptual design using SysML. In *ASME 2009 International Design Engineering Technical Conferences and Computers and Information in Engineering Conference* (pp. 635–645). New York: American Society of Mechanical Engineers.

Chapter 2
Platform-Based Design for Automotive and Transportation Cyber-Physical Systems

Chung-Wei Lin, Bowen Zheng, Hengyi Liang, and Qi Zhu

Acronyms

ADAS	Advanced driver assistance systems
AUTOSAR	Automotive open system architecture
CACC	Cooperative adaptive cruise control
CAN	Controller area network
ECU	Electronic control units
OEM	Original equipment manufacturer
PBD	Platform-based design
TDMA	Time division multiple access
TSN	Time-sensitive networking
V2V	Vehicle-to-vehicle
V2X	Vehicle-to-X

C.-W. Lin (✉)
National Taiwan University, Taipei, Taiwan
e-mail: cwlin@csie.ntu.edu.tw

B. Zheng
University of California, Riverside, CA, USA

H. Liang · Q. Zhu
Northwestern University, Evanston, IL, USA

© Springer Nature Switzerland AG 2019
M. A. Al Faruque, A. Canedo (eds.), *Design Automation of Cyber-Physical Systems*, https://doi.org/10.1007/978-3-030-13050-3_2

2.1 Platform-Based Design Methodology for Connected Vehicles

Automotive design has become more complex than ever due to the rapid development of connected and autonomous technology. This trend affects not only the design of individual vehicles but also the operation of entire vehicular transportation system, through connected vehicle applications such as intelligent traffic signals, collaborative adaptive cruise control (CACC), and vehicle platooning. The safety-critical nature of these systems makes it essential to rigorously ensure functional correctness and to quantitatively evaluate system metrics throughout the design process and across all system layers. In this chapter, we will introduce the application of the platform-based design (PBD) paradigm in connected vehicles. We will present how the principles of the PBD paradigm, in particular the definition of platforms and the mapping between functional and architectural platforms, may be carried out across the system layers, from connected vehicle applications to individual vehicle functionality, and then to in-vehicle software, hardware, and physical layers.

2.1.1 Design Challenges for Connected Vehicles

In the following, we will first introduce some of the major challenges for connected vehicles, and then outline how the PBD paradigm may be applied to their design.

- **Addressing high-volume and dynamic input data:** The size of a signal in conventional control systems is usually not very large. It can be only a binary to indicate "on" or "off" of a component, or several bytes to represent the value of a measurement. However, for advanced driver assistance systems (ADAS) and autonomous functions in modern vehicles, the inputs from lidars, radars, cameras, and other sensors could induce much larger data at a high input data rate. For example, an advanced lidar can have input data rate that is up to 100 Mbps, which far exceeds the capacity of currently prevalent in-vehicle bus protocol, the controller area network (CAN), and the processing capability of current electronic control units (ECUs). Moreover, such input data rate may significantly vary under different road conditions, moving speed, and light intensity, which presents further challenges to the system design, as detailed below.
- **Computation architecture design:** High-volume and dynamic input data has a significant impact on the design of the computation platform. Should system designers add more ECUs or upgrade existing ECUs to more powerful ones for handling the data? What types of new computation elements such as GPUs, FPGAs, or ASIC accelerators are needed? Can the computation architecture be dynamically adapted to handle the changing data rate? Answering these questions requires the development of new design methodologies.

- **Communication architecture design:** To address the high-volume and dynamic input data, original equipment manufacturers (OEMs) have been exploring new in-vehicle communication architectures such as those based on the Ethernet protocol. However, systematic methodologies are still greatly needed to meet the data processing requirements. Furthermore, the new communication protocols, including both in-vehicle protocols and inter-vehicle protocols for vehicle-to-everything (V2X) communication, should be carefully designed and integrated with the conventional protocols that are still important for conventional/legacy components. The integration of different protocols also relies on the design and analysis of gateways, which further increase the design complexity.
- **Topology design:** As there are different protocols and multiple network devices in an automotive system, it is not trivial to decide the connection of sensors, actuators, and ECUs to network devices. The decisions are constrained by design requirements and affected by the trade-offs between performance, cost, and even wiring weight. Furthermore, the topology should follow the harness and routing graph in an automotive system and is often challenging to design.
- **Safety:** Automotive systems are safety-critical systems, and there are many constraints that have to be met for ensuring system safety. For instance, the end-to-end latency from detecting sensor input to applying control often has to meet a strict deadline, which requires rigorous worst-case analysis based on formal mathematical models. However, with the increase of functional and architectural complexity, accurately building those models and conducting worst-case analysis has become increasingly challenging.
- **Reliability:** The reliability of automotive systems relies on many factors, such as the fault-tolerant and redundant architectures for single-point-of-failures. Several protocols such as the time-sensitive networking (TSN) support replications and eliminations (if redundant at destinations) of frames. As shown in Fig. 2.1, these operations can increase the reliability of communication, but they also induce higher costs and more communication traffic. Furthermore, they require multiple routing paths, which makes topology design more challenging. Similarly, redundant ECUs may increase the reliability of computation, but they also lead to higher costs and design complexity.

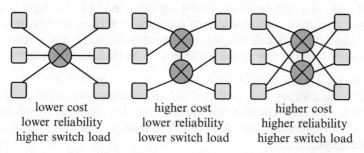

lower cost	higher cost	higher cost
lower reliability	lower reliability	higher reliability
higher switch load	lower switch load	higher switch load

Fig. 2.1 The trade-off between cost, reliability, and switch performance in automotive design

- **Security:** System security is a rising issue for automotive systems. It requires a cross-layer solution that includes security mechanisms compatible with existing V2X communication protocols, lightweight security mechanisms within individual vehicles, and component-level security mechanisms. Due to tight resource constraints and stringent design requirements, security should be considered from the beginning of the design process; otherwise, it is often too late or impossible to add security at late stages.

2.1.2 Mapping Problems for Connected Vehicles

We propose the PDB methodology to address these growing design challenges of connected vehicles. The key idea of PBD is to capture the system with a number of abstraction layers called *platforms*, and divide the complex design process into a series of *mappings* from higher-layer to lower-layer platforms. The mapping between two platform layers is, in fact, a design space exploration process, where different options (abstracted as design variables) for implementing the high-layer platform model (i.e., "functionality") on the lower-layer platform components (i.e., "architecture") are explored with respect to a set of design objectives and constraints.

Figure 2.2 shows how the design of connected vehicles can be addressed with the PBD paradigm as a series of mapping problems across platform layers, including mapping connected vehicle applications to vehicle functionality, mapping vehicle functionality to software tasks, mapping software tasks to hardware components, and mapping hardware components to physical layout.

2.2 Mapping Connected Vehicle Applications to Vehicle Functionality

In the following sections, we will go through some representative problems for each of these mapping problems. At the top layer, the PBD paradigm is applied to the mapping from connected vehicle applications, such as cooperative adaptive cruise control (CACC), lane merging, and autonomous intersection, to functionality of individual vehicles. The mapping problem can be formulated as follows:

- **Platforms**: (1) The higher-layer platform is captured by the models of connected vehicle applications, such as CACC and autonomous intersections; and (2) the lower-layer platform includes the models of individual vehicles in both the cyber domain (computation and communication models) and the physical domain (vehicle dynamics).

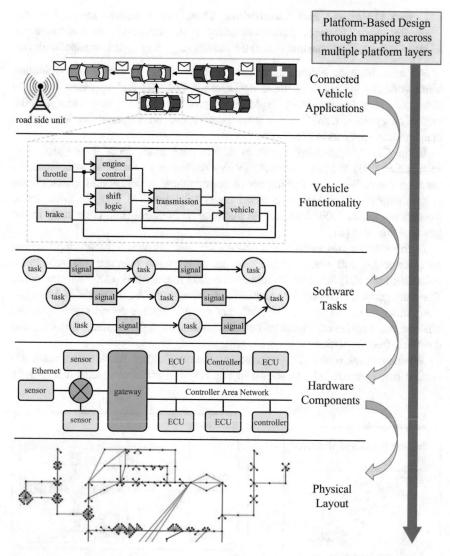

Fig. 2.2 Platform-based design for connected vehicles through mapping across multiple platform layers: (1) mapping connected vehicle applications to vehicle functionality, (2) mapping vehicle functionality to software tasks, (3) mapping software tasks to hardware components, and (4) mapping hardware components to physical layout

- **Design Space**: The design variables to be explored include the setting of con-tracts (constraints) on individual vehicle behavior/functionality—in the physical domain, this means constraints on vehicle's path planning and motion control; in cyber domain, this means constraints on computation latency, communication latency, reliability, etc.

- **Design Objectives and Constraints**: These could include safety, liveness, deadlock-free, fairness, traffic efficiency (e.g., for CACC and autonomous intersections), fuel consumption and emission (e.g., for eco-driving applications).

We have been developing a system-level modeling, synthesis and validation framework for connected vehicle applications [22, 27] and applying it to the above mapping (design space exploration) problem. In particular, we apply the methodology to a CACC application [9, 22] and an autonomous intersection management application [21, 23].

In the CACC application, vehicles inform each other about their speeds and accelerations via vehicle-to-vehicle (V2V) messages to maintain safe distances between them. We study the impact of communication delays and losses on the system safety and performance based on simulations, and then, in turn, derive the constraints for individual vehicle planning and control (i.e., constraints in the physical domain) [22].

In the autonomous intersection application, autonomous vehicles approaching an intersection will communicate with an intersection manager via vehicle-to-infrastructure (V2I) messages to request the right to enter and pass the intersection. The manager will then decide/schedule the entering order for the vehicles. We again study this application with consideration of communication delays and losses, and observe the significant impact of communication on system safety, liveness, and deadlock-free properties.

We then develop and analyze a delay-tolerant protocol for autonomous intersection management [21], as shown in Fig. 2.3. The protocol assures that as long

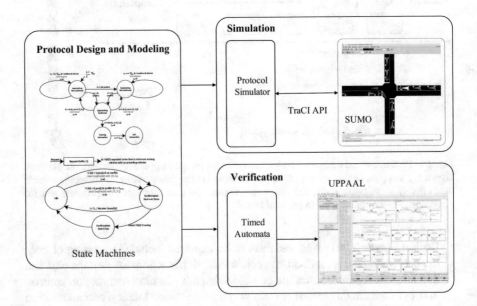

Fig. 2.3 Design and validation of delay-tolerant autonomous intersections [21]

as the communication delays are bounded, every vehicle will eventually cross the intersection (i.e., liveness property) and vehicles with conflicting routes will never enter the intersection at the same time (i.e., safety property). We verify the safety and liveness properties of our protocol by building more abstract timed automata models and leveraging the UPPAAL verification tool [19]. Finally, we implement our protocol in the SUMO traffic simulation suite [18], with the extension of modeling communication delays, to study the system performance. Such analysis allows us to derive the delay constraints on V2I communication in the cyber domain, which includes the delays of in-vehicle processing and the delays of V2I message transmissions, for ensuring system safety, liveness, deadlock-free, and performance.

2.3   Mapping Vehicle Functionality to Software Tasks

Once we have the specifications and constraints of individual vehicle functionality, the PBD paradigm can be applied to conduct the mapping from vehicle functionality to software tasks. This mapping problem can be formulated as follows:

- **Platforms**: (1) The higher-layer platform is captured by the models of vehicle functionality (e.g., Simulink models, timed automata), including in-vehicle sensing, computation and communication models, as well as V2X communication models; and (2) the lower-layer platform includes the models of software tasks and communication protocols.
- **Design Space**: The design variables to be explored include the generation of software tasks from functional models (i.e., mapping from functional blocks to tasks) and the design of communication protocols (including signals) from functional models.
- **Design Objectives and Constraints**: These may include a variety of constraints and optimization objectives on system performance, safety, security, cost, reliability, extensibility, memory size, reusability, modularity, etc.

For the mapping across these two layers, we have developed algorithms for exploring software task generation, allocation, and scheduling from functional models of finite state machines [25] and synchronous block diagrams [4, 6], two main models of computation in synchronous models that are widely used in capturing embedded sensing, control, and computation applications.

In [25], we developed a general partitioned model for multi-task implementations of synchronous finite state machines, and defined two metrics for measuring the quality of task implementations: the breakdown factor and the action extensibility. We then developed a heuristic algorithm to explore robust and extensible task generation and scheduling based on the two metrics. The experimental results demonstrated significant improvement on the two metrics from our algorithm, and showed the importance of exploring task generation options for synchronous finite state machines.

In [6], we developed an algorithm for direct generation of software tasks on single-core platforms from synchronous block diagrams, with respect to modularity, reusability, code size, and latency. This work showed the promise of exploring task generation for synchronous block diagrams.

In [4], we developed a complete model-based synthesis flow for automotive software systems that follow the AUTOSAR standard [1]. The synthesis flow optimizes the generation of AUTOSAR runnables from synchronous block diagrams, the mapping of runnables onto software tasks, and the allocation and scheduling of tasks onto multi-core ECU platforms. A key idea of this flow is to develop a uniformed formalism of firing and execution timing automata (FETA) to capture the periodic timing behavior of functional blocks, runnables, and tasks. Leveraging FETA, the flow can more accurately model and reason about system timing behavior across different layers during the entire mapping process. Finally, the synthesis flow addresses constraints and objectives on a variety of metrics when solving the mapping problems, including software engineering objectives such as runnable modularity, reusability, and code size as well as timing and resource objectives such as system schedulability and memory cost. In particular, the flow focuses on trading off modularity with schedulability during the mapping from functional blocks to runnables, and on minimizing memory cost under schedulability constraints during the mapping from runnables to tasks and from tasks to ECU cores. Similarly as [6, 25], this work showed the importance of exploring task generation options when mapping vehicle functionality to software tasks. Furthermore, it demonstrated the benefits of explicitly considering timing during task generation and having a uniformed formalism such as FETA to capture timing behavior across system layers.

2.4   Mapping Software Tasks to Hardware Components

Once we have a model of software tasks and their communication signals, the PBD paradigm can be further applied to explore the mapping of tasks onto hardware components. We have briefly discussed this above in [4] and will elaborate it more in this section. The mapping formulation for task to hardware platform mapping can be captured as follows:

- **Platforms**: (1) The higher-layer platform is typically modeled as task graphs with communication signals; and (2) the lower-layer platform includes architectural models of hardware components.
- **Design Space**: The design variables include task allocation, task scheduling, signal mapping to memory transactions or bus/wireless messages, message scheduling, etc.
- **Design Objectives and Constraints**: The constraints and objectives address metrics such as latency, schedulability, cost, energy consumption, extensibility, fault tolerance, and security.

In the following, we will demonstrate a few different mapping platforms across these two layers.

2.4.1 Conventional CAN-Bus Systems

The controller area network (CAN) protocol is still the most common in-vehicle network. The mapping problem from a task graph to a CAN-based system can be solved by the PBD paradigm. As shown in Fig. 2.4, the functional model is a task graph that consists of a set of tasks, denoted by $T = \{\tau_1, \tau_2, \ldots, \tau_{|T|}\}$, and a set of signals, denoted by $S = \{\sigma_1, \sigma_2, \ldots, \sigma_{|S|}\}$. Each signal σ_i is between a source task and a destination task. Each task is activated periodically and communicate with each other through signals. The architecture model is a distributed CAN-based platform that consists of a set of ECUs, denoted by $E = \{\varepsilon_1, \varepsilon_2, \ldots, \varepsilon_{n_E}\}$, and a CAN bus that connects all the ECUs. Each ECU ε_k can send a set of messages, denoted by $M_k = \{\mu_{k,1}, \mu_{k,2}, \ldots, \mu_{k,|M_k|}\}$. ECUs are assumed to run AUTOSAR/OSEK-compliant operation systems that support preemptive priority-based task scheduling. The bus uses the standard CAN bus arbitration model that features non-preemptive priority-based message scheduling [2].

A path π is an ordered interleaving sequence of tasks and signals, defined as $\pi = (\tau_{r_1}, \sigma_{r_1}, \tau_{r_2}, \sigma_{r_2}, \ldots, \sigma_{r_{k-1}}, \tau_{r_k})$. $src(\pi) = \tau_{r_1}$ is the path's source and $snk(\pi) = \tau_{r_k}$ is its sink. Sources are activated by external events, while sinks activate actuators. Multiple paths may exist between each source–sink pair. We assume all tasks in a path perform computations that contribute to a distributed function, from the collection of sensor data to the remote actuation. The worst-case end-to-end latency incurred when traveling a path π is denoted as l_π, which

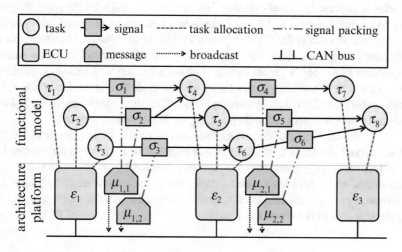

Fig. 2.4 The task mapping problem in a CAN-based system

represents the largest possible time interval that is required for the change of the input (or sensed) value at the source to be propagated and cause a value change (or an actuation response) at the sink.

During mapping, the functional model is mapped onto the architecture platform, as shown in Fig. 2.4. Specifically, the tasks are allocated to ECUs, and the signals are packed into messages and transmitted on the CAN bus in a broadcast fashion. Messages are triggered periodically and each message contains the latest values of the signals that mapped to it. Static priorities are assigned to tasks and messages for priority-based scheduling. The design space of task allocation, signal packing, and priority assignment is explored with respect to a set of design objectives and constraints.

For detailed problem formulations and their corresponding algorithms, please refer to our previous publications on task mapping for the CAN-based platform, with the consideration of end-to-end latency [3, 5, 24, 30], extensibility [10, 26, 28, 29], fault tolerance [20], and security [13, 15].

In the following, we will introduce task mapping onto two different architectural platforms—one replaces the CAN bus by a time division multiple access (TDMA) switch, and the other one utilizes an OS hypervisor to support multiple operating systems running on a hardware component.

2.4.2 Advanced Architecture: TDMA-Based Systems

The TDMA-based protocol is a very representative synchronous protocol and an abstraction of many existing protocols, such as the FlexRay [7], the Time-Triggered Protocol [17], the Time-Triggered Ethernet [16], and the Time-Sensitive Networking [8]. These protocols are likely to be adopted in future intelligent vehicles to support high and dynamic data rate. Compared with Ethernet, they also have more deterministic and predictable timing behavior. Compared with priority-based networks such as the CAN protocol, TDMA-based systems have fundamental differences in system modeling (in particular for latency modeling), on security mechanism selection (a global time is available for security reasons), on design space (network scheduling is the focus of this work but not a factor for CAN-based systems), and on algorithm design. Therefore, the approaches for CAN-based systems in the previous section do not apply to TDMA-based systems.

As shown in Fig. 2.5, similar to the system model in the previous section, the functional model is a task graph that consists of a set of tasks, denoted by $T = \{\tau_1, \tau_2, \ldots, \tau_{|T|}\}$, and a set of signals, denoted by $S = \{\sigma_1, \sigma_2, \ldots, \sigma_{|S|}\}$. Each signal σ_i is between a source task and a destination task, and each task is activated periodically and communicates with each other through signals. The architecture model is a distributed platform that consists of a set of ECUs, denoted

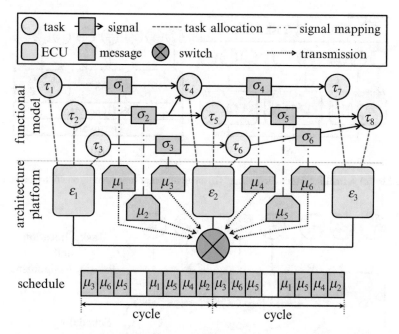

Fig. 2.5 The mapping problem of a TDMA-based system

by $E = \{\varepsilon_1, \varepsilon_2, \ldots, \varepsilon_{n_E}\}$, and ECUs are assumed to support preemptive priority-based task scheduling. The nodes are connected through a TDMA-based switch (we focus on the single-switch case in this chapter, and our formulation can be extended to multi-switches cases). A set of messages is communicated among nodes through the switch, denoted by $M = \{\mu_1, \mu_2, \ldots, \mu_{|M|}\}$. The switch uses a TDMA-based model for scheduling, in which each *time slot* in the schedule can be assigned to one message. Several time slots form a *cycle*, and the network switch repeats the same scheduling sequence after each cycle. It is possible that a time slot is empty (not assigned to any message) in a schedule, and it is also possible that there are more than one time slots assigned to the same message in a cycle.

During mapping, the functional model is mapped onto the architecture platform, as shown in Fig. 2.5. Specifically, the tasks are allocated to ECUs, and the signals are one-to-one mapped onto messages and transmitted on the network. Messages are triggered periodically and each message contains the latest values of the signals that are mapped to the message. Static priorities are assigned to tasks for priority-based scheduling, and the time slots in the schedule are assigned to messages. The design space of task allocation, priority assignment, and switch scheduling is explored with respect to a set of design objectives and constraints.

For detailed problem formulation and its corresponding algorithm, please refer to our previous publications [12, 14] that address security in the mapping process.

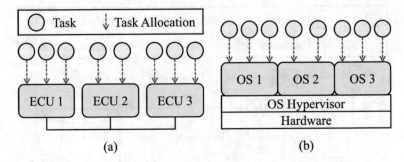

Fig. 2.6 (**a**) A traditional architecture, and (**b**) an architecture supported by an OS hypervisor

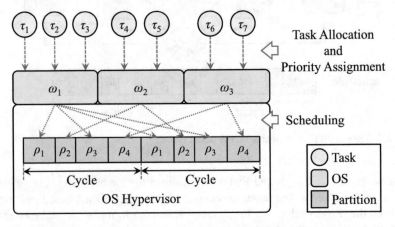

Fig. 2.7 The tasks are allocated to the operating systems, and the operating systems are scheduled on the OS hypervisor

2.4.3 Advanced Architecture: OS-Hypervisor-Based Systems

In this section, we consider mapping onto platforms with OS hypervisor. In Fig. 2.6a, there is a traditional architecture where the tasks are allocated directly on the ECUs. In Fig. 2.6b, an OS hypervisor runs between hardware and operating systems and virtualizes hardware. As a result, tasks and operating systems can be executed in a hardware-independent way. The OS hypervisor in Fig. 2.6b is categorized as a type-1 OS hypervisor which runs directly on hardware, while a type-2 OS hypervisor runs on a host operating system and supports other guest operating systems. In the market, there have been several OS hypervisors available. Although they have different features and specific applications (not only for automotive systems), the fundamental goal is still to virtualize hardware and provide high flexibility and isolation.

As shown in Fig. 2.7, the system model consists of a set of tasks, a set of operating systems, and an OS hypervisor. Each task τ_i is triggered periodically. We assume that all operating systems are identical, and each operating system ω_i supports the preemptive fixed-priority scheduling. The OS hypervisor supports the TDMA scheduling and maintains a *schedule* in which each *partition* ρ_i is assigned to one operating system,[1] and the OS hypervisor repeats the schedule after each *cycle*. It is possible that there is more than one partition assigned to the same operating system in a cycle.

The research on developing the mapping algorithm for this model is still ongoing.

2.4.4   Heterogeneous Communication Architectures

There are still some limitations with those approaches above. First, there is usually only one protocol to be considered, so the design methods cannot be applied to heterogeneous communication architectures. Next, the designs are for conventional functions which do not have very high data rates, and thus they cannot support ADAS and autonomous functions. Lastly, the architectures are usually fixed so that system designers have no flexibility to select appropriate hardware devices and design a topology for them. To address these problems and the challenges in Sect. 2.1, in this section, we propose a design methodology based on the PBD paradigm for heterogeneous communication architectures in automotive systems.

The design methodology is based on the mapping from functional models to architectural models. The notations which will be used in the methodology are listed in Table 2.1. We first define a device and an architectural model as follows:

Definition 2.1 A device δ is either a sensor, an actuator, an ECU, or a network device.

The location of a device is usually fixed according to the floor planning of an automotive system. A network device can be a CAN bus, a TSN switch, or a gateway.

Definition 2.2 An architectural model Δ is a set of devices.

An architectural model can be given by system designers directly or extracted from standardized languages. Each device in an architectural model is only a *candidate*, which means that it is possibly not selected during the mapping.

Definition 2.3 For each device δ, it is associated with a parameter C_δ as the device cost of δ. For each pair of devices δ and δ', it is with a parameter $D_{\delta,\delta'}$ as the connection cost of δ and δ' and another parameter $E_{\delta,\delta'}$ as the compatibility of δ and δ'.

---

[1]Some existing OS hypervisors allow one partition to be assigned to more than one operating system, and those operating systems are scheduled by their priorities. This can be generalized to the system model by defining task priority as a 2-tuple.

Table 2.1 Notations in the design methodology for heterogeneous communication architectures

δ	A device
C_δ	The device cost of δ
$D_{\delta,\delta'}$	The connection cost of δ and δ'
$E_{\delta,\delta'}$	The compatibility of δ and δ'
ι	An implementation
S_ι	The set of devices of ι
T_ι	The set of logical connections between devices of ι
U_ι	The set of reliability and safety constraints on logical paths between devices of U_ι
Δ	An architectural model or a set of devices
I	A functional model or a set of implementations
n	The number of functional models
σ	A sensor
π	An actuator
θ	An ECU
ϕ	A network device
Σ	The set of sensors
Π	The set of actuators
Θ	The set of ECUs
Φ	The set of network devices

The parameter $D_{\delta,\delta'}$ can be pre-computed based on the harness and routing graph in an automotive system, and it can be set as the distance or the wiring weight between δ and δ' which are physically connected. If $D_{\delta,\delta'} = \infty$, it means that there is no physical connection between δ and δ'. On the other hand, $E_{\delta,\delta'} = 1$ if and only if δ and δ' can be selected at the same time. The existence of the parameter $E_{\delta,\delta'}$ is to address the challenge of device selection mentioned in Sect. 2.1, e.g., if both of a regular ECU and an upgraded ECU are the candidates at the same location, only one of them can be selected.

As shown in Fig. 2.8a, the architectural model Δ has five devices including one sensor, one actuator, two ECUs, and one CAN bus. If δ_2 is a regular ECU, δ_3 is an upgraded ECU, and both of them are the candidates at the same location, then only one of them can be selected. Therefore, $E_{\delta_2,\delta_3} = 0$, while $E_{\delta_i,\delta_j} = 1$ for any other pair of devices. On the other hand, all devices except the CAN bus are only connected to the CAN bus, so $D_{\delta_i,\delta_j} = \infty$ for any pair of devices where $i, j \in \{1, 2, 3, 4\}$.

Definition 2.4 Given Δ, Σ is the set of sensors, Π is the set of actuators, Θ is the set of ECUs, Φ is the set of network devices, and thus $\Delta = \Sigma \cup \Pi \cup \Theta \cup \Phi$. Throughout the section, σ is a sensor, π is an actuator, θ is an ECU, ϕ is a network device,

Then, we define an implementation and a functional model as follows:

Definition 2.5 An implementation ι is associated with S_ι as the set of devices, T_ι as the set of logical connections between devices, and U_ι as the set of reliability and safety constraints on logical paths between devices.

Note that T_ι can be represented by a set of subsets in S_ι, and U_ι can be represented by a set of reliability and safety constraints on tuples of elements in S_ι.

Definition 2.6 A functional model I is a set of implementations, and it can be implemented by any $\iota \in I$.

We define a functional model by its possible implementations on devices because we can translate system designers' experience into candidate implementations and significantly reduce the complexity and search space during design space exploration (e.g., we can keep the scenario that both ECUs need to be upgraded at the same time).

It should be mentioned that, in most cases, a functional model has the same sensors and actuators in all of its implementations, e.g., the sensors and actuators that the functions of a blind spot monitor use are fixed. On the other hand, there is usually some flexibility selecting ECUs to execute corresponding functions, no matter they are at the same location or at different locations, so a functional model usually has different sets of ECUs in its implementations. Lastly, there is usually no network device in an implementation, although it may be implied by the harness and routing graph or objective optimization and constraint satisfaction during mapping.

As shown in Fig. 2.8b, the functional model is $\{\iota_1, \iota_2\}$. For ι_1, $S_{\iota_1} = \{\delta_1, \delta_2, \delta_4\}$, $T_{\iota_1} = \{\{\delta_1, \delta_2\}, \{\delta_2, \delta_4\}\}$, and U_{ι_1} consists of the constraints on path $(\delta_1, \delta_2, \delta_4)$, e.g., its end-to-end latency of the functional path of ι_1 must be smaller than its deadline. Similarly, for ι_2, $S_{\iota_2} = \{\delta_1, \delta_3, \delta_4\}$, $T_{\iota_2} = \{\{\delta_1, \delta_3\}, \{\delta_3, \delta_4\}\}$, and U_{ι_2} consists of the constraints on path $(\delta_1, \delta_3, \delta_4)$.

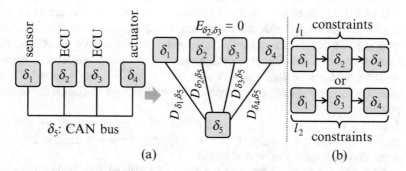

(a)   (b)

Fig. 2.8 (a) An architectural model with one sensor, one actuator, two ECUs, and one CAN bus, where the two ECUs are incompatible, and only one of them can be selected. (b) Two implementations of a functional model. One of them should be selected, depending on objective optimization and constraint satisfaction during mapping

With the definitions of an architectural model and a functional model, the design problem can be defined as follows:

- Given an architectural model Δ and n function model $\{I_1, I_2, \ldots, I_n\}$, select an implementation for each functional model such that all devices of selected implementations are compatible, all reliability and safety constraints of selected implementations are satisfied, and the objective is optimized.

As mentioned in Sect. 2.1, a reliability constraint can be the requirement of multiple routing paths. A safety constraint can be the utilization bound of each device or the requirement that the end-to-end latency of a functional path must be smaller than its deadline. The most typical objective is to minimize total cost which includes all device costs and all connection costs. Some other possible objectives are weight minimization, latency minimization, and performance maximization. As shown in Fig. 2.8, one implementation in Fig. 2.8b should be selected to implement the functional model, depending on objective optimization and constraint satisfaction during mapping.

Here is the summary of how the methodology addresses the design challenges mentioned in Sect. 2.1.

- **Addressing high-volume and dynamic input data**. By objective optimization and constraint satisfaction during mapping, a functional model with high data rate will be served by faster network devices (protocols) after mapping. If nearby ECUs are not powerful enough for dynamic data rate, a functional model will connect its sensors or actuators to further ECUs, and related objectives (e.g., connection cost) and constraints (e.g., end-to-end latency) will also be considered during mapping.
- **Computation architecture design**. In the methodology, different types of ECUs at the same location are all included in an architectural model and marked by the compatibility ($E_{\delta,\delta'}$). As mentioned above, this allows us to translate system designers' experience into candidate implementations and keep the scenario that both ECUs need to be upgraded at the same time. Then, the challenges in Sect. 2.1 can be addressed by objective optimization and constraint satisfaction during mapping. If a device has no load after mapping, it means that it is not selected.
- **Communication architecture design**. Similar to device selection, a network device may have no load on it, which means that it is not selected. On the other hand, gateways are considered in the methodology to composite different protocols.
- **Topology design**. In an architectural model, all possible connections and their costs ($D_{\delta,\delta'}$) are pre-computed. During mapping, those connections are candidates, and their costs can be considered.
- **Safety**. The end-to-end latency of a path can be defined with U_l in a functional model and its implementations. Note that a frame is a special case of a path between two devices. The methodology leaves flexibility for system designers to apply different timing models. If those models are not available or their results are over-pessimistic so that simple bounds on the utilization of network devices

are adopted as safety constraints, the utilization of a device, which is a special case of a path with only one device, can also be defined with U_l.

- **Reliability**. Similarly, all possible connections are candidates so that a function model can construct multiple routing paths from them.
- **Security**. Although security is not the focus of the methodology, other protocols and gateways in heterogeneous communication architecture can provide opportunities for adding security protections. The methodology is a platform for further security considerations during design stages.
- **Optimization objective**. The methodology leaves flexibility for system designers to set total cost, wiring weight, reliability, or performance as their objectives. To deal with different objectives, generalized optimization approaches should be applied.

Heterogeneous communication architectures are expected to be deployed to support ADAS and autonomous functions. In this section, we propose a methodology to address those challenges on heterogeneous communication architectures. Based on the methodology, we can formulate a problem and its corresponding algorithm to solve mapping problems at this level. The corresponding research is still ongoing.

2.5 Mapping Hardware Components to Physical Layouts

Finally, we can apply the PBD paradigm to map hardware components to physical layouts. Hardware components typically have pre-defined places for them. For example, radars should be placed at the front or rear side of a vehicle, not inside the vehicle. These components are connected by wires, which need to go through harnesses as shown in Fig. 2.9. The mapping problem from hardware to physical layouts can be captured as follows and illustrated in Fig. 2.10.

- **Platforms**: (1) The higher-layer platform includes a set of logical connections between hardware components; and (2) the lower-layer platform includes a physical routing graph that consists of wiring harnesses, connections between harnesses, locations (where a wire gets in or out of a harness) of wire harnesses, and hardware components.
- **Design Space**: The design variables to be explored include placement of splices, physical routing paths, and wire sizes.
- **Design Objectives and Constraints**: The metrics to be considered include total wiring length, total wiring weight, fuel efficiency, resistance, signal quality, space, and capacities of locations.

One problem formulation and its corresponding algorithm have been proposed in [11]. The features of the problem are:

- A logical connection can be defined as a hypergraph, i.e., a connection (hyperedge) can connect more than two components (vertices). To physically connect those components, we need to add splices physically (Steiner vertices logically),

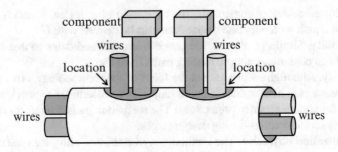

Fig. 2.9 A harness model with its locations [11]. Two components are connected by a wire, and the wire goes through the harness

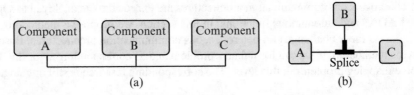

Fig. 2.10 (**a**) The logical connection between three hardware components is mapped to (**b**) the physical routing graph including a splice

which are similar to switches in network routing. A Steiner-tree problem for wire routing is also common in electronic design automation.

- The placement of harnesses is fixed. As a result, the problem is to select routing paths upon the given harnesses, and thus the number of potential routing paths is limited. From this perspective, the problem is closer to network routing rather than wire routing in electronic design automation.
- Similarly, a splice can only be placed at a location of a harness, so the number of potential locations is also limited.
- Considering resistance, the total wiring length and the total wiring weight have a quadratic relation because to maintain the same resistance for a wire, its length and the area of its cross section need to increase or decrease linearly. The total wiring weight is relevant to fuel efficiency as it is up to 30 kg in modern vehicles.

Please refer to [11] for detailed problem formulation and its corresponding algorithm.

2.6 Summary

In this chapter, we introduced the platform-based design (PBD) paradigm for automotive and transportation systems, and the application of PBD to map the high-level specification of connected vehicle applications to individual vehicle functionality, and then to software and hardware implementations, and finally to physical layouts. We believe that the PBD paradigm is a promising methodology to address the rapidly growing complexity of automotive design and improve design quality and productivity.

Acknowledgements We gratefully acknowledge the support from the National Science Foundation of the United States under Awards 1834324, 1834701, and 1839511, the Ministry of Education in Taiwan under Grant Number NTU-107V0901, and the Ministry of Science and Technology in Taiwan under Grant Number MOST-108-2636-E-002-011.

References

1. AUTOSAR. http://www.autosar.org
2. Robert Bosch GmbH. (1991). CAN specification (Version 2.0).
3. Davare, A., Zhu, Q., Di Natale, M., Pinello, C., Kanajan, S., & Sangiovanni-Vincentelli, A. (2007, June). Period optimization for hard real-time distributed automotive systems. In *Design Automation Conference (DAC'07)*.
4. Deng, P., Cremona, F., Zhu, Q., Di Natale, M., & Zeng, H. (2015, April). A model-based synthesis flow for automotive CPS. In *2015 ACM/IEEE International Conference on Cyber-Physical Systems (ICCPS)* (pp. 198–207).
5. Deng, P., Zhu, Q., Davare, A., Mourikis, A., Liu, X., & Natale, M. D. (2016, December). An efficient control-driven period optimization algorithm for distributed real-time systems. *IEEE Transactions on Computers, 65*(12), 3552–3566.
6. Deng, P., Zhu, Q., Di Natale, M., & Zeng, H. (2014, June). Task Synthesis for latency-sensitive synchronous block diagram. In *2014 9th IEEE International Symposium on Industrial Embedded Systems (SIES)* (pp. 112–121).
7. FlexRay Consortium. (2010, October). FlexRay communications system protocol specification (Version 3.0.1).
8. IEEE. (2011, March). IEEE standard for local and metropolitan area networks — timing and synchronization for time-sensitive applications in bridged local area networks. In *IEEE Std 802.1AS-2011* (pp. 1–292).
9. Liang, H., Jagielski, M., Zheng, B., Lin, C., Kang, E., Shiraishi, S., et al. (2018, November). Network and system level security in connected vehicle applications. In *2018 IEEE/ACM International Conference on Computer-Aided Design (ICCAD)*.
10. Liang, H., Wang, Z., Zheng, B., & Zhu, Q. (2017, November). Addressing extensibility and fault tolerance in can-based automotive systems. In *2017 IEEE/ACM International Symposium on Networks-on-Chip (NOCS)*.
11. Lin, C.-W., Rao, L., Giusto, P., D'Ambrosio, J., & Sangiovanni-Vincentelli, A. (2015, November). Efficient wire routing and wire sizing for weight minimization of automotive systems. *IEEE Transactions on Computer-Aided Design of Integrated Circuits and Systems, 34*(11), 1730–1741.
12. Lin, C.-W., Zheng, B., Zhu, Q., & Sangiovanni-Vincentelli, A. (2015, December). Security-aware design methodology and optimization for automotive systems. *ACM Transactions on Design Automation of Electronic Systems, 21*(1), 18:1–18:26.
13. Lin, C.-W., Zhu, Q., Phung, C., & Sangiovanni-Vincentelli, A. (2013). Security-aware mapping for CAN-based real-time distributed automotive systems. In *2013 IEEE/ACM International Conference on Computer-Aided Design* (pp. 115–121)
14. Lin, C.-W., Zhu, Q., & Sangiovanni-Vincentelli, A. (2014, November). Security-aware mapping for TDMA-based real-time distributed systems. In *2014 IEEE/ACM International Conference on Computer-Aided Design* (pp. 24–31)

15. Lin, C.-W., Zhu, Q., & Sangiovanni-Vincentelli, A. (2015, March). Security-aware modeling and efficient mapping for CAN-based real-time distributed automotive systems. *IEEE Embedded Systems Letters, 7*(1), 11–14.

16. SAE. (2011, November). Time-triggered ethernet. *SAE Standard AS6802*.

17. SAE. (2011, February). TTP communication protocol. *SAE Standard AS6003*.

18. SUMO. (2017). http://www.dlr.de/ts/en/desktopdefault.aspx/tabid-9883/16931_read-41000/

19. UPPAAL. (2017). https://www.uppaal.org/

20. Zheng, B., Gao, Y., Zhu, Q., & Gupta, S. (2015, October). Analysis and optimization of soft error tolerance strategies for real-time systems. In *2015 International Conference on Hardware/Software Codesign and System Synthesis (CODES+ISSS)* (pp. 55–64).

21. Zheng, B., Lin, C. W., Liang, H., Shiraishi, S., Li, W., & Zhu, Q. (2017, May). Delay-aware design, analysis and verification of intelligent intersection management. In *2017 IEEE International Conference on Smart Computing (SMARTCOMP)* (pp. 1–8).

22. Zheng, B., Lin, C.-W., Yu, H., Liang, H., & Zhu, Q. (2016, November). CONVINCE: A cross-layer modeling, exploration and validation framework for next-generation connected vehicles. In *2016 IEEE/ACM International Conference on Computer-Aided Design (ICCAD)*.

23. Zheng, B., Sayin, M. O., Lin, C. W., Shiraishi, S., & Zhu, Q. (2017, November). Timing and security analysis of VANET-based intelligent transportation systems: (invited paper). In *2017 IEEE/ACM International Conference on Computer-Aided Design (ICCAD)* (pp. 984–991).

24. Zheng, W., Zhu, Q., Natale, M. D., & Sangiovanni-Vincentelli, A. (2007). Definition of task allocation and priority assignment in hard real-time distributed systems. In *RTSS '07: Proceedings of the 28th IEEE International Real-Time Systems Symposium* (pp. 161–170)

25. Zhu, Q., Deng, P., Di Natale, M., & Zeng, H. (2013, March). Robust and extensible task implementations of synchronous finite state machines. In *Design, Automation Test in Europe Conference Exhibition (DATE), 2013* (pp. 1319–1324)

26. Zhu, Q., Liang, H., Zhang, L., Roy, D., Li, W., & Chakraborty, S. (2017, June). Extensibility-driven automotive in-vehicle architecture design. In *2017 54th ACM/EDAC/IEEE Design Automation Conference (DAC)* (pp. 1–6)

27. Zhu, Q., & Sangiovanni-Vincentelli, A. (2018, Sept). Codesign methodologies and tools for cyber–physical systems. *Proceedings of the IEEE, 106*(9), 1484–1500.

28. Zhu, Q., Yang, Y., Natale, M. D., Scholte, E., & Sangiovanni-Vincentelli, A. (2010). Optimizing the software architecture for extensibility in hard real-time distributed systems. *IEEE Transactions on Industrial Informatics, 6*(4):621–636.

29. Zhu, Q., Yang, Y., Scholte, E., Natale, M. D., & Sangiovanni-Vincentelli, A. (2009). Optimizing extensibility in hard real-time distributed systems. In *RTAS '09: Proceedings of the 2009 15th IEEE Real-Time and Embedded Technology and Applications Symposium* (pp. 275–284).

30. Zhu, Q., Zeng, H., Zheng, W., Di Natale, M., & Sangiovanni-Vincentelli, A. (2012). Optimization of task allocation and priority assignment in hard real-time distributed systems. *ACM Transactions on Embedded Computing Systems, 11*(4), 85:1–85:30.

Chapter 3
An Hourglass-Shaped Architecture for Model-Based Development of Networked Cyber-Physical Systems

Muhammad Umer Tariq and Marilyn Wolf

3.1 Introduction

Many technological achievements have been enabled by the field of *feedback control systems*, which deals with the process of controlling a physical system through a feedback controller. If the feedback controller is implemented as a *real-time computer system*, the resulting configuration of the feedback control system is referred to as *embedded control system*. Some prime examples of embedded control systems are automotive systems, avionics systems, and smart grid. The typical development process of an embedded control system can be partitioned into two distinct stages: controller design and controller implementation. During the controller design stage, a control systems engineer models the physical plant, derives the feedback control law, and validates the controller design through mathematical analysis and simulation. During the controller implementation stage, a computer systems engineer implements the feedback controller as a real-time computer system.

The field of *embedded control systems* brings together the fields of *control theory* and *real-time computer systems*. However, as noted in [15], the fields of *control theory* and *real-time computer systems* typically employ two completely different types of models: analytical models and computational models. As a result, two vastly different design processes are currently popular for the two stages of embedded control system development process: feedback controller design and feedback controller implementation as real-time computer system. Due

M. U. Tariq (✉)
ProsumerGrid, Inc., Atlanta, Georgia
e-mail: mumertariq@prosumergrid.com

M. Wolf
Georgia Institute of Technology, Atlanta, Georgia
e-mail: marilyn.wolf@ece.gatech.edu

© Springer Nature Switzerland AG 2019
M. A. Al Faruque, A. Canedo (eds.), *Design Automation of Cyber-Physical Systems*, https://doi.org/10.1007/978-3-030-13050-3_3

to the inherent differences between the abovementioned two stages, currently popular development methodologies for embedded control systems support very few correct-by-construction properties and depend heavily on testing the final implementation for creating confidence in the correct operation of an embedded control system under various runtime operating conditions. Therefore, current development techniques for embedded control systems are not capable of efficiently handling the ever-increasing complexity of these systems.

These limitations of the traditional embedded control system development techniques have created interest in taking a fresh look at the abstractions used in the traditional embedded control systems development process, resulting in a new field, *cyber-physical systems* (CPS) [39, 40]. The aim of CPS research is to develop an integrated theory as well as an integrated development toolset for controller design and controller implementation phases of the embedded control system development process. The hope is that this CPS research will enable the cost-effective development and maintenance of more complex versions of embedded control systems.

Recent CPS research efforts can be divided into two major categories: *platform-imperfection-aware feedback controller design* and *CPS-friendly computing platform design*. Under the category of *platform-imperfection-aware feedback controller design*, theoretical developments from the fields of hybrid systems [3], switched systems [21], time-delay systems [7], networked control systems [41], multi-agent networked systems [29], and game theory [16] are leveraged to develop a feedback controller design that takes into account the imperfections of the runtime computing platform (such as communication delays or failures caused by communication network congestion or cyber security attacks) at the design time [37]. The resulting "platform-imperfection-aware" feedback controller is either robust against the imperfections of runtime computing platform or possesses the capability to switch between different *control modes* to overcome the imperfections of runtime computing platform. Under the category of *CPS-friendly computing platform design*, CPS research has focused on specialized runtime computing platforms that have more predictable timing performance or provide correct-by-construction composition of software components. Some examples of this approach are provided in [17, 19, 22].

Model-based development (or model-driven development) of cyber-physical systems has the potential to bind the abovementioned CPS research efforts into an integrated, cross-layer CPS development methodology. In model-based development paradigm, high-level or platform-independent models (PIM) are transformed into lower-level or platform-specific models (PSM) through the process of model transformation. Both high-level and lower-level models are described using their own domain-specific modeling languages (DSMLs) [32]. In this chapter, we propose an approach to model-based development of networked cyber-physical systems (CPS) that is centered on the notion of a standardized design specification language. The proposed design specification language can be used to build a CPS design specification model that can serve as a CPS-aware interface between control systems engineer and embedded systems engineer.

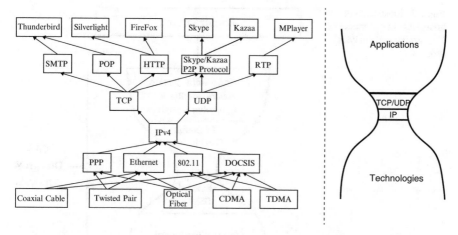

Fig. 3.1 Illustration of hourglass-shaped architecture of Internet; adapted from [2]

The proposed approach is inspired by the hourglass-shaped architecture of Internet, illustrated in Fig. 3.1. The narrow waist of hourglass-shaped architecture suggests that there is less diversity of protocols at this layer of Internet [2]. Any application that can operate based on the services of IP layer can be deployed on the Internet, and any underlying technology that can transport bytes from one point to another according to IP services can be used in the Internet. Similarly, according to the proposed approach to the model-based development of networked CPS (Fig. 3.2), a wide range of DSMLs (and associated analysis tools) can be utilized to develop a platform-imperfection-aware feedback controller design, which is then specified using a standardized CPS design specification language. The proposed feedback controller design can then be analyzed for mapping on to wide range of runtime CPS computing platforms by utilizing their corresponding DSMLs (and associated analysis tools). This approach can support the goals of an integrated CPS theory and development methodology while still taking into account the differences between the domain-specific skillset that control systems engineers and embedded system engineers typically possess.

The rest of the chapter is organized as follows. In Sect. 3.2, we present some related work. In Sect. 3.3, we present the details of the proposed hourglass-shaped architecture for model-based development of networked cyber-physical systems. In Sect. 3.4, we document a number of requirements that any standardized CPS design specification language must satisfy. In Sect. 3.5, we present the overview of a proposed CPS design specification language. In Sects. 3.6–3.8, we discuss the concrete syntax, abstract syntax, and semantics of the proposed CPS design specification language, respectively. In Sect. 3.9, we present the conclusion.

Fig. 3.2 Illustration of
hourglass-shaped model of
CPS design and analysis
process

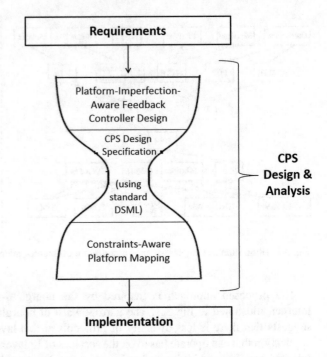

3.2   Related Work

Figure 3.3 presents a summary of specification languages and analysis tools
used in the different stages of a typical embedded control system development
process. Simulink [27] (combined with auxiliary tools such as Stateflow [28] and
Simscape [26]) has become a de facto standard in the field of embedded control
systems for specification and refinement (through simulation) of the feedback
controller design, developed by a control engineer through the application of
various analytical controller design strategies available in the literature for the field
of *control theory* [5]. Once a feedback controller design has shown acceptable
performance in the Simulink-based simulation environment, a computer system
engineer takes on the task of implementing this feedback controller design as
a *real-time computer system*. Various tools have been developed over the years
to help a computer systems engineer in this process of converting a feedback
controller design from a Simulink-based specification to a real-time computer
system implementation. Specialized modeling languages, such as UML (combined
with MARTE profile) [30], SysML [10], and AADL [8], help in the process of
designing the system and software architecture of the required real-time computer
system. Specialized programming languages, such as Lustre [12], Esterel [4],
Signal [20], and Giotto [14], help in the development of real-time computer system
whose timing performance can be formally guaranteed.

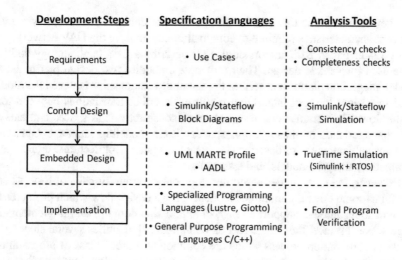

Development Steps	**Specification Languages**	**Analysis Tools**
Requirements	• Use Cases	• Consistency checks • Completeness checks
Control Design	• Simulink/Stateflow Block Diagrams	• Simulink/Stateflow Simulation
Embedded Design	• UML MARTE Profile • AADL	• TrueTime Simulation (Simulink + RTOS)
Implementation	• Specialized Programming Languages (Lustre, Giotto) • General Purpose Programming Languages C/C++)	• Formal Program Verification

Fig. 3.3 Embedded control systems: development steps, specification languages, and analysis tools; adapted from [36]

Model-based development (MBD) paradigm has also been successfully employed in the domain of embedded control system in order to improve the productivity of a computer systems engineer during the process of conversion of a feedback controller design into a real-time computer system. In MBD paradigm, high-level or platform-independent models (PIM) are transformed into lower-level or platform-specific models (PSM) through the process of model transformation. Both high-level and lower-level models are described using their own domain-specific modeling languages (DSMLs) [32]. A DSML is first defined through a meta-modeling step. A meta-model of a DSML defines the basic constructs (along with their relationships and constraints) that can be used in a DSML. Model transformation step of MBD paradigm uses the meta-models of DSMLs to define transformation rules from higher-level (platform-independent) models to lower-level (platform-specific) models. Model-driven architecture (MDA) [9], model integrated computing (MIC) [18], and eclipse modeling framework (EMF) [11, 33] initiatives represent three popular MBD efforts.

In the domain of embedded control systems, various model transformation (code generation) tools have been developed to automatically generate executable code from Simulink models for various real-time computing platforms. Embedded Coder [25], from Mathworks, Inc., is a commercially available example of such a code generation tool. Another example of a Simulink-based MBD toolset for a more specialized real-time computing platform has been reported in [6].

Building on the MBD paradigm, Sztipanovits et al. [35] describe a methodology for cyber-physical system integration and illustrate their methods on the design of a network of quadrotor UAVs. They identify three design layers: physical, platform, and computation/communication. Their methodology emphasizes component-based design and its associated requirement, compositionality. They identify passivity as a key characteristic that enables composition of control systems. They identify network characteristics required to compositionally analyze the UAV network.

In a later paper, Sztipanovits et al. [34] describe a CPS methodology and tool suite used for vehicle design. Their tool suite embodies two design platforms: the model integration platform describes the semantic relationships between the models used in design; the tool integration platform describes translations between tools in the flow. Their framework allows them to construct design spaces and analyze the characteristics of those design spaces. Their modeling language CyPhyML includes sublanguages to describe components, system architectures, architectural parameters, analysis models, and testbenches.

However, the CPS model-based development community has not been as successful as some other communities in identifying a design flow which promotes the reuse of tools and can support a range of application domains and implementation targets. For instance, the classic text on compilers [1] identifies several steps in the classical compilation process which are common to a broad class of programming languages: lexical analysis, syntactic analysis, semantic analysis, intermediate code generation, code optimization, and code generation. In this classical compilation process, the intermediate code (developed in an intermediate language such as three-address code) plays a pivotal role by providing an independent narrow interface between a set of source code languages and a set of target machines. Similarly, as illustrated in Fig. 3.1 and detailed in [2], the IP layer can be considered the narrow waist of an hourglass-shaped architecture of Internet. Any application that can operate based on the services of IP layer can be deployed on the Internet, and any underlying technology that can transport bytes from one point to another according to IP services can be used in the Internet.

While model-based development of networked cyber-physical systems is a challenging problem, we believe that abovementioned observations from the domains of software compilation and Internet architecture can be leveraged to improve the model-based development process for networked cyber-physical systems. Therefore, in this chapter, we propose an approach to model-based development of networked cyber-physical systems (CPS) that is centered on the notion of a standardized CPS design specification language, capable of playing an analogous role to the intermediate language and the IP layer from the domains of software compilation and Internet architecture.

3.3 Hourglass-Shaped Architecture for Model-Based CPS Development

Two major categories of CPS research are *platform-imperfection-aware feedback controller design* and *CPS-friendly computing platform design*. Model-based development of cyber-physical systems has the potential to bind the abovementioned CPS research efforts into an integrated, cross-layer CPS development methodology.

This section presents an approach to model-based development of networked cyber-physical systems (CPS) that is centered on the notion of a standardized design specification language. The proposed design specification language can be used to build a CPS design specification model that can serve as a CPS-aware interface between control systems engineer and embedded systems engineer. The proposed approach is inspired by the hourglass-shaped architecture of Internet, illustrated in Fig. 3.1. The narrow waist of hourglass-shaped architecture suggests that there is less diversity of protocols at this middle layer of Internet [2], while many different protocols can be employed at top and bottom layers of Internet.

According to the proposed hourglass-shaped architecture for model-based networked CPS development, illustrated in Fig. 3.2, a wide range of DSMLs (and associated analysis tools) can be utilized to develop a platform-imperfection-aware feedback controller design, which is then specified using a standardized DSML for CPS design specification. Furthermore, according to the proposed hourglass-shaped architecture, the platform-imperfection-aware feedback controller design (specified using the standardized DSML) can then be analyzed for mapping on to various runtime CPS computing platforms by utilizing corresponding DSMLs (and associated analysis tools).

The proposed hourglass-shaped architecture can enable effective coordination between control systems engineer and embedded systems engineer during model-based development of networked cyber-physical system, while still allowing them to concentrate and specialize in the CPS-aware, model-based tools developed in their respective domains. This approach can support the goals of an integrated CPS theory and development methodology while taking into account the differences between the domain-specific skillset that control systems engineer and embedded system engineer must acquire during their respective academic training.

The proposed hourglass-shaped architecture for model-based development of networked CPS consists of three explicit phases: (1) platform-imperfection-aware feedback controller design, (2) CPS design specification, and (3) constraints-aware platform mapping.

3.3.1 Platform-Imperfection-Aware Feedback Controller Design

In this phase, control systems engineer designs a feedback controller that takes into account the imperfections of the runtime computing platform (such as communication delays or failures caused by communication network congestion) at the design time. The resulting "platform-imperfection-aware" feedback controller is either robust against the imperfections of runtime computing platform or possesses the capability to switch between different *control modes* to overcome the imperfections of runtime computing platform. In this phase, control systems engineer utilizes various results from CPS research [37] that have been achieved over the recent years by leveraging the theoretical advances from the fields of hybrid systems [3], switched systems [21], time-delay systems [7], networked control systems [41], multi-agent networked systems [29], and game theory [16].

During this phase, a control systems engineer can utilize any model-based tool from the following three categories: (a) various DSMLs (and associated analysis tools) that were used in the traditional control system design process [26–28], (b) recently proposed DSMLs (and associated analysis tools) that are employed by the numerous cyber-physical co-design CPS research efforts [13, 31], and (c) any DSMLs (and associated analysis tools) that are proposed by any future CPS research into integrated cyber-physical design.

3.3.2 CPS Design Specification

In this phase, the results of the platform-imperfection-aware feedback controller design process are captured using a standardized DSML for CPS design specification. This CPS design specification must capture the sensed and actuated-upon physical plant parameters as well as the networked controller aspects of a CPS design. However, the networked controller aspects of CPS design should not be described by specifying the runtime computing infrastructure, instead networked controller aspects of CPS design should be described at an abstract level by specifying various *control nodes* and *sensor ports*, *actuator ports*, *input message ports*, and *output message ports* associated with these *control nodes*.

This CPS design specification must also capture the feedback control adaptation strategy to handle the imperfect performance of runtime computing and communication platform. This element of CPS design can also be captured at an abstract level by specifying various *controller modes* of a *control node* and a mode switching logic based on QoS violations associated with *sensor ports*, *actuator ports*, *input message ports*, and *output message ports* of the *control node*. A CPS design specification can also declare some QoS constraints of *sensor ports*, *actuator ports*, *input message ports*, and *output message ports* to be *hard*. This will indicate that these QoS properties must be satisfied by runtime computing platform, because there is no safe backup mode of operation in case of violation of these QoS properties.

3.3.3 Constraints-Aware Platform Mapping

In this phase, the mapping of the CPS design specification (described using standardized DSML) onto various runtime computing platform is analyzed to either choose the most appropriate mapping or figure out the appropriate parameter settings for a runtime computing platform so that the platform can meet the QoS constraints of CPS design (and minimize the time that the system has to spend in a backup mode of operation). During this process, various model transformations can also be applied to translate the CPS design specification model into appropriate models that can be used as input for corresponding analysis tools (simulation or formal verification) associated with each of the candidate runtime computing platform technologies. Some specialized examples of these runtime computing platforms are Lustre [12], Esterel [4], Signal [20], and Giotto [14] with their own formal computing semantics. More traditional RTOS-based computing platforms can be captured and analyzed through UML (MARTE Profile) or AADL-based models and analysis tools [8, 30].

3.4   Requirements for Standardized CPS Design Specification Language

Following are some of the major requirements that a CPS design specification language (CPS-DSL) must meet:

3.4.1   Physical Plant Parameter Specification

A CPS-DSL must clearly identify the physical plant parameters that are sensed or actuated upon by the feedback controller.

3.4.2   Networked Controller Specification

An appropriate CPS-DSL must also describe the various elements of a networked controller design. These elements include topology of sensors, actuators, and control nodes, local control law for each control node, and information exchanged between different control nodes.

3.4.3   Specification of Controller Adaptation Strategies

For the emerging wide-area CPS application domains, such as smart grid, the performance of communication subsystem cannot be guaranteed. Therefore, CPS-DSL must also define the timing constraints on the information exchange among different control nodes and the control adaptation strategies in case of violation of these timing constraints.

3.4.4 Interface Between Control Systems Engineer and Real-Time Computer Systems Engineer

A CPS design specification captures the output of platform-imperfection-aware feedback controller design process, and it also serves as input to the process of developing a functionally equivalent embedded implementation of the feedback controller design. Therefore, the CPS-DSL should be designed in such a way that it can serve as an effective communication interface between control systems engineer and real-time computer systems engineer.

3.4.5 Formal Semantics

A CPS design specification language must support formal semantics. The existence of formal semantics of a CPS design specification language (CPS-DSL) opens up the possibility to prove formal equivalence properties between a CPS-DSL-based CPS design specification and the corresponding CPS deployment on a computing platform.

3.5 A Proposed CPS Design Specification Language: Overview

This section presents the summary of a proposed CPS-DSL that can meet the requirements identified in Sect. 3.4. Various aspects (such as concrete syntax, abstract syntax, and semantics) of the definition of proposed CPS-DSL are described in detail in Sects. 3.6–3.8.

The individual language elements of the proposed CPS-DSL can be divided into three categories: *physical system elements*, *cyber system elements*, and *cyber-physical interface elements*. Table 3.1 provides a list of the language elements in each of the abovementioned three categories.

Table 3.1 Language elements of the proposed CPS-DSL

Category	Language elements
Physical system elements	CompoundPhysicalPlant, PhysicalSystemParameter
Cyber-physical interface elements	Sensor, Actuator
Cyber system elements	ComputingNode, CommunicationNetwork, ControlApp, SensorPort, ActuatorPort, InputMsgPort, OutputMsgPort, Mode, ModeSwitchLogic, ControllerFunction, ControllerFunctionMemory, PeriodicControllerInput, PeriodicControllerOutput

3.5.1 Physical System Elements

CompoundPhysicalPlant and *PhysicalSystemParameter* elements belong to the category of *physical system elements*. *CompoundPhysicalPlant* element is used to represent the physical plant of a CPS. A *CompoundPhysicalPlant* element contains a set of *PhysicalSystemParameter* elements. *PhysicalSystemParameter* elements of the proposed CPS-DSL are used to identify the parameters of a physical plant that are to be sensed and actuated upon by the cyber subsystem of a CPS.

3.5.2 Cyber-Physical Interface Elements

Sensor and *Actuator* elements make up the category of *cyber-physical interface elements*. Cyber-physical interface of a CPS design is captured by a set of *Sensor* and *Actuator* elements. Each *Sensor* and *Actuator* element is associated with a corresponding *PhysicalSystemParameter* element.

3.5.3 Cyber System Elements

ComputingNode, CommunicationNetwork, ControlApp, SensorPort, ActuatorPort, InputMsgPort, OutputMsgPort, Mode, ModeSwitchLogic, ControllerFunction, ControllerFunctionMemory, PeriodicControllerInput, and *PeriodicController-Output* make up the category of *cyber system elements*. Cyber aspects of a CPS design include the topology of computing nodes, the controller application executing on each computing node, and the message exchange among computing nodes. The topology of controller computing nodes is captured by connecting a set of *ComputingNode* elements to a *CommunicationNetwork* element. Each *ComputingNode* element includes a *ControlApp* element and a set of *SensorPort, ActuatorPort, InputMsgPort,* and *OutputMsgPort* elements. *SensorPort, ActuatorPort,* and *ControlApp* elements combine to capture the local control application executing on a computing node.

    InputMsgPort and *OutputMsgPort* elements of proposed CPS-DSL are intended to capture the message exchange among computing nodes of a CPS. However, in a generic cyber-physical system, perfect behavior of communication subsystem cannot be guaranteed. As a result, a CPS design must specify the timing constraints on information exchange among computing nodes and different modes of operation for local feedback control law that are used in case of violation of these timing constraints. In the proposed CPS-DSL, *InputMsgPort* and *OutputMsgPort* elements capture the timing constraints on the information exchange among computing node.

    Each *ControlApp* element includes a *ModeSwitchLogic* element and a set of *Mode* elements to capture the different modes of operation of feedback control law for handling QoS fault scenarios. Each *Mode* element specifies the control action taken by the feedback controller in that mode of operation through a set of *ControllerFunction*, *PeriodicControllerInput*, and *PeriodicControllerOutput* elements.

3.6 Proposed CPS Design Specification Language: Concrete Syntax

Since Simulink [27] (combined with auxiliary Stateflow [28] and Simscape [26] blocks) has become a de facto standard in the domain of embedded control systems, concrete syntax of the proposed CPS-DSL has been implemented as an extension to standard blocks available in Simulink. In particular, a new Simulink library [36] has been developed that provides a Simulink block for each element of the proposed CPS-DSL, described in Sect. 3.5. Moreover, Simulink's mask interface capability has been used to provide each new Simulink block with a custom look, and a dialog box for entering element-specific parameters, such as the timing constraints associated with an *InputMsgPort* element.

Figure 3.4 shows a Simulink model that specifies a CPS design using the Simulink-based concrete syntax of the proposed CPS-DSL. Figure 3.5 shows the internal details of a *ComputingNode* block, which contains a *ControlApp* block and a set of *SensorPort*, *ActuatorPort*, *InputMsgPort*, and *OutputMsgPort* blocks. Figure 3.6 shows the internal details of *ControlApp* block, which consists of a set of *Mode* blocks and a *ModeSwitchLogic* block. Figure 3.7 shows the internal details of *Mode* block, which contains a set of *ControllerFunction*, *PeriodicControllerInput*, and *PeriodicControllerOutput* blocks. Figure 3.8 shows the internal details of *ControllerFuncton* block, which contains a description of feedback control law using standard Simulink computation blocks.

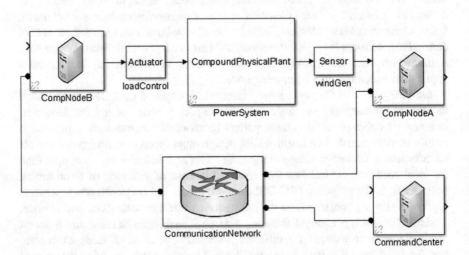

Fig. 3.4 A CPS design, specified as Simulink model with the proposed CPS-DSL

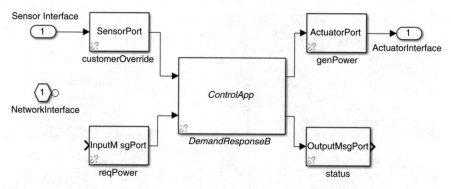

Fig. 3.5 Internal details of *ComputingNode* block, named CompNodeB, in Fig. 3.4

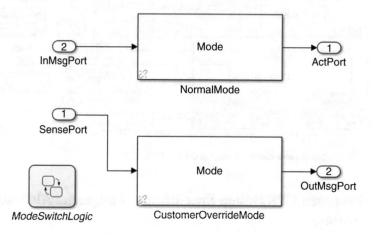

Fig. 3.6 Internal details of *ControlApp* block, named DemandResponseB, in Fig. 3.5

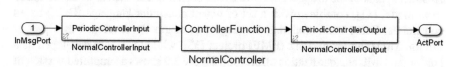

Fig. 3.7 Internal details of *Mode* block, named NormalMode, in Fig. 3.6

Fig. 3.8 Internal details of *ControllerFunction* block, named NormalControllerFunction, in Fig. 3.7

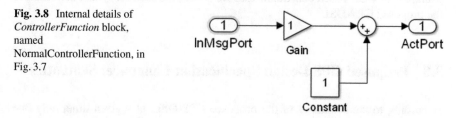

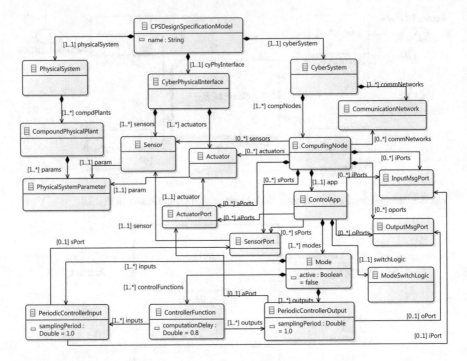

Fig. 3.9 Ecore-based meta-model of proposed CPS-DSL

3.7 Proposed CPS Design Specification Language: Abstract Syntax

Abstract syntax of the proposed CPS-DSL has been implemented as an Ecore-based meta-model [11], combined with a set of object constraint language (OCL)-based constraints. Ecore meta-modeling language was originally developed as a part of Eclipse Modeling Framework (EMF) project [33], while OCL was developed as a part of the UML standardization effort [38]. Figure 3.9 shows a simplified version of the Ecore-based meta-model for the proposed CPS-DSL. Table 3.2 provides some examples of OCL-based constraints that are part of the abstract syntax definition of the proposed CPS-DSL.

3.8 Proposed CPS Design Specification Language: Semantics

According to the semantics of the proposed CPS-DSL, at a given time, only one *Mode* element inside a *ControlApp* is active. *ModeSwitchLogic* element is evaluated at specific time instants, defined by the following two properties of the currently-active mode: mode period and switch frequency from active mode to mode j (the

Table 3.2 Abstract syntax definition of proposed CPS-DSL: examples of OCL-based constraints

context ControlApp
inv numOfSimultanoeousActiveModes:
modes— >select(active = true)— >size() = 1
context ControllerFunction
inv equalityOfSamplingPeriods:
inputs— >any(true).samplingPeriod = outputs— >any(true).samplingPeriod
context ControllerFunction
inv limitOnComputationDelay:
self.computationDelay < outputs— >any(true).samplingPeriod
context CPSDesignSpecificationModel
inv noUnusedSensor:
cyPhyInterface.sensors— >asSet() = cyberSystem.compNodes.sPorts.sensor— >asSet()
context CPSDesignSpecificationModel
inv noUnusedActuator:
cyPhyInterface.actuators— >asSet() = cyberSystem.compNodes.aPorts.actuator— >asSet()

number of equally-distant time instants in a single mode period at which the mode switch condition from active mode to mode j is evaluated).

As long as a certain *Mode* element is active, its constituent *PeriodicControllerInput* and *PeriodicControllerOutput* elements periodically sample the values at their inputs and store them at the output until the next sampling time instant. A *ControllerFunction* element contains the specification of feedback control law computation and is always sandwiched between a pair of *PeriodicControllerInput* and *PeriodicControllerOutput* elements with same sampling period T and synchronized sampling instants. The sampling period T, associated with a *ControllerFunction*, is defined in terms of the following two properties: mode period and controller function frequency (the number of equally-distant time instants in a single mode period at which the controller function is evaluated). Moreover, a *ControllerFunction* element takes time Δt to transfer any change in its input to its output where $0 < \Delta t < T$. A *ControllerFunction* element may also contain one or more *ControllerFunctionMemory* elements.

By design, the proposed CPS-DSL leaves its exact semantics dependent on the language used to define the control law computation inside a *ControllerFunction* element. This capability makes the proposed CPS-DSL more flexible. However, for the rest of this chapter, it will be assumed that Simulink computation blocks are used to define the control law computation inside a *ControllerFunction* element.

As outlined in Sect. 3.4.5, semantics of the proposed CPS-DSL should ideally be formally defined. In their seminal work on the application of linear temporal logic (LTL) for formal verification of reactive computer systems, Manna and Pnueli [23, 24] presented a generic model of a reactive computer system in the form of a *transition system*. (This transition system will be referred to as *Manna–Pnueli Transition System* in the rest of this chapter.) They showed that various existing programming languages and specification formalisms for reactive computer systems can be mapped into this generic model. They also observed that their generic model of reactive computer systems is designed to be capable of capturing any programming language or specification formalism for reactive computer system, proposed in the future. In Sect. 3.8.1, we summarize the abovementioned *Manna–Pnueli Transition System*. In Sect. 3.8.2, we describe the semantics of the proposed CPS-DSL in terms of *Manna–Pnueli Transition System*.

3.8.1 Manna-Pnueli Transition System

Manna–Pnueli Transition System $< \Pi, \Sigma, T, \Theta >$, intended to serve as a generic model for reactive computer systems, consists of the following components:

- $\Pi = \{u_1, \ldots, u_n\}$—A finite set of *state variables*.

  Each state variable is a typed variable, whose *type* indicates the domain from which the values of that variable can be assigned. Some of these state variables are *data variables*, which represent the data elements that are declared and manipulated by the program of a reactive computer system. Other state variables are *control variables*, which keep track of the progress in the execution of a reactive computer system's program.
- Σ—A set of states.

  Each *state s* in Σ is an *interpretation* of Π. An *interpretation* of a set of typed variables is a mapping that assigns to each variable a value in its domain. Therefore, each *state s* in Σ assigns each variable u in Π a value over its domain, which is denoted by $s[u]$.
- T—A finite set of transitions.

  Each transition τ in T represents a state-changing action of the reactive computer system and is defined as a function $\tau : \Sigma \to 2^{\Sigma}$ that maps a state s in Σ into the (possibly empty) set of states $\tau(s)$ that can be obtained by applying action τ to state s. Each state s' in $\tau(s)$ is defined to be a τ-*successor* of s. A transition τ is said to be *enabled* on s if $\tau(s) \neq \phi$, that is, s has a τ-successor. It is required that one of the transitions, τ_I, called the *idling transition*, is an identity transition, i.e., $\tau_I(s) = \{s\}$ for every state s. The transitions other than the idling transition are called *diligent transitions*.
- Θ—An *initial condition*.

  Initial condition is an assertion (Boolean expression) that characterizes the states at which the execution of reactive computer system's program can begin. A state s satisfying Θ is called an *initial state*.

Each transition τ can be characterized by an assertion $\rho_\tau(\Pi, \Pi')$, called the *transition relation*, of the following form:

$$\rho_\tau(\Pi, \Pi') : C_\tau(\Pi) \wedge (y'_1 = e_1) \wedge \cdots \wedge (y'_k = e_k)$$

This transition relation consists of the following elements:

- An *enabling condition* $C_\tau(\Pi)$, which is an assertion, describing the condition under which the state s may have a τ-successor.
- A conjunction of *modification statements*

$$(y_1' = e_1) \wedge \cdots \wedge (y_k' = e_k),$$

  which relate the values of the state variables in a state s to their values in a successor state s' obtained by applying τ to s. Each modification statement $y_i = e_i$ describes the value of a state variable in state s' as an expression consisting of the state variable values in state s.

As an example, for a transition system with $\Pi = \{x, y, z\}$,

$$\rho_\tau : (x > 0) \wedge (z' = x - y)$$

describes a transition τ that is enabled only when x is positive and this transition assigns the value of z in state s' equal to the value of $x - y$ in state s.

3.8.1.1 Computations

A *computation* of Manna–Pnueli Transition System $< \Pi, \Sigma, T, \Theta >$ is defined to be an infinite sequence of states

$$\sigma : s_0, s_1, s_2, \ldots$$

satisfying the following requirements:

- *Initiation*: The first state s_0 is an initial state, i.e., it satisfies the initial condition of the transition system.
- *Consecution*: For each pair of consecutive states s_i, s_{i+1} in σ, $s_{i+1} \in \tau(s_i)$ for some transition τ in T. The pair s_i, s_{i+1} is referred to as a τ-*step*. It is possible for a given pair to be both a τ-step and a τ'-step for $\tau \neq \tau'$.
- *Diligence*: Either the sequence contains infinitely many diligent steps or it contains a terminal state (defined as a state to which only idling transitions can be applied). This requirement excludes the sequences in which, even though some diligent transition is enabled, only idling steps are taken beyond some point. A computation that contains a terminal state is called a *terminating computation*.

Indices i of states in a computation σ are referred to as *positions*. If $\tau(s_i) \neq \phi$ (τ enabled on s_i), it is said that the transition τ is *enabled* at position i of computation σ. If $s_{i+1} \in \tau(s_i)$, it is said that transition τ is *taken* at position i. Several transitions may be enabled at a single position. Moreover, one or more transitions may be considered to be taken at the same position. A state s is called *reachable* in a transition system if it appears in some computation of the system.

3.8.2 Manna–Pnueli Transition System-Based Representation of CPS-DSL

According to the proposed CPS design specification language (CPS-DSL), a *ComputingNode* block contains a *ConrolApp* block and a set of *SensorPort*, *ActuatorPort*, *InputMsgPort*, and *OutputMsgPort* blocks. Furthermore, the *ControlApp* block contains a set of *Mode* blocks and a *ModeSwitchLogic* block. Based on these constituent blocks, a *ComputingNode* block, *CompNode*1, of CPS-DSL can be represented as the Manna–Pnueli Transition System, $P_{CompNode} < \Pi_{P_{CompNode}}, \Sigma_{P_{CompNode}}, T_{P_{CompNode}}, \Theta_{P_{CompNode}} >$, outlined below, where:

- $\Pi_{P_{CompNode}}$—A finite set of *state variables*.

$$
\begin{aligned}
\Pi_{P_{CompNode1}} = \{&t, t^{switch}_{CompNode1}, mode_{CompNode1}, t^{next}_{CompNode1}, \\
&sensePort^1_{CompNode1}, sensePort^2_{CompNode1}, \\
&\dots, sensePort^p_{CompNode1}, \\
&inMsgPort^1_{CompNode1}, inMsgPort^2_{CompNode1}, \\
&\dots, inMsgPort^r_{CompNode1}, \\
&actPort^1_{CompNode1}, actPort^2_{CompNode1}, \dots, actPort^q_{CompNode1}, \\
&outMsgPort^1_{CompNode1}, outMsgPort^2_{CompNode1}, \\
&\dots, outMsgPort^l_{CompNode1}, \\
&periodicControllerIn^1_{CompNode1}, \\
&periodicControllerIn^2_{CompNode1}, \\
&\dots, periodicControllerIn^a_{CompNode1}, \\
&periodicControllerOut^1_{CompNode1}, \\
&periodicControllerOut^2_{CompNode1}, \\
&\dots, periodicControllerOut^b_{CompNode1}, \\
&controllerFunctionMemory^1_{CompNode1}, \\
&controllerFunctionMemory^2_{CompNode1}, \\
&\dots, controllerFunctionMemory^c_{CompNode1}\}
\end{aligned}
$$

where

t = time,

$t^{switch}_{CompNode1}$ = latest mode switch time of *ControlApp* block, associated with *ComputingNode* block *CompNode*1,

$mode_{CompNode1}$ = current mode of *ControlApp* block, associated with *ComputingNode* block *CompNode*1,

$t^{next}_{CompNode1}$ = next relevant time instant (actuator update, output message update) during the current mode of operation of *ControlApp* block, associated with *ComputingNode* block *CompNode*1,

$sensePort^i_{CompNode1}$ = A *SensorPort* block, contained in the *ComputingNode* block *CompNode*1,

$inMsgPort^i_{CompNode1}$ = An *InputMsgPort* block, contained in the *ComputingNode* block *CompNode*1,

$actPort^i_{CompNode1}$ = An *ActuatorPort* block, contained in the *ComputingNode* block *CompNode*1,,

$outMsgPort^i_{CompNode1}$ = An *OutputMsgPort* block, contained in the *ComputingNode* block *CompNode*1,

$peridoicControllerIn^i_{CompNode1}$ = A *PeriodicControllerInput* block that is contained in a mode of the *ControlApp* block, associated with *ComputingNode* block *CompNode*1,

$peridoicControllerOut^i_{CompNode1}$ = A *PeriodicControllerOutput* block that is contained in a mode of the *ControlApp* block, associated with *ComputingNode* block *CompNode*1,

$controllerFunctionMemory^i_{CompNode1}$ = A *ControllerFunctionMemory* block that is contained in the *ControllerFuction* block of a mode of the *ControlApp* block, associated with *ComputingNode* block *CompNode*1,

- $\Sigma_{P_{CompNode}}$—A set of states.

Each *state* s in Σ is an *interpretation* of Π. An *interpretation* of a set of typed variables is a mapping that assigns to each variable a value in its domain. The domain of state variables $t, t^{switch}_{CompNode1}$, and $t^{next}_{CompNode1}$ is $\mathbf{R}_{\geq 0}$. The domain of state variable $mode_{CompNode1}$ is $Modes_{CompNode1} = \{$Set of modes of *ControlApp* block, contained in the *ComputingNode* block *CompNode*1$\}$. Given the following definitions of Π_α and $\mathbf{D}$, all the state variables in Π_α have the domain $\mathbf{D}$:

$$\Pi_\alpha = \{sensePort^i_{CompNode1}, actPort^i_{CompNode1}, outMsgPort^i_{CompNode1},$$

$$periodicControllerIn^i_{CompNode1}, periodicControllerOut^i_{CompNode1},$$

$$controllerFunctionMemory^i_{CompNode1}\}$$

$$\mathbf{D} = \{x \mid (x \in \mathbf{R})$$

$$\wedge \ (x \text{ can be represented by type } double \text{ of computer system})\}$$

The state variable $inMsgPort^i_{CompNode1}$ has the following domain:

$$P = \{(x, y) \mid (x \in \mathbf{R}) \wedge (y \in \mathbf{D})\}$$

- $TP_{CompNode}$—A finite set of transitions.

$$TP_{CompNode1} = \tau_I \cup T^{ModeSwitches}_{CompNode1} \cup T^{TimeIncrement}_{CompNode1}$$

where

τ_I = Idling Transition

$T^{ModeSwitches}_{CompNode1} = \{\tau^{mode_i mode_j}_{CompNode1} \mid \exists$ a *mode switch* from $mode_i$ to $mode_j$ in the *ModeSwitchLogic* block of *ControlApp* block, associated with *ComputingNode* block *CompNode1*$\}$

$T^{TimeIncrement}_{CompNode1} = \{\tau^{mode_1}_{CompNode1}, \tau^{mode_2}_{CompNode1}, \ldots, \tau^{mode_M}_{CompNode1}\}$

As outlined in the summary of Manna–Pnueli Transition System approach, presented in Sect. 3.8.1, each transition τ can be characterized by an *enabling condition* and a *set of modification statements*. Based on the abovementioned set of transitions $TP_{CompNode1}$ of $P_{CompNode1}$, all the diligent transitions of $P_{CompNode1}$ can be completely described through the enabling conditions and modification statements of the following generic transitions: $\tau^{mode_i mode_j}_{CompNode1}$ and $\tau^{mode_i}_{CompNode1}$.

(a) $\tau^{mode_i mode_j}_{CompNode1}$: *Enabling Condition*

$$C_{\tau^{mode_i mode_j}_{CompNode1}} = (mode_{CompNode1} == mode_i)$$

$$\wedge\ ModeSwitchCondition_{CompNode1}(t, mode_i, mode_j)$$

$$\wedge\ ModeSwitchCheckTime_{CompNode1}$$

$$(t, t^{switch}_{CompNode1}, mode_i, mode_j)$$

where

$ModeSwitchCondition_{CompNode1}(t, mode_i, mode_j)$ = An assertion that returns true if the *mode switch condition* associated with *mode switch* from $mode_i$ to $mode_j$ in the *ModeSwitchLogic* block, contained in the *ComputingNode* block *CompNode1*, is true at time t.

$ModeSwitchCheckTime_{CompNode1}(t, t^{switch}_{CompNode1}, mode_i, mode_j)$ = An assertion that returns true if $t - t^{switch}_{CompNode1} = a\{\frac{Period_{mode_i}}{SwitchFreq_{mode_i mode_j}}\}$, for some $a \in \{1, 2, \ldots, SwitchFreq_{mode_i mode_j}\}$.

(b) $\tau^{mode_i mode_j}_{CompNode1}$: *Modification Statements*

1. $mode_{CompNode1}' = mode_j$
2. $t^{switch}_{CompNode1}' = t$
3. $t^{next}_{CompNode1}' = t + t_{jump}$
   where

$$t_{jump} = \min \left\{ t_j \mid (t_j > 0) \wedge (t + t_j = t^{switch}_{CompNode1}' \right.$$

$$+ a\{ \frac{Period_{mode_j}}{ControllerFunctionFreq_{controllerFucntion_d}} \}),$$

for some

$$a \in \{1, 2, \ldots, ControllerFunctionFreq_{controllerFunction_d}\}$$

and for some

$$\left. controllerFunction_d \in ControllerFunctions^{mode_j}_{CompNode1} \right\}$$

4.

$$periodicControllerOuts^{mode_j}_{CompNode1}' = ModeSwitchFunction^{mode_i mode_j}_{CompNode1}$$

$$(periodicControllerOuts^{mode_i}_{CompNode1})$$

where
$ModeSwitchFunction^{mode_i mode_j}_{CompNode1}$ = A function that produces the values to
which $periodicControllerOuts^{mode_j}_{CompNode1}$ are initialized after the *mode switch* from $mode_i$ to $mode_j$ of *ControlApp*, associated with *CompNode1*

5.

$$actPorts^{mode_j}_{CompNode1}' = ControllerOutsToActs^{mode_j}_{CompNode1}$$

$$(periodicControllerOuts^{mode_j}_{CompNode1}')$$

where
$ControllerOutsToActs^{mode_j}_{CompNode1}$ = A function that captures the input–output relationship (produced by the combined effect) of all the connections between *PeriodicControllerOutput* blocks and *ActuatorPort* blocks in $mode_j$ of *CompNode1*.

6.

$$outMsgPorts^{mode_j}_{CompNode1}' = ControllerOutsToOutMsgs^{mode_j}_{CompNode1}$$

$$(periodicControllerOuts^{mode_j}_{CompNode1}')$$

where

$ControllerOutsToOutMsgs_{CompNode1}^{mode_j}$ = A function that captures the input–output relationship (produced by the combined effect) of all the connections between *PeriodicControllerOutput* blocks and *OutputMsgPort* blocks in $mode_j$ of $CompNode1$.

7.

$$periodicControllerIns_{controllerFucntion_b}'$$

$$= LoadControllerInputs_{controllerFunction_b}^{mode_j}(sensePorts_{CompNode1}^{mode_j}{}',$$

$$inMsgPorts_{CompNode1}^{mode_j}{}', periodicControllerOuts_{CompNode1}^{mode_j}{}')$$

for every $controllerFunction_b \in ControllerFunctions_{CompNode1}^{mode_j}$

where

$LoadControllerInputs_{controllerFunction_b}^{mode_j}$ = A function that captures the input–output relationship (produced by the combined effect) of all the connections between *PeriodicControllerInput* blocks, associated with *ControllerFunction* block $controllerFunction_b$ in $mode_j$, and *SensorPorts*, *InputMsgPorts*, and *PeriodicControllerOutput* blocks in $mode_j$ of $CompNode1$.

(c) $\tau_{CompNode1}^{mode_i}$: *Enabling Condition*

$$C_{\tau_{CompNode1}^{mode_i}} = (mode_{CompNode1} == mode_i)$$

$$\wedge \neg(ModeSwitchCondition_{CompNode1}(t, mode_i, mode_c)$$

$$\wedge ModeSwitchCheckTime_{CompNode1}(t, t_{CompNode1}^{switch}, mode_i, mode_c))$$

$$\forall mode_c \in \{mode_c \mid \exists \text{ a mode switch from } mode_i \text{ to } mode_c \text{ of } ControlApp$$

$$\text{associated with } ComputingNode \text{ block } CompNode1\}$$

(d) $\tau_{CompNode1}^{mode_i}$: *Modification Statements*

1. $t' = t_{CompNode1}^{next}$
2. $t_{CompNode1}^{next}{}' = t' + t_{jump}$
where

$$t_{jump} = \min\left\{t_j \mid (t_j > 0) \wedge (t' + t_j = t_{CompNode1}^{switch}\right.$$

$$\left. + a\{\frac{Period_{mode_i}}{ControllerFucntionFreq_{controllerFunction_d}}\})\right)$$

for some $a \in \{1, 2, \ldots, ControllerFunctionFreq_{controllerFunction_d}\}$
and
for some $controllerFunction_d \in ControllerFunctions_{CompNode1}^{mode_i}$ $\Big\}$

3.

$$(periodicControllerOuts_{controllerFunction_e}{}',$$

$$controllerFunctionMemory_{controllerFunction_e}{}') =$$

$$f^{controllerFunction_e}(periodicControllerIns_{controllerFunction_e},$$

$$controllerFuctionMemory_{controllerFunction_e})$$

$$\forall controllerFunction_e \in \Big\{ controllerFunction_e \mid$$

$$(controllerFunction_e \in ControllerFunctions_{CompNode1}^{mode_i})$$

$$\wedge (t' = t_{CompNode1}^{switch} + a\{\frac{Period_{mode_i}}{ControllerFunctionFreq_{controllerFunction_e}}\})$$

$$\text{for some } a \in \{1, 2, \ldots, ControllerFunctionFreq_{controllerFunction_e}\} \Big\}$$

where
$f^{controllerFunction_e}$ = The function implemented by the internal components (Simulink blocks) of *ControllerFunction* block *controller Fucntion_e*.

4.

$$periodicControllerIns_{controllerFunction_f}{}' =$$

$$LoadControllerInputs_{controllerFunction_f}^{mode_i}(sensePorts_{CompNode1}^{mode_i}{}',$$

$$inMsgPorts_{CompNode1}^{mode_i}{}', periodicControllerOuts_{CompNode1}^{mode_i}{}')$$

$$\forall controllerFunction_f \in \Big\{ controllerFunction_f \mid$$

$$(controllerFunction_f \in ControllerFunctions_{CompNode1}^{mode_i})$$

$$\wedge (t' = t_{CompNode1}^{switch} + a\{\frac{Period_{mode_i}}{ControllerFunctionFreq_{controllerFunction_f}}\})$$

$$\text{for some } a \in \{1, 2, \ldots, ControllerFunctionFreq_{controllerFunction_f}\} \Big\}$$

5.

$$act Ports_{CompNode1}^{mode_i}{}' =$$

$$Controller OutsToActs_{CompNode1}^{mode_i}(periodicController Outs_{CompNode1}^{mode_i}{}')$$

6.

$$out Msg Ports_{CompNode1}^{mode_i}{}' =$$

$$Controller OutsToOut Msgs_{CompNode1}^{mode_i}$$

$$(periodicController Outs_{CompNode1}^{mode_i}{}')$$

- $\Theta_{P_{CompNode}}$—An *initial condition*. Any initial state s of transition system $P_{CompNode}$ must satisfy the following initial conditions:

$t = 0$

$t_{CompNode1}^{switch} = 0$

$mode_{CompNode1} = mode_1$

$t_{CompNode1}^{next} = \min \left\{ t_j \mid (t_j > 0) \wedge (t_j = a\{\dfrac{Period_{mode_1}}{Controller Function Freq_{controller Function_d}}\}) \right.$

for some $a \in \{1, 2, \ldots, Controller Function Freq_{controller Function_d}\}$ and for

some $controller Function_d \in Controller Functions_{CompNode1}^{mode_1} \left. \right\}$

3.9   Conclusion

Taking inspiration from the hourglass-shaped architecture of the Internet, this chapter has proposed an hourglass-shaped architecture for model-based development of networked cyber-physical systems. Similar to the central role played by TCP/IP protocols in the Internet architecture, the proposed architecture for model-based networked CPS development is centered on the notion of a standardized CPS design specification language.

The proposed hourglass-shaped architecture can enable effective coordination between control systems engineers and embedded systems engineers during a model-based CPS development process, while still acknowledging the differences between the domain-specific skillset that control systems engineer and embedded system engineer typically possess. The chapter has also proposed a version of the abovementioned CPS design specification language and discussed its various aspects such as concrete syntax, abstract syntax, and semantics.

References

1. Aho, A. V., Lam, M. S., Sethi, R., & Ullman, J. D. (2006). *Compilers: Principles, techniques, and tools* (2nd ed.). Boston: Addison-Wesley Longman Publishing Co., Inc.
2. Akhshabi, S., & Dovrolis, C. (2013). The evolution of layered protocol stacks leads to an hourglass-shaped architecture. In *Dynamics on and of complex networks* (Vol. 2, pp. 55–88). New York: Springer.
3. Antsaklis, P. J. (1998). Hybrid control systems: An introductory discussion to the special issue. *IEEE Transactions on Automatic Control, 43*(4), 457–460.
4. Berry, G., & Gonthier, G. (1992). The esterel synchronous programming language: Design, semantics, implementation. *Science of Computer Programming, 19*(2), 87–152.
5. Brogan, W. L. (1991). *Modern control theory*. Upper Saddle River: Prentice-Hall.
6. Caspi, P., Curic, A., Maignan, A., Sofronis, C., Tripakis, S., & Niebert, P. (2003). From simulink to scade/lustre to TTA: A layered approach for distributed embedded applications. In *ACM sigplan notices* (Vol. 38, pp. 153–162). New York: ACM.
7. Dugard, L., & Verriet, E. (1998). *Stability and control of time-delay systems. Lecture notes in control and information sciences*. Berlin: Springer.
8. Feiler, P. H., & Gluch, D. P. (2012). *Model-based engineering with AADL: An introduction to the SAE architecture analysis & design language*. Boston: Addison-Wesley.
9. Frankel, D. S. (2003). *Model driven architecture: Applying MDA to enterprise computing*. Hoboken: Wiley.
10. Friedenthal, S., Moore, A., & Steiner, R. (2014). *A practical guide to SysML: The systems modeling language*. Burlington: Morgan Kaufmann.
11. Gronback, R. C. (2009). *Eclipse modeling project: A domain-specific language toolkit*. Boston: Addison-Wesley Professional.
12. Halbwachs, N., Caspi, P., Raymond, P., & Pilaud, D. (1991). The synchronous data flow programming language lustre. *Proceedings of the IEEE, 79*(9), 1305–1320.
13. Henriksson, D., & Elmqvist, H. (2011). Cyber-physical systems modeling and simulation with Modelica. In *International Modelica Conference* (Vol. 9). Linköping: Modelica Association.
14. Henzinger, T., Horowitz, B., & Kirsch, C. (2003). Giotto: A time-triggered language for embedded programming. *Proceedings of the IEEE, 91*(1), 84–99.
15. Henzinger, T. A., & Sifakis, J. (2006). The embedded systems design challenge. In *FM 2006: Formal Methods* (pp. 1–15). Berlin: Springer.
16. Jones, M., Kotsalis, G., & Shamma, J. S. (2013). Cyber-attack forecast modeling and complexity reduction using a game-theoretic framework. In *Control of cyber-physical systems* (pp. 65–84). Heidelberg: Springer.
17. Kang, W., Kapitanova, K., & Son, S. H. (2012). Rdds: a real-time data distribution service for cyber-physical systems. *IEEE Transactions on Industrial Informatics, 8*(2), 393–405.
18. Karsai, G., Sztipanovits, J., Ledeczi, A., & Bapty, T. (2003). Model-integrated development of embedded software. *Proceedings of the IEEE, 91*(1), 145–164.
19. Lee, E. A. (2009). Computing needs time. *Communications of the ACM, 52*(5), 70–79. https://doi.org/10.1145/1506409.1506426
20. LeGuernic, P., Gautier, T., Le Borgne, M., & Le Maire, C. (1991). Programming real-time applications with signal. *Proceedings of the IEEE, 79*(9), 1321–1336.
21. Liberzon, D., & Morse, A. S. (1999). Basic problems in stability and design of switched systems. *IEEE Control Systems, 19*(5), 59–70.
22. Liu, I., Reineke, J., Broman, D., Zimmer, M., & Lee, E. A. (2012). A PRET microarchitecture implementation with repeatable timing and competitive performance. In *IEEE 30th International Conference on Computer Design (ICCD), 2012* (pp. 87–93). https://doi.org/10.1109/ICCD.2012.6378622
23. Manna, Z., & Pnueli, A. (1991). *The temporal logic of reactive and concurrent systems: Specification*. New York: Springer.

24. Manna, Z., & Pnueli, A. (1995). *Temporal verification of reactive systems: Safety*. New York: Springer.
25. Mathworks inc. (2016). Embedded coder r2015b. http://www.mathworks.com/products/embedded-coder/
26. Mathworks inc. (2016). Simscape r2015b. http://www.mathworks.com/products/simscape/
27. Mathworks inc. (2016). Simulink r2015b. http://www.mathworks.com/products/simulink/
28. Mathworks inc. (2016). Stateflow r2015b. http://www.mathworks.com/products/stateflow/
29. Mesbahi, M., & Egerstedt, M. (2010). *Graph theoretic methods in multiagent networks*. Princeton: Princeton University Press.
30. Selic, B., & Gérard, S. (2013). *Modeling and analysis of real-time and embedded systems with UML and MARTE: Developing cyber-physical systems*. New York: Elsevier.
31. Simko, G., Lindecker, D., Levendovszky, T., Neema, S., & Sztipanovits, J. (2013). Specification of cyber-physical components with formal semantics–integration and composition. In *Model-driven engineering languages and systems* (pp. 471–487). Berlin: Springer.
32. Stahl, T., Völter, M., Bettin, J., Haase, A., & Helsen, S. (2006). *Model-driven software development: Technology, engineering, management*. Hoboken: Wiley.
33. Steinberg, D., Budinsky, F., Paternostro, M., & Merks, E. (2008). *EMF: Eclipse modeling framework*. Boston: Addison-Wesley Professional.
34. Sztipanovits, J., Bapty, T., Koutsoukos, X., Lattmann, Z., Neema, S., & Jackson, E. (2018). Model and tool integration platforms for cyber-physical system design. *Proceedings of the IEEE, 106*, 1–26. https://doi.org/10.1109/JPROC.2018.2838530
35. Sztipanovits, J., Koutsoukos, X., Karsai, G., Kottenstette, N., Antsaklis, P., & Gupta, V. (2012). Toward a science of cyber–physical system integration. *Proceedings of the IEEE, 100*(1), 29–44. https://doi.org/10.1109/JPROC.2011.2161529
36. Tariq, M. U., Florence, J., & Wolf, M. (2014). Design specification of cyber-physical systems: Towards a domain-specific modeling language based on simulink, eclipse modeling framework, and giotto. In: *ACESMB@ MoDELS* (pp. 6–15).
37. Tarraf, D. C. (2013). Control of cyber-physical systems. In *Proceedings of Lecture Notes in Control and Information Sciences* (Vol. 449).
38. Warmer, J. B., & Kleppe, A. G. (2003). The object constraint language: Getting your models ready for MDA. Boston: Addison-Wesley Professional.
39. Wolf, W. (2009). Cyber-physical systems. *Computer, 42*(3), 88–89. https://doi.org/10.1109/MC.2009.81
40. Wolf, M., & Serpanos, D. (2017). Safety and security of cyber-physical and internet of things systems [point of view]. *Proceedings of the IEEE, 105*(6), 983–984. https://doi.org/10.1109/JPROC.2017.2699401
41. Zhang, W., Branicky, M. S., & Phillips, S. M. (2001). Stability of networked control systems. *IEEE Control Systems, 21*(1), 84–99.

Part II
Testing and Operation

Chapter 4
Formal Techniques for Verification and Testing of Cyber-Physical Systems

Jyotirmoy V. Deshmukh and Sriram Sankaranarayanan

4.1 Introduction

Cyber-physical systems (CPS) involve the tight coupling of physical components such as electrical, mechanical, hydraulic, and biological with software systems that are primarily involved in tasks such as sensing, communication, control, and interfacing with human operators. Software components in CPS are often designed using the model-based development (MBD) paradigm [113]. The MBD process proceeds in many steps: (1) First, the designer specifies the *plant model*, i.e., the dynamical characteristics of the physical parts of the system using differential, logical, and algebraic equations. Examples of plant models include the rotational dynamics model of the camshaft in an automobile engine, the thermodynamic model of an internal combustion engine, kinematic and dynamic models for ground and air vehicles, and pharmacokinetic models of human physiology. (2) The next step is to design control software to regulate the behavior of the physical system. This step often involves the use of techniques from control theory to design embedded controllers, techniques from distributed systems to achieve communication and coordination, and more recently, techniques from artificial intelligence to allow learning and adaptation. (3) The final step is to define an environment model which encapsulates physical assumptions on the exogenous quantities that affect the system (such as atmospheric turbulence, driver behavior, or meal intake by a patient). The composition of these three types of models (plant, software, and environment) constitutes the overall *closed-loop* system.

J. V. Deshmukh (✉)
University of Southern California, Los Angeles, CA, USA
e-mail: jyotirmoy.deshmukh@usc.edu

S. Sankaranarayanan
University of Colorado, Boulder, CO, USA
e-mail: srirams@colorado.edu

© Springer Nature Switzerland AG 2019
M. A. Al Faruque, A. Canedo (eds.), *Design Automation of Cyber-Physical Systems*, https://doi.org/10.1007/978-3-030-13050-3_4

Typically, plant models in an MBD process are deterministic. Any uncertainty is encoded in the environment model as either a nondeterministic choice on inputs to the plant model (subject to an appropriate set of constraints) or a random choice on the inputs subject to an appropriate probability distribution. Though it is also possible to model certain phenomena such as manufacturing variations, uncertainties in physics-based modeling, and sensor/actuator noise using a *stochastic dynamical* plant model, industrial MBD frameworks rarely use stochastic models during the control design process. The controller models are typically deterministic, as they represent a software implementation. In this chapter, we focus on plant and controller models that are deterministic, and environment models that are nondeterministic (not stochastic[1]).

Mathematical models for CPS applications help us analyze the system in multiple ways: (1) models are simulated under various input conditions to predict how the system as a whole would behave. Often these input conditions may be hard and expensive to recreate in the physical world. For systems involving human operators, models serve as an important alternative to real physical tests that may be dangerous or even unethical; and (2) models can expose latent/hidden system variables that are hard to measure, and thus allow us to examine their presumed behavior. In Sect. 4.2, we summarize various kinds of mathematical models that are used in the CPS domain, and typical applications for each model type.

Next, we describe *behavioral specifications*. Note that many industrial settings use the term *requirements* to mean behavioral specifications. The term specification is instead used to designate *a specification model*—a high-level programmatic description of the embedded software code. Behavioral specifications go hand-in-hand with models and describe desirable properties of the system as a whole. The specifications can be high level ("end-to-end"), describing a desired property of the system as a whole (e.g., the car will not be physically damaged by the action of the adaptive cruise control subsystem) or at the modular level, focusing on an individual module of the system (e.g., when the input to the controller is within $[-2, 2]$, the output must be within $[-1, 1]$). In Sect. 4.2, we also discuss a formalism used for behavioral specifications of CPS models.

Given a mathematical model of the system M, and a behavioral specification φ, there are two main kinds of analysis problems that focus on ensuring correctness of the CPS design: *formal verification* and *falsification*. The main purpose of verification is to prove the absence of failures in a given CPS model, where a failure is defined as the violation of a given formal specification. Many verification procedures perform a best-effort search for a proof of system correctness, wherein a failure to find one may lead to an inconclusive result. On the other hand, test generation or *falsification* focuses on providing evidence of the *presence of failures*

---

[1] Allowing stochasticity in the plant or environment model necessitates treating the closed-loop CPS model as a stochastic dynamical system. The techniques for verification and testing of such systems are quite different. As we wish to focus on techniques that are closer to industrial use of MBD for CPS applications, we refer the reader to [36, 71] for excellent surveys.

in the form of counterexamples. Falsification procedures perform a best effort search for a counterexample to the property of interest, with a failure to find a counterexample leading to an inconclusive result.

We now formalize these problems. A typical abstraction for a mathematical model of a CPS, M, is as a *stateful* system that maps timed input behaviors (i.e., input signals) to output signals. A *signal* is defined as a function mapping a *time domain*—a finite or infinite subset of positive real numbers—to some value in a *signal domain*. For simplicity, we consider signal domains that are compact subsets of the real numbers. For ease of exposition, we assume that the time domain for the input and output signals is the same set $\mathbb{T}$, and the input and output signal domains are respectively $\mathscr{U}$ and $\mathscr{Y}$. Let the initial set of states for M be the set $\mathscr{X}_0$. Let $u \in \mathbb{T}^{\mathscr{U}}$ be an input signal and let $y \in \mathbb{T}^{\mathscr{Y}}$ be an output signal. Thus, M defines a function that maps a state $x_0 \in \mathscr{X}_0$, and an input signal u to an output signal y, i.e., $y = M(x_0, u)$. Finally, assume that we are given a specification φ, which maps every pair (u, y) to *true* or *false*.

Definition 4.1 (Verification) Given a model M, with initial states $\mathscr{X}_0$, a time domain $\mathbb{T}$, input domain $\mathscr{U}$ and output domain $\mathscr{Y}$, and a specification φ, the formal verification problem provides a proof that for all $x_0 \in \mathscr{X}_0$, and for all $u \in \mathbb{T}^{\mathscr{U}}$, if $y = M(x_0, u)$, then $\varphi(u, y)$ is *true*.

There are several techniques that have been proposed to solve the verification problem for CPS models. The most popular among these are *reachability analysis* techniques that are based on computing the set of states reachable (usually within a given finite-time horizon) from a given set of initial conditions and for a given set of input signals. In such techniques, a common assumption is that the system state is fully observable (i.e., the output signals are simply the state trajectories of the system). Further, the specification is typically provided as a set of *unsafe states* that should not be reached by the system. We discuss these techniques in Sect. 4.3.

The advantage of techniques based on reachability is that they are highly automatic; however, for systems with nonlinearities and switching behaviors, these techniques may suffer from imprecision. An alternative approach is to use manual insight to propose an *invariant* for the given CPS model. An invariant is a set that is guaranteed to contain the system behaviors for all time. The computational effort is then to automate the invariant generation process (as much as possible) and verify the validity of the system invariant. We discuss such techniques in Sect. 4.4.

In Sect. 4.5, we discuss various specification-driven falsification techniques for CPS models. A falsification problem attempts to provide a refutation to a verification question for a system. Formally, we define falsification as follows.

Definition 4.2 (Falsification) Given a model M, with initial states $\mathscr{X}_0$, a time domain $\mathbb{T}$, input domain $\mathscr{U}$ and output domain $\mathscr{Y}$, and a specification φ, the falsification problem provides a proof that there is some $x_0 \in \mathscr{X}_0$, and some $u \in \mathbb{T}^{\mathscr{U}}$, such that $y = M(x_0, u)$, and $\varphi(u, y)$ is *false*.

Falsification approaches are based on systematically searching for a counterexample to a specification. In Sect. 4.5, we present robustness-guided falsification approaches that use a robustness metric to map properties $\varphi(u, y)$ that provide *true/false* interpretation to signals to real-valued interpretations that measure how close a trace comes to satisfying or violating a property.

Finally, in Sect. 4.6, we highlight a significant challenge on the horizon for CPS applications that aspire to become autonomous or semi-autonomous. Developers for such applications are increasingly using AI-based software such as artificial (and deep) neural networks for various aspects such as perception, planning/decision-making, and control. We review some of the key challenges in this domain and summarize some of the recent work seeking to address these challenges.

4.1.1 Motivating Examples

In this section, we describe two motivating examples that illustrate the need for model-based design supported by formal design verification tools.

4.1.1.1 Autonomous Driving

There has been significant recent interest in the ability of vehicles to drive autonomously, i.e., without any intervention by a human driver [28, 77, 100]. The typical software stack for an autonomous vehicle consists of several components: (1) a perception component that processes data about the environment coming through sensors such as a Radar, forward-facing cameras, and LiDAR (light detection and ranging), (2) a decision/planning component that uses the environment models created by the perception component to plan the motion of the vehicle, and (3) a low-level control component that interfaces with the actuators of the vehicle to physically realize the motion plan determined by the planning component. There is ample scope for model-based design of the interfaces between each of these components. In particular, we consider one of the simplest problems for an autonomous vehicle, which is that of regulating its speed. This is based either on a desired speed determined by the high-level motion planner in accordance with the current weather conditions and speed-limit regulations, or based on the speed of the vehicle in front (whichever is lesser). The objective is twofold, if there is a lead car, then the ego car should always maintain a safe following distance from the lead car; otherwise, it should maintain a speed close to that suggested by the high-level motion plan.

An *adaptive cruise controller* (ACC) is a control scheme that seeks to automate the task of choosing the right acceleration for the ego vehicle so as to maintain its safety and performance objectives. Radar-based ACC systems have been implemented in several commercial cars, but continue to be of relevance in the autonomous-driving space, where the sensor inputs are not restricted to Radar. Furthermore, a typical autonomous vehicle has several subsystems that may try to control the longitudinal acceleration of the car (e.g., a controller that attempts to execute a lane-change maneuver, or a controller to execute an emergency stopping

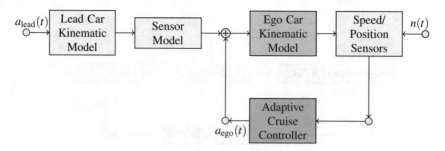

Fig. 4.1 Schematic diagram for an adaptive cruise control system

maneuver). In such cases, it is important that the ACC system is not designed in isolation, but is cognizant of other systems around it.

A schematic model of a typical ACC system is shown in Fig. 4.1. The typical model of the environment is to construct a kinematic model of the lead car (based on Newton's laws of motion), while assuming that the lead car can dynamically change its acceleration (denoted by $a_{\text{lead}}(t)$). The sensor model then captures the quantities in the lead car's motion that can be measured by the ego car. For example, for a Radar-based sensor, this would be the relative distance between the cars and the velocity of the lead car. The kinematic model of the ego car models the effect of the controller and environment inputs on the ego car's motion. We assume that the adaptive cruise controller estimates the ego car's motion through speed sensors (possibly coupled with an odometry-based position computation model). These sensors could have an associated measurement noise (modeled by $n(t)$). Finally, the ACC outputs a control signal (typically the ego car's acceleration, shown as $a_{\text{ego}}(t)$).

Recent work has focused on formal verification and correct-by-construction synthesis of ACC systems. In [102], the authors use quantified dynamic logic to verify the local lane control problem which uses an invariant-based theorem-proving approach. In [104], the authors use reachability analysis for proving safety of ACC systems. On the other hand, in [17, 114], the authors use correct-by-construction approaches using Lyapunov theory and control barrier certificates to automatically obtain safe implementations of ACC systems. While these studies have demonstrated the power of formal verification, more work can be done in formalizing behavioral specifications for an ACC system, and then applying the different techniques considered in this chapter to prove correctness of such a system.

4.1.1.2   Artificial Pancreas

Type-1 diabetes is characterized by the inability to regulate the blood glucose (BG) levels within an *euglycemic range* [70, 180] mg/dl in the human body due to the absence of insulin, a hormone that is responsible for reducing BG levels. The treatment is to externally replace the lost insulin. However, this insulin must

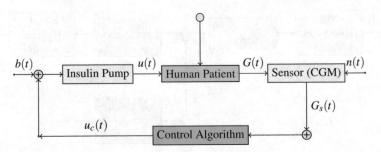

Fig. 4.2 Overview of the key components of an artificial pancreas control system. $b(t)$: external insulin, $u(t)$: insulin infused, $G(t)$: BG level, $n(t)$: measurement error, $G_s(t)$: sensed glucose level, $u_c(t)$: insulin infusion commanded

be delivered to compensate increases in blood glucose levels due to meals or endogenous glucose production by the liver. Too much insulin can expose the patient to the risk of hypoglycemia wherein the blood glucose levels fall below 70 mg/dl, whereas too little insulin causes high BG levels due to *hyperglycemia* wherein BG levels rise above 180 mg/dl leading to long-term damage to kidneys, eyes, heart, and the peripheral nerves. In order for insulin to be delivered, it is often infused subcutaneously through an insulin pump—a device that is programmed to deliver a constant low rate of insulin, known as *basal insulin*, or a larger *bolus* of insulin in advance of a meal or to treat high BG values [38, 138].

The artificial pancreas project seeks to partially or fully automate the delivery of insulin by combining a continuous glucose sensor which periodically senses BG levels subcutaneously, an insulin pump that delivers insulin, and a closed-loop control algorithm that uses inputs from the CGM and the user to control BG levels to a target value [48, 86, 143]. A schematic diagram is shown in Fig. 4.2.

Because of the severe risks posed by hypo- and hyperglycemia, AP devices are safety critical. They need to be used by patients 24/7/365 without expert supervision, though they are capable of serious harm to the patient. As a result, their design and implementation require careful consideration and thus form an ideal target for formal methods/automated reasoning approaches.

Notable attempts to verify medical devices include work on pacemakers and implantable cardiac defibrillators (ICDs). This started with physiological models of excitable cells in the heart [119], leading to approaches that employ these models to test closed-loop systems [88, 117].

Lee and collaborators studied a PID-based closed-loop system meant for intra-operative use in patients [39], using the dReal SMT solver [76] to prove safety for a range of parameters and controller gains. Other approaches to verifying artificial pancreas systems have relied on *falsification*, using temporal logic robust-ness [56, 68], and incorporated in tools such as S-TaLiRo [1, 111] and Breach [55]. Sankaranarayanan et al. have studied the use of falsification techniques for verifying closed-loop control systems for the AP [34]. Their initial work investigated a PID controller proposed by Steil et al. [140, 141] based on published descriptions of the control system available. Another recent study by Sankaranarayanan et al. [135] was performed to test a predictive pump shutoff controller designed by Cameron et al. [35] that has undergone outpatient clinical trials, recently [103]. Recently, Kushner et al. studied a personalized approach to analyzing controller parameters using data-driven models [96]. These studies have demonstrated the ability of formal approaches to verification and falsification to provide important behavioral specifications, combine a variety of models for every aspect of the artificial pancreas, and prove/falsify important properties.

4.2   Mathematical Models and Specifications

> *"All models are wrong but some are useful"*–George E. Box [32].

Verifying properties of a system requires mathematical models and formal specifications. In this section, we briefly describe the varieties of mathematical models and specification formalisms that are used in cyber-physical systems (CPS). As mentioned earlier, CPS combine a variety of heterogeneous components, including physical (mechanical, electrical, chemical, and biological) systems, electronic (analog and digital circuits), and software components. Furthermore, they are subject to a wide variety of input stimuli from the environment that can range from disturbances such as wind to inputs from human operators. As a result, mathematical modeling is a key first step in order to provide a framework wherein we can define key properties of the system in a formal manner. A variety of mathematical models are employed in CPS, including ordinary and partial differential equation models for physical and biological components, automata-based models for digital electronic components, and software. Finally, stochastic models capture the behavior of disturbances such as the wind, noise, measurement errors, component failures, or mistakes made by human operators.

4.2.1   Mathematical Models

Table 4.1 lists some commonly employed mathematical models and the type of components that they are used to model. These models range from continuous-time models such as ODEs and SDEs to discrete time models such as finite and extended state machines. Each of these models have been well studied by communities of mathematicians, physicists, and engineers.

However, the challenge of CPS applications arises in the combination of multiple modeling paradigms within the same system. Due to this combination, the modeling

Table 4.1 Commonly employed mathematical models for various aspects of a CPS

Model type	Component type	Examples
Ordinary differential equation (ODE) [107]	Physical/analog	Vehicle body, engine speed, drug pharmacokinetics
Partial differential equation (PDE)	Physical continuum	Fluid flow, electromagnetic field, fabric, paper
Finite state automata [137]	Software/electronics	Switching logic, relays, digital circuits, software
Extended state automata	Software	Software controllers
Timed automata [12]	Real-time software	Schedulers, watchdog timers
Markov chains [115]	Disturbances/failures	Component failures, job arrivals
Stochastic differential equation (SDE) [116]	Disturbances	Wind disturbances, measurement noise

of CPS has focused on the combination of discrete-time models such as automata and continuous models such as ODEs to yield *hybrid dynamical systems* that are capable of continuous-time evolution in conjunction with discrete mode transitions.

4.2.1.1 Hybrid Systems

Hybrid systems model processes that combine the continuous evolution of state over time with discrete jumps that can instantaneously change the state as well as the future course of the dynamics. Such systems arise from a variety of sources: physical systems involving contact forces, biological systems, controlled systems with switched or periodically updated control action, and in general, software-driven control systems. The field of hybrid systems evolved historically from two complementary sources that included *computer scientists* studying languages and formalisms defined by the interaction of automata with physical process [9, 83]; *control theorists* extending previously well studied continuous models to include discrete switching actions [33, 139]. Labinaz et al. present an early survey that touches upon the historical development of hybrid systems [97].

The *hybrid automaton model* was proposed to provide a conceptual model for expressing hybrid systems [10]. Figure 4.3 shows an example of a hybrid automaton model expressing a temperature controller for a house that is heated by turning on/off a source of heated/cooled air. The automaton has two modes ON: representing the dynamics of the room temperature when the heater is turned on and OFF: representing the dynamics when the heat is turned off.

Definition 4.3 (Hybrid Automaton) Given a vector of system variables $\mathbf{x} \in X$, control inputs $\mathbf{u} \in U$, and disturbances $\mathbf{w} \in W$, a hybrid automaton $\mathcal{H}$: $\langle L, E, I, F, G, R \rangle$ consists of the following components:

1. A finite set of *modes* $L : \{\ell_1, \ldots, \ell_n\}$ and *transitions* that form edges between locations $E \subseteq L \times L$,

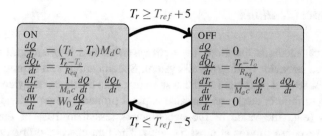

Fig. 4.3 Hybrid automaton model for the house heating demo example. The state variables include Q, the heat flowing in to the room, Q_L, the heat lost to the outside, T_r, the room temperature, and W, the total heating cost. The parameters are shown in blue and include M_a the mass of air inside the house, R_{eq} the "thermal resistance" equivalent of the house, M_d, the air flow rate through the heater, c the heat capacity of air at constant pressure, W_0 is cost per unit heat, and T_{ref} the desired reference temperature. The disturbance input is T_o the outside air temperature, shown in red

2. A map I that associates each location $\ell \in L$, a location invariant $I_\ell \subseteq X$,
3. A map F that associates each mode $\ell_i \in L$ with a vector field $F_l : X \times U \times W \mapsto$ $(\mathscr{T}X)$ that forms the RHS of the ODE: $\frac{dx}{dt} = F_{\ell_i}(\mathbf{x}, \mathbf{u}, \mathbf{w})$. The function F_ℓ is assumed to be Lipschitz continuous over $\mathbf{x}$ and continuous over the remaining inputs for all $\ell \in L$.
4. A guard map G that associates with a guard set $G_{(\ell_1, \ell_2)}$ with each transition, and
5. A reset map R that associates each transition with an update function $R_{(\ell_1, \ell_2)} :$ $I_{\ell_1} \mapsto I_{\ell_2}$.

The initial condition of a hybrid automaton is given by a location $\ell_0 \in L$ and an initial state $\mathbf{x}_0 \in I_{\ell_0}$. Let $\mathbf{u} : [0, T] \mapsto U$ be a control input signal and $\mathbf{w} : [0, T] \mapsto W$ be a disturbance input. The state of a hybrid automaton is given by a pair $(\ell, \mathbf{x})$ where $\ell \in L$ and $\mathbf{x}$ is a state belonging to the invariant set I_ℓ associated with the mode ℓ. The execution of a hybrid automaton over a time horizon T (can be finite or infinite $T = \infty$) is given by a sequence of *flows* and *jumps*:

- A flow $(\ell, \mathbf{x}, \tau) \rightsquigarrow (\ell, \mathbf{x}', \tau + \delta)$ for $\delta \geq 0$ is a solution to the ODE $\frac{dx}{dt} = F_\ell(\mathbf{x}, \mathbf{u}, \mathbf{w})$ starting from the initial condition $t_0 = \tau, \mathbf{x}(\tau) = \mathbf{x}$ with $\mathbf{u}(\cdot)$ as the signal $\mathbf{u}(t)$ with $t \in [\tau, \tau + \delta)$ and likewise, $\mathbf{w}$ as the signal $\mathbf{w}(t)$ over $t \in [\tau, \tau + \delta)$. This trajectory is uniquely defined since F_ℓ is Lipschitz. Finally, $\mathbf{x}'$ is the state $\mathbf{x}(\tau + \delta)$.
- A jump $(\ell, \mathbf{x}, \tau) \rightarrow (\ell', \mathbf{x}', \tau)$ is an instantaneous transition from mode ℓ to ℓ' wherein $(\ell, \ell') \in E$, and $\mathbf{x} \in G_{(\ell, \ell')}$ must belong to the guard set of the transition. The state $\mathbf{x}' = R_{(\ell, \ell')}(\mathbf{x})$ is obtained by applying the reset map corresponding to the transition (ℓ, ℓ') to the state $\mathbf{x}$.

An execution trace of the hybrid automaton yields a hybrid time trajectory comprised of flows and jumps starting from the initial state $(\ell_0, \mathbf{x}_0)$ at time 0.

$$(\ell_0, \mathbf{x}_0, 0) \rightsquigarrow (\ell_1, \mathbf{x}_1, t_1) \rightarrow (\ell_1', \mathbf{x}_1', t_1) \rightsquigarrow (\ell_2, \mathbf{x}_2, t_2) \rightarrow \cdots$$

4.2.2 Specifications

In the formal methods literature, the term *specifications* is often used to describe the expected behavior of the overall system. Specifications can express properties defined over several behaviors of the system (e.g., the average energy consumption, mean time to failure, etc.), and can also express properties over individual system executions (e.g., the value of the overshoot is less than 10% of the reference value, the response time is at most 5 s, etc.). The first class of properties (that are defined over several system behaviors) are called *hyperproperties* [47]. The second class of properties are *trace* properties, i.e., given a (discrete or continuous) trace representing a system behavior, we can check the satisfaction or violation of such a property on this trace.

Types of Properties In hyperproperties, we can further make a distinction between *statistical* hyperproperties, i.e., properties that reason about statistical aspects of the system (such as average energy consumption, mean time to failure, etc.), and *relational* hyperproperties. There has been limited work on estimating statistical properties of CPS models [2], but not much work has been done to verify or falsify statistical hyperproperties. Relational hyperproperties are gaining popularity for expressing security and privacy properties such as information leakage, robust I/O behavior, noninterference, noninference, etc. [47, 112]. For example, consider a potential side-channel power attack: there exists a system behavior where for the input u the signal representing the magnitude of power (say y) that exceeds the value c for τ seconds, but for all other inputs u' *near* u, the corresponding y' is always below some value d s.t. $d < c$. There has not been much work on verification of relational hyperproperties for CPS models. Thus, as verification or testing for hyperproperties is a nascent field with limited results for narrow subproblems [29, 30, 53, 70]. Hence, we do not discuss this aspect in detail in this chapter, but instead focus on verification and falsification for trace properties.

4.2.2.1 Temporal Logics for Trace Properties

There are several possible ways in which trace-level properties can be expressed and checked. Many industrial practitioners often write custom programs in their preferred programming language to check a trace-level property. These programs are also known as *property monitors*. An *offline* monitor checks the satisfaction of a *finite-time* trace-level property by a given finite-time system execution after the execution has terminated. On the other hand, an *online* monitor continuously checks the satisfaction or violation of the property as the system runs. In an MBD framework, the same terminology applies to simulations of system behavior: offline monitoring requires the simulation to have terminated.

Having customized programs for property monitors can pose challenges in terms of interpretability and maintainability, and is prone to manual programming errors. An elegant alternative is to use a suitable logical formalism to describe the desired trace-level property. One such formalism is that of linear temporal logic (LTL). LTL was introduced in the late 1970s [123] to reason about the temporal behaviors of reactive systems, i.e., input-output systems with Boolean, discrete-time signals. CPS rarely have discrete-valued, discrete-time behaviors, as the physical components in a CPS have real-valued behaviors that evolve continuously in time. To reason about such systems, later, temporal logics such as timed propositional temporal logic (TPTL) [14], the duration calculus [37], and metric temporal logic (MTL) [94] were introduced to deal with dense-time system executions. These logics required first creating a set of atomic Boolean predicates over signals, and then introduced formulas that contained temporal operators that could be interpreted over dense time.

Signal Temporal Logic (STL) STL [106] was proposed in the context of analog and mixed-signal circuits as a specification language for expressing constraints on real-valued signals directly in the formula expressing the property of interest. Let $\mathbf{x}$ be an n-dimensional signal representing the system execution over some finite time, and for simplicity, let the codomain of this variable be $\mathbb{R}^n$. Without loss of generality, these predicates can be reduced to the form $\mu = f(\mathbf{x}) \sim c$, where f is a scalar-valued function from $\mathbb{R}^n$ to $\mathbb{R}$.

Temporal formulas are formed using temporal operators, "always" (denoted as $\mathbf{G}$), "eventually" (denoted as $\mathbf{F}$) and "until" (denoted as $\mathbf{U}$). Each temporal operator is indexed by intervals of the form (a, b), $(a, b]$, $[a, b)$, $[a, b]$, (a, ∞) or $[a, \infty)$ where each of a, b is a nonnegative real-valued constant. If I is an interval, then an STL formula is written using the following grammar:

$$\varphi := true$$
$$\begin{array}{lll} & \mid & \mu & \text{atomicproposition} \\ & \mid & \neg\varphi & \text{negation} \\ & \mid & \varphi_1 \wedge \varphi_2 & \text{conjunction} \\ & \mid & \varphi_1 \, \mathbf{U}_I \, \varphi_2 & \text{untiloperator} \end{array}$$

The always and eventually operators are defined as special cases of the until operator as follows: $\mathbf{G}_I\varphi \triangleq \neg\mathbf{F}_I\neg\varphi$, $\mathbf{F}_I\varphi \triangleq true \, \mathbf{U}_I \, \varphi$. When the interval I is omitted for the until operator, we take it as the default interval of $[0, +\infty)$. The semantics of STL formulas are defined informally through examples as follows.

Example 4.1 The signal $\mathbf{x}$ satisfies an atomic predicate $f(\mathbf{x}) > 10$ at time t (where $t \geq 0$) if the value of $f(\mathbf{x}(t))$ at time t is greater than 10.

The signal x satisfies $\varphi = \mathbf{G}_{[0,2)} (x > -1)$ if for all time $0 \leq t < 2$, $x(t) > -1$.

The signal x_1 satisfies $\varphi = \mathbf{F}_{[1,2)} \, x_1 > 0.4$ iff there exists time t such that $1 \leq t < 2$ and $x_1(t) > 0.4$.

The signal $\mathbf{x} = (x_1, x_2)$ over two-dimensional space satisfies the formula $\varphi = (x_1 > 10) \, \mathbf{U}_{[2.3,4.5]} \, (x_2 < 1)$ iff there is some time u where $2.3 \leq u \leq 4.5$ and $x_2(u) < 1$, and for all time v in $[2.3, u)$, $x_1(u)$ is greater than 10.

We formally define the semantics of STL as follows:

Definition 4.4 (STL Semantics) STL semantics are defined in terms of the *satisfaction* operator $\models$, for a given signal $\mathbf{x}$ at each time t as follows:

$$
\begin{aligned}
(\mathbf{x}, t) &\models \mu & &\Longleftrightarrow \mathbf{x}(t) \models \mu \\
(\mathbf{x}, t) &\models \neg\varphi & &\Longleftrightarrow (\mathbf{x}, t) \not\models \varphi \\
(\mathbf{x}, t) &\models \varphi_1 \wedge \varphi_2 & &\Longleftrightarrow (\mathbf{x}, t) \models \varphi_1 \text{ and } (\mathbf{x}, t) \models \varphi_2 \\
(\mathbf{x}, t) &\models \mathbf{G}_{[a,b]}\varphi & &\Longleftrightarrow \forall t' \in [t+a, t+b](\mathbf{x}, t') \models \varphi \\
(\mathbf{x}, t) &\models \mathbf{F}_{[a,b]}\varphi & &\Longleftrightarrow \exists t' \in [t+a, t+b](\mathbf{x}, t') \models \varphi \\
(\mathbf{x}, t) &\models \varphi_1 \mathbf{U}_{[a,b]} \varphi_2 & &\Longleftrightarrow \exists t' \in [t+a, t+b] \ s.t. \\
& & &\quad (\mathbf{x}, t') \models \varphi_2 \text{ and} \\
& & &\quad \forall t'' \in (t, t'), (\mathbf{x}, t'') \models \varphi_1
\end{aligned}
$$

Beyond STL Recently, there have been several efforts to consider alternatives to STL to address specific properties that may be cumbersome to express in STL, or inexpressible in STL. Timed regular expressions (TRE) first introduced in 2002 [21] allow expressing localized patterns in CPS behaviors. An efficient monitoring procedure has been proposed for TREs in [150], and an implementation of this procedure is available in the Montre tool [149]. Quantitative regular expressions (QREs) [13, 16] is yet another modeling and programming abstraction for specifying complex numerical queries over data streams. These have been used for analyzing complex behaviors such as arrhythmia in cardiac signals [4].

Finally, differential dynamic logic [120] is a logic for specifying and verifying correctness of hybrid systems. The language allows specifying hybrid systems operationally as hybrid programs and uses automated deduction-based theorem proving tools (such as KeyMaera and its extensions [74, 122]) to verifying program correctness. A key difference between deductive techniques and those that we consider in this chapter is that deductive techniques often require manual intervention in the form of lemmas and proof strategy selection when the automated theorem prover fails to prove program correctness. We omit such techniques from this chapter, and the interested reader can find an extensive treatment in [121].

4.3 Reachability Analysis

Reachability analysis asks whether a hybrid system starting from a set of initial states X_0 can reach any state in a given target set U. The problem is of fundamental importance to hybrid systems since the target set U often describes dangerous states which we wish to avoid reaching during an execution of the system.

Example 4.2 Consider the house heating system shown in Fig. 4.3. It is considered dangerous if the temperature of the house falls below 10 centigrade, while the system continues to be operational and the outside temperature behaves "reasonably":

that is, it must be in the range $[-20, 50]\,°C$ and cannot increase/decrease more than $5\,°C/h$. Let us assume an initial state with $T_r = 27$. Is there a scenario in which the value of $T_r \leq 10$ is possibly under the constraints on the behavior of the outside temperature? Here, the target unsafe set is $U : \{(\ell, Q, Q_L, T_r, W) \mid T_r \leq 10\}$.

Another safety property asks whether it is possible for $T \leq T_{\mathrm{ref}} - 5$ and simultaneously, the heater is in the "OFF" state. Here, the unsafe set is $V :$ $\{(\ell, Q, Q_L, T_r, W) \mid T \leq T_{\mathrm{ref}} - 5 \wedge \ell = \text{OFF}\}$.

Reachability analysis has been studied using a variety of approaches, and for various restrictions on the hybrid automaton model.

4.3.1 Decidability of Reachability

First, it is well known that the reachability problem is undecidable even for simple cases. For instance, in the absence of hybrid dynamics, reachability is undecidable for polynomial ODEs involving 3 or more state variables [82]. Furthermore, for linear dynamical systems, it is known that reachability of a single target state $\mathbf{y}$ from a single given initial state $\mathbf{x}_0$ is decidable. However, the reachability problem of a hyperplane target from a single state initial set (known as the Skolem–Pisot problem) is open. Recent result by Chonev connects the undecidability of this problem to a well-known and open number theoretic conjecture called the *Schaunel* conjecture [44]. Alur and Dill showed that the reachability problem is decidable for timed automata that can be seen as hybrid automata whose continuous variables are all *clocks* with dynamics $\frac{dT}{dt} = 1$. Furthermore, the guard conditions are restricted to comparing clocks with fixed constants, and resets are limited to setting clocks to fixed constant values. The result relies on an *untiming* construction through the region abstraction that produces a finite state automaton which is bisimulation equivalent to the original infinite state timed automaton [12]. However, Henzinger et al. show that if we allow "stopwatches," i.e., clocks that can be stopped by setting $\frac{dT}{dt} = 0$ in certain modes, even the presence of a single stopwatch in a timed automaton model renders the reachability problem undecidable [84]. The timed automaton model can be generalized to rectangular hybrid dynamics that allows the derivative of each variable x_i to lie within an interval $\frac{dx_i}{dt} \in [l_i^{(m)}, u_i^{(m)}]$ in each mode m. Henzinger et al. [84] show that the reachability problem is decidable for *initialized* rectangular hybrid automata that adds the following constraint: for every transition τ from mode m to $\hat{m}$, if the dynamics for $\frac{dx_i}{dt}$ changes going from m to $\hat{m}$, then the variable x_i must be reset to constant value by τ. However, failing this condition, the problem is undecidable, in general. Asarin et al. consider polyhedral hybrid systems that are defined by partitioning the state space into polyhedral regions defining modes and associating each polyhedral region with a mode m and a corresponding constant differential equation $\frac{dx_i}{dt} = c_i^{(m)}$ [23]. Transitions happen when the system moves from one polyhedral region to another in this model. While the reachability problem is decidable for 2D (planar) systems, it is undecidable for

Table 4.2 Summary of a few results establishing decidability/undecidability of reachability for hybrid systems

System	Outcome	Description
Timed automata [12]	Decidable	$\frac{dx_i}{dt} = 1$ for all x_i and all modes, guards $x_i\{\leq, \geq, =\}c$ and resets $x_i := c$
Stopwatch automata [84]	Undecidable	Timed automata + at least one stopwatch with $\frac{dx_i}{dt} = 0$ allowed in some modes
Initialized rectangular automata [83, 84]	Decidable	Rectangular dynamics $\frac{dx_i}{dt} \in [l_i, u_i]$ for each x_i and mode, guards + resets as in timed automata. Transition between different dynamics should reinitialize a variable
Polyhedral hybrid automata [23]	Undecidable	Decidable for 2 or fewer state variables
O-minimal hybrid automata [98]	Decidable	Automata whose guards, reset maps, and flows can be defined in an O-minimal logical theory

systems involving 3 or more variables. Table 4.2 summarizes some of the significant results on decidability/undecidability of reachability for various classes of hybrid systems.

Understanding the boundary between decidable and undecidable subclasses has been an active area of investigation with some open problems. However, early results showed that seemingly simple hybrid automata models can exhibit a high degree of complexity in terms of their behaviors. As a result, the focus has gradually shifted from finding new decidable classes to finding practical algorithms that can be useful to analyze models of interest to practitioners, even if the overall problem is known to be undecidable.

4.3.2 Reachability Using Over-Approximations

As discussed previously, the problem of deciding questions of reachability is undecidable. However, for many practical systems, the problem of reachability analysis can be resolved by computing over-approximations of the reachable set of states starting from the initial set X_0, or alternatively, by computing over-approximations of the backward reachable set starting from the unsafe/target set U. This is pictorially illustrated in Fig. 4.4. Over-approximations can be obtained for a *finite* time horizon if the value of T is finite, or an *infinite* time horizon if $T = \infty$. Naturally, infinite-time horizon approximations are more complicated and approached using deductive methods discussed in subsequent sections. The rest of this section focuses, for the most part, on finite-time horizon reachability analysis.

Let $S \subseteq X$ be a subset of states of a system and X_0 be the initial set. We say that S is a (forward) over-approximation for a time interval $[0, T)$ iff for any initial state $\mathbf{x}(0) \in X_0$, any state $\mathbf{x}(t)$ reachable from $\mathbf{x}(0)$ at time $t \in [0, T)$ belongs to S.

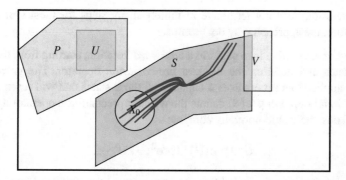

Fig. 4.4 S: Over-approximation of reachable set of states includes the initial condition X_0 and all states reached by trajectories starting from X_0. P: Backward-over approximation containing all states that can reach U. The set of states U is proven unreachable since $U \cap S$ is empty, or alternatively, However, the set of states V may or may not be reachable since $V \cap S$ is not empty

Using the forward over-approximation S, we may conclude that U is unreachable if $U \cap S = \emptyset$.

Alternatively, we can prove unreachability by computing a set of backward reachable states $P \subseteq X$ such that $U \subseteq P$ for the target set and every trajectory of the system $\mathbf{x}(\cdot)$ such that $\mathbf{x}(t) \in U$ at time $t \in [0, T]$ must satisfy $\mathbf{x}(0) \in P$. If $P \cap X_0 = \emptyset$, we may now conclude that no run of the system starting from X_0 may reach U within the given time horizon. Figure 4.4 illustrates how a backward reachable set can be used to prove unreachability, as well.

4.3.2.1   Approximate Reachability: Overview

We will now discuss how reachability analysis works at the high level, focusing first on computing over-approximations of forward reachable states starting from the initial state X_0 and an initial mode ℓ_0 of the hybrid system. The approach is based on *symbolic model checking*, wherein a set of reachable states is iteratively computed by repeatedly applying the *post-condition* operator to the initial set of states. The post-condition operator applied to a set of states S captures all the states reachable from S in a single "computational" step. Let $\mathsf{post}(S)$ denote the post-condition of S. Thus, we would normally compute

$$X_0 \cup \mathsf{post}(X_0) \cup \mathsf{post}^2(X_0) \cup \cdots.$$

However, there are three core problems with this approach:

1. Hybrid systems combine the continuous evolution of state variables with discrete transitions. There is no natural notion of a single discrete computational step.
2. The sets $\mathsf{post}^k(X_0)$ become increasingly complicated to represent in a computer, making the process prohibitively expensive.

3. The iteration does not terminate in finitely many steps for most systems, and therefore, the approach may not terminate.

The other alternative is to perform a *backward* iteration, starting from the unsafe set of states and iterating the weakest *pre-condition* operator. The precondition operator applied to a set of states S captures all those states that will reach S in one computational step. Let $\mathsf{pre}(S)$ denote the weakest precondition operator applied to a set S. Thus, we would normally compute

$$U \cup \mathsf{pre}(U) \cup \mathsf{pre}^2(U) \cup \cdots.$$

Once again, the same three problems we encountered for post-conditions arise for preconditions as well.

Reachability algorithms overcome the three key problems mentioned above through two important and closely intertwined ideas: (1) *abstraction* of the hybrid system by a simpler model; and (2) *abstract* (over-approximate) representation of sets of states by geometric primitives such as rectangles, polyhedra, ellipsoids, zonotopes, and Taylor models.

4.3.2.2 Abstractions

A system abstraction seeks to replace a given hybrid automaton $\mathscr{S}$ by another finite or infinite state system $\mathscr{T}$ over the same state space and set of modes as $\mathscr{S}$, such that *every trajectory of $\mathscr{S}$ is also a trajectory of $\mathscr{T}$*. In this case, we will write $\mathscr{S} \succeq \mathscr{T}$. Note, however, that $\mathscr{T}$ may have more trajectories that are not trajectories of $\mathscr{S}$. It is easy to show that any reachable state of $\mathscr{S}$ starting from a given state X_0 is also reachable in $\mathscr{T}$ (starting from a suitable superset of X_0).

Early approaches considered *finite state* abstractions that transform a given hybrid automaton into a finite state machine which simulates the original system, or in special cases, such as timed automata or initialized rectangular automata, exhibits a stronger connection through bisimulation relations [15, 24]. However, most systems of interest have been observed not to have finite bisimulation quotients. To circumvent this, Girard and Pappas consider the notion of an *approximate* bisimulation relation that is defined by means of a comparison metric between states of the two systems so that as the systems evolve in time starting from related initial states, the distance decays over time [79]. The notion of approximate bisimulation relations expands the class of systems for which we may find suitable finite state abstractions with some property preservation guarantees. Nevertheless, it remains the case that finding finite (approximate) bisimulation quotients is rare and seldom feasible for practical systems. Other approaches for finding finite abstractions have employed the use of predicate abstractions with counterexample guided refinements [11]. While the approach can perform well if the right set of predicates were to be provided, the problem of deriving such a set of predicates is often hard in practice. Furthermore, the refinement loop may often generate a large number of predicates making the finite state abstraction prohibitively expensive. More recently, hybridization approaches have investigated the abstraction of more complex dynamics such as nonlinear ODEs, linear hybrid systems by simpler dynamics such as rectangular automata [22, 50, 125, 132]. On one hand, these approaches can provide tradeoffs between the accuracy of the abstraction and its size. On the other hand, these approaches can also suffer from the curse of dimensionality since they rely on decomposing the state space into small compact regions in order to bound the error between the original system and the abstraction.

A related class of abstractions seeks to eliminate continuous dynamics by replacing the ODEs by relational models that relate a state $\mathbf{x}$ and a future reachable state $\mathbf{x}'$. Building such relational models can then allow off-the-shelf tools for model checking infinite state discrete systems to tackle the verification problem. The idea of relationalization, though implicit in earlier works such as Podelski and Wagner [124], was first formalized by Tiwari and Sankaranarayanan under the notion of relations that abstracted time away as well as relations that captured change in state over a fixed time step [136, 154]. Subsequent work studied various ways of constructing these relations that tracked time explicitly as "time-aware" relations [109]. Recently, Chen et al. explored the construction of these relations for nonlinear systems [42]. One of the key drawbacks for existing methods lies in the lack of approaches to refine these relations once they are constructed. A related issue lies in the tradeoff between constructing a coarse but simple relation versus a more complex and less conservative approximation. Approaches that can construct "multi-scale" relations that selectively refine interesting parts of the relation remain unexplored at the time of writing.

4.3.2.3 Flowpipe Computation

Flowpipe computation approaches rely primarily on computing reachable sets by approximating the time trajectories of the system rather than abstracting the system itself. A large variety of flowpipe computation approaches have been proposed in the literature, and many proposed techniques are supported by tools for experimental validation. Some of these tools are specialized to linear hybrid systems, while others tackle a larger class of nonlinear systems. Most flowpipe construction methods are instances of the forward reachability computation using the post-condition operator presented previously. However, in order to extend this scheme to hybrid systems, it is important to consider four important aspects (illustrated in Fig. 4.5).

1. A systematic way to represent sets of continuous states. Since not all sets are representable inside a computer, common representations include intervals, convex polyhedra, ellipsoids, zonotopes, support vectors, and Taylor models.
2. Once a representation is chosen, we need to compute sets of reachable states for given nonlinear dynamics inside a mode. This operation has been variously termed "time elapse," "continuous post-condition," or "continuous image computation" in the literature.

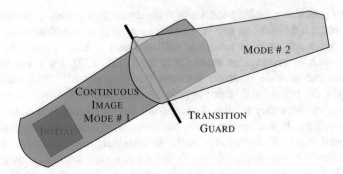

Fig. 4.5 Illustration of basic primitives for flowpipe computation. Starting from initial set in mode 1, we perform a continuous image computation for a given time horizon. Next, we compute states from which a transition to a different mode (mode 2) is possible. From these states, we compute the reachable states for mode 2 shown in orange

3. The effect of a discrete transition must be computed. This operation is called "discrete post" or "discrete image" computation.
4. Finally, the primitives mentioned above must be integrated into a model checking scheme that employs them in order to compute the reachable set estimation for the system as a whole. To this end, operations such as subsumption checks, aggregation, simplification, and extrapolation are often used.

The fundamental scheme of performing forward reachability using a combination of continuous and discrete image computation with specialized operators has been carried out through a variety of approaches, which are summarized in Table 4.3.

Table 4.3 presents an overview of selected approaches-based classified in terms of the representations used for sets, and the type of models handled by the approach. As we note in the table, there has been significant recent work in scaling up the reachability analysis of linear ODEs to millions of variables [25], linear hybrid systems to many hundreds of variables [73], and nonlinear systems up to a few tens of variables (assuming nonchaotic and nonstiff ODEs) [7, 41]. Furthermore, a variety of recent tricks including decomposition of a monolithic model into smaller submodels that can be exploited by the reachability analysis [43]. However, significant variability in performance is seen across models. Furthermore, many of the approaches have numerous tunable parameters that need to be carefully adjusted for each model to obtain optimal performance. Another important drawback lies in the lack of support for richer models of hybrid systems that can incorporate features such as lookup tables, gain scheduling, predictive models, and learning-enabled loops involving neural networks. Supporting these features remains the subject of ongoing research at the time of writing this survey.

Table 4.3 Reachability analysis approaches using flowpipe construction at a glance

Reference, *Representation*, and Dynamics	Remarks
Krogh et al. [45, 46], *Polyhedra*, NLHybrid	Precise flowpipes for linear systems. Uses numerical optimization for nonlinear systems. Builds abstract finite state model for checking
Dang et al. [31, 49], *Orthogonal Polyhedra*, LHybrid	Introduced face lifting algorithm for computing reachable sets
Kurzhanski and Varaiya [95], *Ellipsoids*, LODE	Uses ellipsoidal calculus and introduced the idea of support vectors. Handling of discrete transitions requires approximations due to ellipsoid–hyperplane intersections
Mitchell and Tomlin [108], Level Sets, *NLHybrid*	Uses Hamilton–Jacobi PDEs solved using state-space discretization. Solves viability problems (computation of control and reachability)
Girard [78], *Zonotopes*, LHybrid	Efficient image computation for continuous dynamics. Handling of discrete transitions remains problematic similar to ellipsoids. Available as part of Spaceex tool
Frehse et al. [73], *Support Functions*, LHybrid	Efficient image computation and handling of discrete transitions. Implemented in tool SpaceEx
Berz and Makino [27, 105], *Taylor Models*, NLODE	No handling of discrete transitions. Introduced higher-order interval methods for guaranteed ODE integration
Chen et al. [40, 41], *Taylor Models*, NLHybrid	Extends techniques from Berz et al. with handling of discrete transitions
Althoff et al. [7, 8], *Multiple*, NLHybrid	Combination of multiple set valued representations including nonlinear zonotopes, matrix zonotopes, and Taylor models for nonlinear hybrid systems reachability analysis
Bak and Duggirala [25], *Polyhedron*, LODE	Using simulations to implicitly compute reachable sets and resolve safety properties. Shown to scale beyond hundreds of thousands of state variables

NLHybrid: Nonlinear hybrid, NLODE: Nonlinear ODEs with continuous RHS, LODE: Linear ODEs, LHybrid: Linear hybrid

4.3.2.4 Constraint Solvers and Reachability

Another approach relies on using constraint solvers for estimating reachable sets that can be used to prove properties of interest. This approach essentially integrates many of the ideas summarized thus far naturally into a constraint-solving framework. The approach has been termed the SAT-modulo ODE approach, originated in the work by Ratschan and She [129, 130], Ratschan et al. [72], incorporated into tools such as HySAT [85]. More recently, the approach was formalized by Gao et al. into *delta-decision* procedures for proving properties of hybrid systems [76]. The key idea is to provide procedures that can either conclude that a system does not satisfy a property or that the system under a bounded perturbation violates a perturbed property, under a well-defined perturbation model. A similar idea is presented independently by Ratschan wherein termination of the reachability analysis is guaranteed under the condition of robust safety wherein a bounded perturbation of the system continues to satisfy the safety property in question [130].

4.3.2.5 Simulation-Guided Reachability Analysis

A significantly different approach for estimating reachable states relies on using simulations coupled with user-provided annotations. The main idea is to obtain a simulation trajectory and to bloat the trajectory in such a way that for each initial state included in the bloated trajectory, the trajectory beginning at this initial state is also included in the bloated trajectory. Such a bloated trajectory is also known as a *reach tube*. The first idea to compute reach tubes was by exploiting the sensitivity of the numerical solutions of an ODE to perturbations in its initial conditions [57]. A similar idea was also explored in [91] for continuous dynamical systems with inputs. Recent advances in simulation-based reachability have shown promise in being able to handle models with industrial-scale complexity [62, 63, 69]. These techniques rely on a user-provided annotation in the form of a discrepancy function. Essentially, a discrepancy function provides a mechanism of bounding the distance between adjacent trajectories as a function of the distance between the initial states for the trajectories. Thus, with a reasonably tight discrepancy function, an over-approximation of the reachable state space can be obtained by performing a (potentially) small number of simulations.

4.4   Techniques Based on Safety Invariants

Techniques based on reachability are highly automated and have shown remarkable progress. However, when faced with highly nonlinear plant models, and especially in the presence of discrete switching, these techniques can suffer from loss of precision.

A different approach is offered by semi-automated techniques based on *invariants*. The simplest definition of an invariant is that it is a set such that starting from an element of this set, the time evolution of the system trajectories remains within this set at all times. Typically, we consider the forward time evolution of the system trajectories (i.e., time increases along a trajectory), and thus focus on *forward invariants*. Given a set of *safe* states S, an *invariant* set $\mathscr{I}$ is called a *safety invariant*, if $\mathscr{I} \cap \overline{S} = \emptyset$. Various kinds of invariants have been proposed in the literature to help automate proofs of safety. The prime challenge in invariant-based verification is that it is typically very difficult to find invariants in an automated fashion, and may require human insight.

A key body of work in invariant-based verification is with the use of the KeyMaera and KeyMaeraX theorem proving tools [74, 122]. These tools allow a user to systematically construct the proof of safety of a hybrid system (modeled as a hybrid program). The user has the choice of introducing various kinds of invariants to automate safety proofs. An important class considered is that of *differential invariants* [120]. These are invariants that allow proving the properties of a differential equation without having to solve the equation itself. See [120] for a comprehensive survey. There are certain specializations of invariant-based reasoning that we discuss now.

Control Envelopes. Arechiga et al. [20] present the problem of safety verification for embedded control systems. Here, given a model of the continuous dynamics of a plant, the technique postulates the computation of an *envelope-invariant* pair. The technique assumes that the plant dynamics are given by an ODE of the form:

$$\dot{\mathbf{x}} = f(\mathbf{x}, \mathbf{u}),$$

where $\mathbf{x}$ is the state of the plant, and $\mathbf{u}$ is the control input from some set U. We assume that we are given an *invariant* set N (a subset of the plant state space X). We then compute a *control envelope* E that is a function from X to $\mathscr{P}(U)$.[2] The pair (N, E) satisfy the property that for all times t, for any given state $\mathbf{x}(t) \in N$, if the input provided by the controller $\mathbf{u}(t)$ is in the set $E(\mathbf{x})$, then for all $t' > t$, $\mathbf{x}(t') \in N$. Further, if the intersection of N and the set of unsafe states is empty, this gives us a proof of the safety of the closed-loop control system. They also provide specific examples of control envelope-invariant pairs, but does not provide a procedure to compute such pairs for general systems. Computing such control envelopes remains an interesting problem that has attracted recent interest due to applications to runtime monitoring.

Barrier Certificates. A barrier certificate is a type of a safety certificate. Let X be the state space of a system specified by the ODE $\dot{\mathbf{x}} = f(\mathbf{x})$, let I be the set of initial states for the system, and let S be the set of safe states for the system. Then, a barrier certificate is defined as a differentiable function B, which has the following properties:

1. $\forall \mathbf{x} \in I : B(\mathbf{x}) \leq 0,$
2. $\forall \mathbf{x} \in \overline{S} : B(\mathbf{x}) > 0,$
3. $\forall \mathbf{x} \in X : (B(\mathbf{x}) = 0) \implies \frac{\partial B}{\partial \mathbf{x}} \cdot f(\mathbf{x}) < 0$

The intuitive idea is that the set $B(\mathbf{x}) = 0$ serves as a barrier preventing the trajectories of the system that originate in the set I from reaching the set $\overline{S}$. As B is a continuous and differentiable function, every trajectory that starts at a point where B is negative must pass a point where B is zero before reaching a point where B is positive. However, because the Lie derivative of B along the manifold where $B(\mathbf{x}) = 0$ is negative, at each point, the system dynamics forces the B function from not increasing. Barrier certificates were first proposed in [126, 127], and a procedure

---

[2]For a set X, let $\mathscr{P}(X)$ denote its power set.

based on Sum-of-Squares programming was proposed for finding barrier certificates for systems with polynomial dynamics. These techniques were extended for systems with certain nonpolynomial dynamics [81, 118]. However, the problem of finding barrier certificates for general nonlinear systems remains open.

Simulation-Guided Search for Invariants. Though invariant-based techniques show a lot of promise to prove safety of systems with highly nonlinear and hybrid dynamics, finding the required invariants remains a hard problem. In [145, 146], the authors suggested a simulation-guided technique to estimate the region-of-attraction (ROA) for a given dynamical system. The main idea in this work was to convert a set of bilinear matrix inequalities encountered in estimating the ROA (which are computationally expensive to solve) into linear matrix inequalities, which are computationally less expensive.

In [90], the authors propose a technique to iteratively compute an invariant using simulations, based on the idea of estimating *Lyapunov functions*. Given a system of the form $\dot{\mathbf{x}} = f(\mathbf{x})$, where $f(\mathbf{0}) = 0$, a Lyapunov function $V(\mathbf{x})$ is a function that is positive everywhere except when $\mathbf{x} = \mathbf{0}$, its Lie derivative $\frac{\partial V}{\partial \mathbf{x}}$ is negative everywhere except at $\mathbf{0}$, and at $\mathbf{x} = \mathbf{0}$, both the value of V and its Lie derivative is 0. A Lyapunov function is a tool that can be used to prove stability of a system to the point $\mathbf{x} = \mathbf{0}$. Furthermore, any level set of the Lyapunov function, i.e., $L(\mathbf{x}) = \{\mathbf{x}|V(\mathbf{x}) = \ell\}$ is an invariant for the system. The iterative procedure in the technique proposed in [90] is as follows: (1) the technique fixes the form of a candidate Lyapunov function as some polynomial $P(\mathbf{c}, \mathbf{x})$, where $\mathbf{c}$ is a vector of coefficients of the polynomial function, (2) it uses a set of discrete-time trajectories of the system from a given set of initial states, and uses these to impose constraints on $\mathbf{c}$, (3) it solves the constraints using an appropriate solver to obtain a candidate Lyapunov function, (4) it searches for counterexample for the candidate using an SMT solver, and (5) if a counterexample exists, it is added to the set of initial conditions used in step 2, and the method repeats; else, it terminates with an answer.

The key step is in the formulation of constraints in step 2. For a fixed polynomial form with unknown coefficients, imposing positivity of V at each point in a system trajectory results in a linear constraint. Suppose we are given two points in a discrete-time trajectory of the system (say $\mathbf{x}_n = \mathbf{x}(t_n)$, and $\mathbf{x}_{n+1} = \mathbf{x}(t_{n+1})$), where $t_{n+1} > t_n$. Then, a sufficient condition for the negativity of the Lie derivative is to impose that $V(\mathbf{x}_n) - V(\mathbf{x}_{n+1}) > 0$. Note that this is again a linear constraint in the coefficients of V as $\mathbf{x}_n$ and $\mathbf{x}_{n+1}$ are known. Thus, solving the constraints in step 3 can be done using a standard linear-programming solver.

Step 4 also merits a remark. A candidate Lyapunov function (or by extension a candidate invariant that is the level set of the candidate Lyapunov function) obtained in Step 4 satisfies the required conditions for being a valid Lyapunov function (resp., invariant) on the selected set of system trajectories, but there is no guarantee that these conditions are met globally in the state space. Thus, the method uses a satisfiability modulo theories (SMT) solver that is equipped to reason about satisfiability of arbitrary nonlinear queries; δ-sat solver dReal is such a solver [76]. It returns an answer **unsat** if the query is unsatisfiable, otherwise returns a interval of width δ in the state space where the query may be satisfiable. As checking validity of a condition is equivalent to checking the satisfiability of its negation, an **unsat** answer from dReal helps us establish the conditions required for a given set to be an invariant.

4.5   Falsification Techniques

In this section, we will review techniques to perform *requirement-driven* test generation of CPS models. There are several automated test generation procedures and heuristics that attempt to tackle this problem by viewing it as a special case of software testing. Commercial tools such as the Simulink Design Verifier™ (SLDV) toolbox from the Mathworks [75, 99], the Reactis® [131] tool, and the TestWeaver tool from QTronic [89] are notable for their adoption within industrial MBD practice.

The Reactis Tester tool evaluates open-loop controller models with a patented technique to generate test inputs using a combination of random and targeted methods. The targeted phase of the tool uses data structures to store intermediate states, and constraint-solving algorithms to search for previously uncovered coverage targets. SLDV uses techniques based on SAT modulo theories (SMT) in conjunction with the Prover tool to automatically generate test inputs to maximize coverage criteria. SLDV is intended for open-loop (discrete-time) controller models, as it cannot process closed-loop (hybrid) models.

The TestWeaver tool does test generation with the goal to maximize state coverage of the underlying system (where coverage is defined in a specific fashion). The test generation algorithm itself is based on proprietary heuristics. The tool relies on the user to quantize the inputs to the model-under-test, discretize the time domain, and also to manually identify system variables that are most sensitive to the inputs. This user intervention may require an understanding of the system dynamics and engineering intuition to use the tool effectively.

With the exception of certain features in TestWeaver, the above tools are primarily focused on testing the controller models for CPS systems, while unable to effectively reason about the plant/environment model. Furthermore, the properties that these tools check are typically hand-coded by the user and tend to be simpler static properties (such as the bounds on a signal value over a specified time interval).

We now discuss falsification techniques that overcome some of the shortcomings of the existing commercial techniques in various ways:

1. They allow specifications expressed in formal specification languages such as those based on signal temporal logic (STL). This allows complex temporal properties over continuous-valued, continuous-time signal to be seamlessly specified.
2. They can effectively search both continuous and hybrid state spaces that arise from closed-loop models.

3. They can be augmented with metrics to measure coverage of continuous and hybrid state spaces.
4. They can combine search for bugs in the software controller with corner case behaviors in the continuous plant model.

In this chapter, we discuss two main classes of such techniques. The first class of techniques allows falsifying closed-loop specifications of temporal behavior with the help of black-box optimization tools. The second class of techniques combines a novel exploration of plant model behaviors with a technique inspired by multiple shooting methods found in numerical ODE solving with symbolic execution techniques for analyzing controller code.

4.5.1   Falsifying Temporal Specifications Using Optimization

A key technology that enables falsification techniques is *quantitative satisfaction semantics* for real-time temporal logics. Robust satisfaction semantics were proposed for metric temporal logic by Fainekos and Pappas in their seminal paper [68], while quantitative semantics for STL were proposed by Donzé and Maler [58], which we now explain.

Quantitative Semantics of STL. For a formula φ in a given logical formalism and a signal trace $\mathbf{x}$, Boolean satisfaction semantics for the logic provide a *true/false* answer for whether $\mathbf{x}$ satisfies φ. Quantitative semantics extend this notion to *robust satisfaction*, i.e., they define a *robust satisfaction degree* (abbreviated as *robustness*) of φ by $\mathbf{x}$. The intuition is that if the robustness value is a positive number, then $\mathbf{x}$ satisfies φ; if it is negative, it does not satisfy φ, and the magnitude of the robustness degree indicates how *strongly* φ is satisfied (or violated).

We provide the formal robustness semantics for STL below in terms of a function ρ that maps a given trace $\mathbf{x}$, a formula φ, and a time t to a real number. This function for a predicate of the form $f(\mathbf{x}) > 0$ at time t is simply the value of $f(\mathbf{x})$ at time t, i.e., $\rho(f(\mathbf{x}) > 0, \mathbf{x}, t) = f(\mathbf{x}(t))$. Then, ρ is defined inductively for every STL formula using the following rules:

$$\rho(\neg\varphi, \mathbf{x}, t) = -\rho(\varphi, \mathbf{x}) \tag{4.1}$$

$$\rho(\varphi_1 \wedge \varphi_2, \mathbf{x}, t) = \min(\rho(\varphi_1, \mathbf{x}, t), \rho(\varphi_2, \mathbf{x}, t)) \tag{4.2}$$

$$\rho(\mathbf{F}_I\varphi, \mathbf{x}, t) = \sup_{t' \in t+I} \rho(\varphi, \mathbf{x}, t') \tag{4.3}$$

$$\rho(\mathbf{G}_I\varphi, \mathbf{x}, t) = \inf_{t' \in t+I} \rho(\varphi, \mathbf{x}, t') \tag{4.4}$$

$$\rho(\varphi_1\mathbf{U}_I\varphi_2, \mathbf{x}, t) = \sup_{t' \in t+I} \left(\min\left(\rho(\varphi_2, \mathbf{x}, t'), \inf_{t'' \in [t,t']} \rho(\varphi_1, \mathbf{x}, t'') \right) \right) \tag{4.5}$$

By convention, the robustness of φ by $\mathbf{x}$ is then simply $\rho(\varphi, \mathbf{x}, 0)$. If we omit the time argument, the implicit assumption is that we are computing the robustness at time 0, i.e., $\rho(\varphi, \mathbf{x}) = \rho(\varphi, \mathbf{x}, 0)$.

We recall that a closed-loop model M can be viewed as a function mapping finite-time input signals $\mathbf{u}$ (defined over time $[0, T]$) to output signals $\mathbf{y}$. For simplicity, we assume that the specification φ is an appropriate STL formula over output signals. Then, the falsification problem can be restated as a search for an input signal $\mathbf{u}$ such that $\rho(\varphi, \mathbf{y}) < 0$. The central idea in most falsification tools is to solve this problem by solving the following optimization problem:

$$\mathbf{u}^* = \underset{\mathbf{u} \text{ s.t. } \mathbf{y}=M(\mathbf{u})}{\arg\min} \rho(\varphi, \mathbf{y}) \qquad (4.6)$$

If we find a $\mathbf{u}^*$ such that $\rho(\varphi, M(\mathbf{u}^*)) < 0$, then we have effectively found a violation of the specification, or successfully falsified the model. While the above setup seems straightforward, there are several caveats.

Input Signal Parameterization. The first is that optimizing over a dense-time input signal is an infinite-dimensional optimization problem. A common approach is to make the search space finite by assuming a *finite parameterization* of the input signal space. For example, one of the approaches adopted by tools such as S-TaLiRo [19] and Breach [55] is to introduce n uniformly spaced discrete time points $t_0, \ldots, t_{n-1}$ along the time axis, also known as *control points*. Here, $t_0 = 0$, and $t_{n-1} = T$. Then, the input signal $\mathbf{u}$ is defined in terms of $(\mathbf{u}_0, \ldots, \mathbf{u}_{n-1})$, as follows: for all $i \in [0, n-1]$: $\mathbf{u}(t) = \mathbf{u}_i$ if $\frac{i}{n-1}T \leq t < \frac{i+1}{n-1}T$. In simpler terms, the signal $\mathbf{u}(t)$ is obtained by *constant interpolation* over values $(\mathbf{u}_0, \ldots, \mathbf{u}_{n-1})$ equally spaced in time. This notion can be generalized by introducing variably spaced time points, and user-defined interpolation functions (such as piecewise linear, splines, etc.).

Another approach is to define a finite grid over the input signal space, i.e., in addition to discretization of the time axis, we also quantize the value axis of the signal. The input signal is ultimately constructed using interpolation over points over this finite grid. (See Fig. 4.6a for an illustration.) Such a grid can then be refined iteratively by the optimization algorithm. This is the approach explored in [52].

Nonconvex Search Space. Most optimization tools critically rely on the optimization problem being defined over a convex space, which enables gradient descent-like optimization methods. Further, such approaches may also require the exact analytic gradient to be available. The optimization problem set up in Eq. (4.6) almost never has such nice properties. First, the method M can be an arbitrary hybrid dynamical system with a high-dimensional state space. Further, the cost function ρ is itself not a smooth function of its input. Thus, most falsification tools rely on *black-box* optimization techniques such as the derivative-free Nelder–Mead technique used in Breach [55], heuristic search techniques such as genetic algorithms [19], Ant Colony optimization [18], the Cross Entropy method [134], or stochastic gradient descent combined with discrete Tabu search [52]. A common theme in these

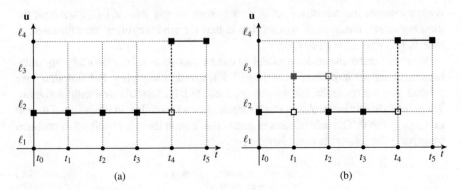

Fig. 4.6 (**a**) Example of using a finite grid to approximate an input signal. The input signal $\mathbf{u}(t)$ is obtained by constant interpolation over the sequence $(\ell_2, \ell_2, \ell_2, \ell_2, \ell_4, \ell_4)$ over the time domain $(t_0, t_1, t_2, t_3, t_4, t_5)$. (**b**) Example of a grid neighbor of the input signal shown in (**a**)

methods is to evaluate the cost function, i.e., the robustness value for a heuristically sequence of points in the input space, and generally choose input points with lower costs. The exact heuristics of how the sequence of inputs is chosen depends on the specific algorithm in question. For example, in Fig. 4.6b, we show how an input signal corresponding to the grid neighbor of the input signal in Fig. 4.6a is chosen for cost function evaluation.

Recently, given the immense success of machine learning techniques in learning and approximating black-box functions, there have been efforts to apply such methods to the falsification problem [6, 51, 92].

Yet another class of methods focuses on simultaneously trying to maximize coverage of the hybrid state space and find a violation of the property of interest. The technique in [59] iteratively computes the input signal incrementally using the rapidly exploring random trees (RRT) algorithm used for motion planning. The RRT algorithm is tuned to pick goal states that maximize a weighted combination of the (incremental) robustness of the output signal, and a coverage metric over the continuous state space of the closed-loop model. In [5], the authors combine a coverage metric on the input signal space with a machine learning technique to classify already covered regions in the input space. In [54], the authors define a hybrid distance metric to obtain coverage over discrete mode switches in the closed-loop model.

4.5.2 Falsification Using Trajectory Splicing

Thus far, the approaches to falsification are *single shooting* approaches that search over a single trajectory starting from some initial condition that falsifies the specification. An alternative approach is to use *multiple shooting*, wherein the approach

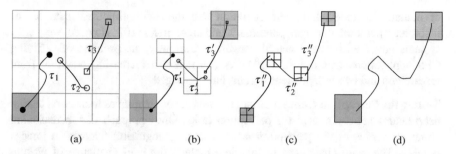

Fig. 4.7 An illustration of the trajectory splicing approach: (**a**) segmented trajectory reaching unsafe states (shaded red) starting from initial states (shaded blue), (**b**) refining an abstract counterexample and narrowing the inter-segment gap, (**c**) further narrowing the gap by refinement, and (**d**) a concrete trajectory with no gaps

splices a collection of trace segments that take us from one state to another in the state space. An approximate trajectory takes a sequence of such trace segments with possible gaps between the ith trace segment and the $(i+1)$th segment. The approach then iteratively narrows the gap through a suitable optimization procedure, leading from an initial sequence of segments to a trajectory of the system obtained when the gaps are reduced to zero. The trajectory splicing approach using local gradient descent was first proposed by Zutshi et al., inspired in turn by collocation-based approaches to integrating systems of differential equations and similar multiple shooting approaches to optimal control (see [152]). Subsequently, this was extended to a larger class of systems using graph-based search and iterative refinement [153]. See Fig. 4.7 for an illustration of the iterative refinement procedure used in the tool S3CAM that performs trajectory splicing for arbitrary hybrid systems.

Trajectory splicing is essentially a state-space exploration technique for hybrid systems. Recall that in many CPS applications, the closed-loop system model is often expressed as a hybrid or continuous plant model composed with a discrete software controller. It is possible to enhance the efficacy of splicing-based falsification techniques by combining trajectory segments explored in the plant's state space by symbolic execution of the controller. This approach was explored in [151], and uses symbolic path exploration tools based on SMT solvers. The scalability of this technique is currently limited by that of existing SMT solvers.

4.6 Challenge Problem: Verification of AI-Based Systems

AI-based systems, especially those based on artificial neural networks (ANNs) and by extension, deep neural networks (DNNs) have gained increasing prominence in CPS applications where they support perception tasks from rich image, LIDAR, and other sensor data [80], and the design of control using ideas such as reinforcement learning [142]. However, a key drawback of neural networks lies in the inability of humans to understand their operation and the well-publicized instances of incorrect operation that can potentially endanger life [110]. How do we verify systems governed by deep neural networks? Currently, the problem of verifying CPS applications that use ANNs/DNNs has received increasing attention from researchers and two independent streams have emerged.

Testing for Perception Components. The first set of techniques focuses on testing deep neural networks used for perception tasks. One approach lies in reasoning about properties of the perception tasks such as recognizing features in images reliably. The main challenges in this area include the hard challenge of writing behavioral specifications for perception tasks that involve feature rich input sources such as images, videos, and LIDAR data streams. Another challenge lies in the sheer size of the network in terms of the number of neurons and the depth of the network, which makes existing verification tools hard to apply directly. *Adversarial test generation* is a popular paradigm which has spawned a number of research papers, focused on identifying mild perturbations to images that result in failed object recognition. Typical approaches use gradient search over the network, or a mixed integer linear programming problem to analyze the robustness of classification tasks to a set of changes to pixels in the images [133, 144]. Another related direction of research is framed as a search for "adversarial" inputs that expose problems with the current network. A linear programming-based approach for finding adversarial inputs is presented by Bastani et al. [26]. A related approach for finding adversarial inputs using SMT solvers that relies on a layer-by-layer analysis is presented by Huang et al. [87]. Currently, falsification-based approaches have proven advantageous for these tasks given the sheer size and complexity of the neural networks involved. Yet, the number of simulations needed, and time taken for each simulations remain astronomically high. Currently, it is important to derive approaches that can significantly reduce both these bottlenecks for falsification.

Dreossi et al. present an approach that uses falsification to test neural network-based perception systems used in autonomous driving by manually generating scenes with known ground truth data [60, 61]. A more elaborate end-to-end approach has been proposed by Abbas et al. using falsification tools to drive the process of testing various scenarios and popular gaming engines to recreate the driving scenarios in order to provide visual inputs to the cameras [3], or the use of robotic simulators to create visual inputs to the perception algorithm (in concert with a closed-loop vehicle dynamics model and a controller) [147].

Testing/Verification of AI-Based Control. The second stream of work considers the safe learning of control laws that take the form of neural network, starting from high level behavioral end-to-end specifications. Here, verification approaches have reported more initial successes due to the much smaller size of neural networks involved in these tasks, when compared to perception tasks. A fundamental primitive that arises in such verification involves the propagation of interval uncertainties over a neural network. Recently, there has been a surge of interest in this problem starting from an approach that linearizes the nonlinear activation function [128], the Reluplex solver by Katz et al. that modifies the Simplex approach to handle piecewise linear constraints posed by the nonlinear rectified linear units [93], an approach using a reduction to mixed integer solvers [101], a combination of local and global search [65], and an integration of convexification with conflict clauses driven by a SAT solver [67]. Whereas these works have considered the neural network in isolation, recent work by Dutta et al. has focused on integrating the learning and verification in a systematic manner using both plant and controller models [64, 66]. The work in [148] uses a closed-loop model of a plant and a neural network-based controller (trained using reinforcement learning) and obtains a barrier certificate for the system. The technique relies on using simulations to find an appropriate barrier certificate and uses the interval constraint propagation-based SMT solver dReal [76] to provide the ultimate proof of safety.

While most of the above approaches have initiated the work of tackling the hard problem of verifying AI-based systems, there is more work to be done. Scaling current approaches to real-world DNNs is a significant challenge, as is the challenge of expressing verification goals for such algorithms in a clean mathematical formalism.

4.7 Conclusion

In this chapter, we reviewed some of the main topics in the formal verification and falsification of cyber-physical systems. The key challenge for such systems is the coupling of the continuous-time behaviors of a physical component with discrete-time control software in the presence of an uncertain environment. Such systems can be mathematically modeled as *hybrid dynamical systems*. Proving safety of such systems over a bounded time horizon can be addressed by solving the *reachability analysis* problem for such hybrid systems, which involves over-approximating the set of behaviors of the system, and proving that this set does not include the unsafe behaviors. An alternate approach is to use *falsification* techniques that seek to find incorrect system behaviors through systematic search procedures. A key assumption for verification or falsification is the ability to express safe behaviors of a system in a formal specification language. We review signal temporal logic, which is a formal logic capable of expressing several interesting properties for CPS applications. We conclude the chapter with a challenge problem that will test the limits of existing verification and falsification techniques.

Acknowledgements We dedicate this chapter to the memory of Dr. Oded Maler, a great friend and collaborator, who shaped our knowledge and perspectives on this vast topic through numerous insightful discussions over the years. The authors also acknowledge contributions from numerous collaborators with special thanks to Xin Chen, Georgios Fainekos, James Kapinski, Nikos Aréchiga, Xiaoqing Jin, and Aditya Zutshi.

This work was funded in part by the US National Science Foundation (NSF) under award numbers CAREER 0953941, CNS 1319457, CPS 1446900, SHF 1527075, CPS 1646556, CCF 1837131, and the Air Force Research Laboratory (AFRL). All opinions expressed are those of the authors and not necessarily of the US NSF or AFRL.

References

1. Abbas, H., Fainekos, G., Sankaranarayanan, S., Ivancic, F., & Gupta, A. (2013). Probabilistic temporal logic falsification of cyber-physical systems. *ACM Transactions on Embedded Computing Systems, 12*, 95.
2. Abbas, H., Hoxha, B., Fainekos, G., & Ueda, K. (2014). Robustness-guided temporal logic testing and verification for stochastic cyber-physical systems. In *2014 IEEE 4th Annual International Conference on Cyber Technology in Automation, Control, and Intelligent Systems (CYBER)* (pp. 1–6). Piscataway: IEEE.
3. Abbas, H., O'Kelly, M., Rodionova, A., & Mangharam, R. (2017). Safe at any speed: A simulation-based test harness for autonomous vehicles. In *7th Workshop on Design, Modeling and Evaluation of Cyber Physical Systems (CyPhy'17)*.
4. Abbas, H., Rodionova, A., Bartocci, E., Smolka, S. A., & Grosu, R. (2017). Quantitative regular expressions for arrhythmia detection algorithms. In *Proceedings of the International Conference on Computational Methods in Systems Biology* (pp. 23–39). Berlin: Springer.
5. Adimoolam, A., Dang, T., Donzé, A., Kapinski, J., & Jin, X. (2017). Classification and coverage-based falsification for embedded control systems. In *International Conference on Computer Aided Verification* (pp. 483–503). Berlin: Springer.
6. Akazaki, T., Liu, S., Yamagata, Y., Duan, Y., & Hao, J. (2018). Falsification of cyber-physical systems using deep reinforcement learning. arXiv preprint arXiv:1805.00200.
7. Althoff, M. (2015). An introduction to CORA 2015. In *Proceedings of the Workshop on Applied Verification for Continuous and Hybrid Systems* (pp. 120–151).
8. Althoff, M., & Grebenyuk, D. (2016). Implementation of interval arithmetic in CORA 2016. In *Proceedings of the 3rd International Workshop on Applied Verification for Continuous and Hybrid Systems* (pp. 91–105).
9. Alur, R., Courcoubetis, C., Halbwachs, N., Henzinger, T. A., Ho, P. H., Nicollin, X., et al. (1995). The algorithmic analysis of hybrid systems. *Theoretical Computer Science, 138*(1), 3–34.
10. Alur, R., Courcoubetis, C., Henzinger, T. A., & Ho, P. H. (1993). Hybrid automata: An algorithmic approach to the specification and verification of hybrid systems. In *Workshop on International Hybrid Systems* (pp. 209–229). Berlin: Springer.
11. Alur, R., Dang, T., & Ivančić, F. (2003). Counter-example guided predicate abstraction of hybrid systems. In *International Conference on Tools and Algorithms for the Construction and Analysis of Systems. Lecture Notes in Computer Science* (Vol. 2619, pp. 208–223). Berlin: Springer
12. Alur, R., & Dill, D.L. (1994). A theory of timed automata. *Theoretical Computer Science, 126*(2), 183–235.
13. Alur, R., Fisman, D., & Raghothaman, M. (2016). Regular programming for quantitative properties of data streams. In *Proceedings of the European Symposium on Programming Languages and Systems* (pp. 15–40). Berlin: Springer.
14. Alur, R., & Henzinger, T. A. (1989). A really temporal logic. In *Proceedings of the Symposium on Foundations of Computer Science* (pp. 164–169).
15. Alur, R., Henzinger, T. A., Lafferriere, G., & Pappas, G.J. (2000). Discrete abstractions of hybrid systems. *Proceedings of the IEEE, 88*(7), 971–984.
16. Alur, R., Mamouras, K., & Ulus, D. (2017). Derivatives of quantitative regular expressions. In *Models, algorithms, logics and tools* (pp. 75–95). Cham: Springer.
17. Ames, A. D., Grizzle, J. W., & Tabuada, P. (2014). Control barrier function based quadratic programs with application to adaptive cruise control. In *2014 IEEE 53rd Annual Conference on Decision and Control (CDC)* (pp. 6271–6278). Piscataway: IEEE.
18. Annapureddy, Y. S. R., & Fainekos, G. E. (2010). Ant colonies for temporal logic falsification of hybrid systems. In *Proceedings of the 36th Annual Conference of IEEE Industrial Electronics* (pp. 91–96). Piscataway: IEEE.

19. Annpureddy, Y., Liu, C., Fainekos, G. E., & Sankaranarayanan, S. (2011). S-TaLiRo: A tool for temporal logic falsification for hybrid systems. In *International Conference on Tools and Algorithms for the Construction and Analysis of Systems* (pp. 254–257). Berlin: Springer.
20. Aréchiga, N., & Krogh, B. (2014). Using verified control envelopes for safe controller design. In *2014 American Control Conference (ACC)* (pp. 2918–2923). Piscataway: IEEE.
21. Asarin, E., Caspi, P., & Maler, O. (2002). Timed regular expressions. *Journal of the ACM, 49*(2), 172–206.
22. Asarin, E., Dang, T., & Girard, A. (2007). Hybridization methods for the analysis of nonlinear systems. *Acta Informatica, 43*(7), 451–476.
23. Asarin, E., Maler, O., & Pnueli, A. (1995). Reachability analysis of dynamical systems having piecewise-constant derivatives. *Theoretical Computer Science, 138*, 35–65.
24. Baier, C., & Katoen, J. P. (2008). *Principles of model checking*. Cambridge, MA: MIT Press.
25. Bak, S., & Duggirala, P. S. (2017). HyLAA: A tool for computing simulation-equivalent reachability for linear systems. In *Proceedings of the 20th International Conference on Hybrid Systems: Computation and Control* (pp. 173–178). New York: ACM.
26. Bastani, O., Ioannou, Y., Lampropoulos, L., Vytiniotis, D., Nori, A., & Criminisi, A. (2016). Measuring neural net robustness with constraints. In *Advances in Neural Information Processing Systems* (pp. 2613–2621).
27. Berz, M. (1999). *Modern map methods in particle beam physics*. Advances in Imaging and Electron Physics (Vol. 108). London: Academic.
28. Bojarski, M., Del Testa, D., Dworakowski, D., Firner, B., Flepp, B., Goyal, P., et al. (2016) End to end learning for self-driving cars. arXiv preprint arXiv:1604.07316.
29. Bonakdarpour, B., & Finkbeiner, B. (2016). Runtime verification for HyperLTL. In *International Conference on Runtime Verification* (pp. 41–45). Cham: Springer.
30. Bonakdarpour, B., Sanchez, C., & Schneider, G. (2018). Monitoring hyperproperties by combining static analysis and runtime verification. In *International Symposium on Leveraging Applications of Formal Methods* (pp. 8–27). Berlin: Springer.
31. Bournez, O., Maler, O., & Pnueli, A. (1999). Orthogonal polyhedra: Representation and computation. In *Hybrid systems: Computation and control. Lecture Notes in Computer Science* (Vol. 1569, pp. 46–60). Berlin: Springer.
32. Box, G. E. P. (1979). Robustness in the strategy of scientific model building. In *Robustness in Statistics* (pp. 201–236). London: Academic.
33. Brockett, R. (1993). Hybrid models for motion control systems. In *Essays on control: Perspectives in the theory and its applications* (pp. 29 –53). Boston: Birkhäuser.
34. Cameron, F., Fainekos, G., Maahs, D. M., & Sankaranarayanan, S. (2015). Towards a verified artificial pancreas: Challenges and solutions for runtime verification. In *Proceedings of Runtime Verification (RV'15). Lecture Notes in Computer Science* (Vol. 9333, pp. 3–17). Cham: Springer.
35. Cameron, F., Wilson, D. M., Buckingham, B. A., Arzumanyan, H., Clinton, P., Chase, H. P., et al. (2012). Inpatient studies of a Kalman-filter-based predictive pump shutoff algorithm. *Journal of Diabetes Science and Technology, 6*(5), 1142–1147.
36. Cassandras, C. G., & Lygeros, J. (2006). *Stochastic hybrid systems*. Boca Raton: CRC Press.
37. Chaochen, Z., Hoare, C. A. R., & Ravn, A. P. (1991). A calculus of durations. *Information Processing Letters, 40*(5), 269–276.
38. Chee, F., & Fernando, T. (2007). *Closed-loop control of blood glucose*. Berlin: Springer.
39. Chen, S., O'Kelly, M., Weimer, J., Sokolsky, O., & Lee, I. (2015). An intraoperative glucose control benchmark for formal verification. In *5th IFAC conference on Analysis and Design of Hybrid Systems (ADHS)* (2015)
40. Chen, X., Ábrahám, E., & Sankaranarayanan, S. (2012). Taylor model flowpipe construction for non-linear hybrid systems. In *Proceedings of the 2012 IEEE 33rd Real-Time Systems Symposium (RTSS'12)* (pp. 183–192). Piscataway: IEEE.
41. Chen, X., Ábrahám, E., & Sankaranarayanan, S. (2013). Flow*: An analyzer for non-linear hybrid systems. In *International Conference on Computer Aided Verification. Lecture Notes in Computer Science* (Vol. 8044, pp. 258–263). Berlin: Springer.

42. Chen, X., Mover, S., & Sankaranarayanan, S. (2017). Compositional relational abstraction for nonlinear systems. *ACM Transactions on Embedded Computing Systems, 16*(5s), 187.

43. Chen, X., & Sankaranarayanan, S. (2016). Decomposed reachability analysis for nonlinear systems. In *2016 IEEE Real-Time Systems Symposium (RTSS)* (pp. 13–24). Piscataway: IEEE.

44. Chonev, V., Ouaknine, J., & Worrell, J. (2016). On the Skolem problem for continuous linear dynamical systems. In *43rd International Colloquium on Automata, Languages, and Programming (ICALP 2016). Leibniz International Proceedings in Informatics* (Vol. 55, pp. 100:1–100:13). Wadern: Schloss Dagstuhl–Leibniz-Zentrum fuer Informatik.

45. Chutinan, A., & Krogh, B. (1998). Computing polyhedral approximations to flow pipes for dynamic systems. In *Proceedings of the 37th IEEE Conference on Decision and Control.* Piscataway: IEEE.

46. Chutinan, A., & Krogh, B. H. (2003). Computational techniques for hybrid system verification. *IEEE Transactions on Automatic Control, 48*(1), 64–75. https://doi.org/10.1109/TAC. 2002.806655

47. Clarkson, M. R., & Schneider, F. B. (2010). Hyperproperties. *Journal of Computer Security, 18*(6), 1157–1210.

48. Cobelli, C., Man, C. D., Sparacino, G., Magni, L., Nicolao, G. D., & Kovatchev, B. P. (2009). Diabetes: Models, signals and control (methodological review). *IEEE Reviews in Biomedical Engineering, 2,* 54–95.

49. Dang, T., & Maler, O. (1998). Reachability via face lifting. In *Hybrid Systems: Computation and Control. Lecture Notes in Computer Science* (Vol. 1386, pp. 96–109). Berlin: Springer

50. Dang, T., Maler, O., & Testylier, R. (2010). Accurate hybridization of nonlinear systems. In *Hybrid Systems: Computation and Control (HSCC '10)* (pp. 11–20). New York: ACM.

51. Deshmukh, J., Horvat, M., Jin, X., Majumdar, R., & Prabhu, V. S. (2017). Testing cyber-physical systems through Bayesian optimization. *ACM Transactions on Embedded Computing Systems, 16*(5s), 170.

52. Deshmukh, J., Jin, X., Kapinski, J., & Maler, O. (2015). Stochastic local earch for falsification of hybrid ystems. In *International Symposium on Automated Technology for Verification and Analysis* (pp. 500–517). Berlin: Springer.

53. Dimitrova, R., Finkbeiner, B., Kovács, M., Rabe, M. N., & Seidl, H. (2012). Model checking information flow in reactive systems. In *International Workshop on Verification, Model Checking, and Abstract Interpretation* (pp. 169–185). Berlin: Springer.

54. Dokhanchi, A., Zutshi, A., Srinivas, R. T., Sankaranarayanan, S., & Fainekos, G. E. (2015). Requirements driven falsification with coverage metrics. In *2015 International Conference on Embedded Software (EMSOFT'15)* (pp. 31–40). Piscataway: IEEE.

55. Donzé, A. (2010). Breach, a toolbox for verification and parameter synthesis of hybrid systems. In *International Conference on Computer Aided Verification* (pp. 167–170). Berlin: Springer.

56. Donzé, A., Ferrère, T., & Maler, O. (2013). Efficient robust monitoring for STL. In *Computer Aided Verification* (pp. 264–279). Berlin: Springer.

57. Donzé, A., & Maler, O. (2007). Systematic simulation using sensitivity analysis. In *International Workshop on Hybrid Systems: Computation and Control* (pp. 174–189). Berlin: Springer.

58. Donzé, A., & Maler, O. (2010). Robust satisfaction of temporal logic over real-valued signals. In *Formal Modeling and Analysis of Timed Systems* (pp. 92–106). Berlin: Springer.

59. Dreossi, T., Dang, T., Donzé, A., Kapinski, J., Deshmukh, J., & Jin, X. (2015). Efficient guiding strategies for testing of temporal properties of hybrid systems. In *NASA Formal Methods Symposium* (pp. 127–142). Berlin: Springer.

60. Dreossi, T., Donzé, A., & Seshia, S. A. (2017). Compositional falsification of cyber-physical systems with machine learning components. In *NASA Formal Methods. Lecture Notes in Computer Science* (Vol. 10227). Berlin: Springer.

61. Dreossi, T., Ghosh, S., Sangiovanni-Vincentelli, A., & Seshia, S.A. (2017). Systematic testing of convolutional neural networks for autonomous driving. In *Reliable Machine Learning in*

the Wild (RMLW) Workshop, Cf. https://people.eecs.berkeley.edu/~tommasodreossi/papers/rmlw2017.pdf

62. Duggirala, P. S., Fan, C., Mitra, S., & Viswanathan, M. (2015). Meeting a powertrain verification challenge. In *Proceedings of the 27th International Conference on Computer Aided Verification. Part I* (pp. 536–543). Cham: Springer.

63. Duggirala, P. S., Potok, M., Mitra, S., & Viswanathan, M. (2015). C2E2: A tool for verifying annotated hybrid systems. In *Proceedings of the 18th International Conference on Hybrid Systems: Computation and Control (HSCC'15)* (pp. 307–308). New York: ACM.

64. Dutta, S., Jha, S., Sankaranarayanan, S., & Tiwari, A. (2018). Learning and verification of feedback control systems using feedforward neural networks. *IFAC-PapersOnLine, 51*(16), 151–156.

65. Dutta, S., Jha, S., Sankaranarayanan, S., & Tiwari, A. (2018). Output range analysis for deep feedforward neural networks. In *Proceedings of NASA Formal Methods Symposium (NFM). Lecture Notes in Computer Science* (Vol. 10811, pp. 121–138). Berlin: Springer.

66. Dutta, S., Kushner, T., & Sankaranarayanan, S. (2018). Robust data-driven control of artificial pancreas systems using neural networks. In M. Češka, & D. Šafránek (Eds.), *Computational methods in systems biology* (pp. 183–202). Cham: Springer.

67. Ehlers, R. (2017). Formal verification of piece-wise linear feed-forward neural networks. In *International Symposium on Automated Technology for Verification and Analysis. Lecture Notes in Computer Science* (Vol. 10482, pp. 269–286). Berlin: Springer.

68. Fainekos, G. E., & Pappas, G. J. (2009). Robustness of temporal logic specifications for continuous-time signals. *Theoretical Computer Science, 410*(42), 4262–4291.

69. Fan, C., Kapinski, J., Jin, X., & Mitra, S. (2018). Simulation-driven reachability using matrix measures. *ACM Transactions on Embedded Computing Systems, 17*(1), 21:1–21:28.

70. Finkbeiner, B., Rabe, M. N., & Sánchez, C. (2015). Algorithms for model checking HyperLTL and HyperCTL*. In *International Conference on Computer Aided Verification* (pp. 30–48). Berlin: Springer.

71. Forejt, V., Kwiatkowska, M., Norman, G., & Parker, D. (2011). Automated verification techniques for probabilistic systems. In *International School on Formal Methods for the Design of Computer, Communication and Software Systems* (pp. 53–113). Berlin: Springer.

72. Fränzle, M., Herde, C., Ratschan, S., Schubert, T., & Teige, T. (2007). Efficient solving of large non-linear arithmetic constraint systems with complex Boolean structure. *Journal on Satisfiability, Boolean Modeling and Computation, 1*, 209–236.

73. Frehse, G., Le Guernic, C., Donzé, A., Cotton, S., Ray, R., Lebeltel, O., et al. (2011). SpaceEx: Scalable verification of hybrid systems. In *International Conference on Computer Aided Verification (CAV'11). Lecture Notes in Computer Science* (Vol. 6806, pp. 379–395). Berlin: Springer.

74. Fulton, N., Mitsch, S., Quesel, J.D., Völp, M., & Platzer, A. (2015). KeYmaera X: An axiomatic tactical theorem prover for hybrid systems. In *Proceedings of International Conference on Automated Deduction* (Vol. 9195, pp. 527–538). Cham: Springer. https://doi.org/10.1007/978-3-319-21401-6_36

75. Gadkari, A., Yeolekar, A., Suresh, J., Ramesh, S., Mohalik, S., & Shashidhar, K. (2008). Automotgen: Automatic model oriented test generator for embedded control systems. In A. Gupta & S. Malik (Eds.), *Computer aided verification. Lecture Notes in Computer Science* (Vol. 5123, pp. 204–208). Berlin: Springer.

76. Gao, S., Kong, S., & Clarke, E. M. (2013). dReal: An SMT solver for nonlinear theories over the reals. In *International Conference on Automated Deduction (CADE'13). Lecture Notes in Computer Science* (Vol. 7898, pp. 208–214). Berlin: Springer.

77. Geiger, A., Lenz, P., & Urtasun, R. (2012) Are we ready for autonomous driving? The Kitti vision benchmark suite. In *2012 IEEE Conference on Computer Vision and Pattern Recognition (CVPR)* (pp. 3354–3361). Piscataway: IEEE.

78. Girard, A. (2005). Reachability of uncertain linear systems using zonotopes. In *International Workshop on Hybrid Systems: Computation and Control. Lecture Notes in Computer Science* (Vol. 3414, pp. 291–305). Berlin: Springer.

79. Girard, A., & Pappas, G. J. (2005). Approximate bisimulations for nonlinear dynamical systems. In *Proceedings of the 44th IEEE Conference on Decision and Control* (pp. 684–689). Piscataway: IEEE.
80. Goodfellow, I., Bengio, Y., & Courville, A. (2016). *Deep learning*. Cambridge, MA: MIT Press. http://www.deeplearningbook.org
81. Goubault, E., Jourdan, J. H., Putot, S., & Sankaranarayanan, S. (2014). Finding non-polynomial positive invariants and lyapunov functions for polynomial systems through darboux polynomials. In *Proceedings of the American Control Conference (ACC)* (pp. 3571–3578). New York: IEEE Press.
82. Hainry, E. (2008). Reachability in linear dynamical systems. In *Logic and theory of algorithms* (pp. 241–250). Berlin: Springer.
83. Henzinger, T. A. (1996). The theory of hybrid automata. In *Proceedings of the Logic in Computer Science* (pp. 278–292). Piscataway: IEEE.
84. Henzinger, T. A., Kopke, P. W., Puri, A., & Varaiya, P. (1998). What's decidable about hybrid automata? *Journal of Computer and System Sciences, 57*(1), 94–124.
85. Herde, C., Eggers, A., Franzle M., & Teige, T. (2008). Analysis of hybrid systems using HySAT. In *Third International Conference on Systems, 2008* (pp. 13–18). Piscataway: IEEE.
86. Hovorka, R. (2005). Continuous glucose monitoring and closed-loop systems. *Diabetic Medicine, 23*(1), 1–12.
87. Huang, X., Kwiatkowska, M., Wang, S., & Wu, M. (2017). Safety verification of deep neural networks. In *Proceedings of the Computer Aided Verification* (pp. 3–29). Cham: Springer.
88. Jiang, Z., Pajic, M., Moarref, S., Alur, R., & Mangharam, R. (2012). Modeling and verification of a dual chamber implantable pacemaker. In *Tools and Algorithms for the Construction and Analysis of Systems (TACAS). Lecture Notes in Computer Science* (Vol. 7214, pp. 188–203). Berlin: Springer.
89. Junghanns, A., Mauss, J., & Tatar, M. (2008). Tatar: Testweaver—a tool for simulation-based test of mechatronic designs. In *6th International Modelica Conference*, Bielefeld, March 3. Citeseer
90. Kapinski, J., Deshmukh, J.V., Sankaranarayanan, S., & Aréchiga, N. (2014). Simulation-guided lyapunov analysis for hybrid dynamical systems. In *Proceedings of the 17th International Conference on Hybrid Systems: Computation and Control* (pp. 133–142). New York: ACM.
91. Kapinski, J., Krogh, B. H., Maler, O., & Stursberg, O. (2003). On systematic simulation of open continuous systems. In *International Workshop on Hybrid Systems: Computation and Control* (pp. 283–297). Berlin: Springer.
92. Kato, K., Ishikawa, F., & Honiden, S. (2018). Falsification of cyber-physical systems with reinforcement learning. In *2018 IEEE Workshop on Monitoring and Testing of Cyber-Physical Systems (MT-CPS)* (pp. 5–6). Piscataway: IEEE.
93. Katz, G., Barrett, C., Dill, D., Julian, K., & Kochenderfer, M. (2017). Reluplex: An efficient smt solver for verifying deep neural networks. In *International Conference on Computer Aided Verification* (pp. 97–117). Berlin: Springer.
94. Koymans, R. (1990). Specifying real-time properties with metric temporal logic. *Real-Time System, 2*(4), 255–299.
95. Kurzhanski, A. B., & Varaiya, P. (2000). Ellipsoidal techniques for reachability analysis. In *International Workshop on Hybrid Systems: Computation and Control. Lecture Notes in Computer Science* (Vol. 1790, pp. 202–214). Berlin: Springer.
96. Kushner, T., Bortz, D., Maahs, D., & Sankaranarayanan, S. (2018). A data-driven approach to artificial pancreas verification and synthesis. In *International Conference on Cyber-Physical Systems (ICCPS'18)*. New York: IEEE Press.
97. Labinaz, G., Bayoumi, M. M., & Rudie, K. (1997). A survey of modeling and control of hybrid systems. *Annual Reviews in Control, 21*, 79–92.
98. Lafferriere, G., Pappas, G. J., & Sastry, S. (2000). O-minimal hybrid systems. *Mathematics of Control, Signals and Systems, 13*(1), 1–21.

99. Leitner, F., & Leue, S. (2008). Simulink design verifier vs. SPIN a comparative case study. In *Proceedings of the 13th International Workshop on Formal Methods for Industrial Critical Systems*.
100. Levinson, J., Askeland, J., Becker, J., Dolson, J., Held, D., Kammel, S., et al. (2011). Towards fully autonomous driving: Systems and algorithms. In *2011 IEEE Intelligent Vehicles Symposium (IV)* (pp. 163–168). Piscataway: IEEE.
101. Lomuscio, A., & Maganti, L. (2017). An approach to reachability analysis for feed-forward ReLU neural networks. http://arxiv.org/abs/1706.07351
102. Loos, S. M., Platzer, A., & Nistor, L. (2011). Adaptive cruise control: Hybrid, distributed, and now formally verified. In *International Symposium on Formal Methods* (pp. 42–56). Berlin: Springer.
103. Maahs, D. M., Calhoun, P., Buckingham, B. A., Chase, H. P., Hramiak, I., Lum, J., et al. (2014). A randomized trial of a home system to reduce nocturnal hypoglycemia in type 1 diabetes. *Diabetes Care, 37*(7), 1885–1891.
104. Magdici, S., & Althoff, M. (2017). Adaptive cruise control with safety guarantees for autonomous vehicles. *IFAC-PapersOnLine, 50*(1), 5774–5781.
105. Makino, K., & Berz, M. (2003). Taylor models and other validated functional inclusion methods. *Journal of Pure and Applied Mathematics, 4*(4), 379–456.
106. Maler, O., & Nickovic, D. (2004). Monitoring temporal properties of continuous signals. In *Proceedings of Formal Modeling and Analysis of Timed Systems* (pp. 152–166). Berlin: Springer.
107. Meiss, J. D. (2007). *Differential dynamical systems*. Philadelphia: SIAM.
108. Mitchell, I., & Tomlin, C. (2000). Level set methods for computation in hybrid systems. In *International Workshop on Hybrid Systems: Computation and Control. Lecture Notes in Computer Science* (Vol. 1790, pp. 310–323). Berlin: Springer.
109. Mover, S., Cimatti, A., Tiwari, A., & Tonetta, S. (2013). Time-aware relational abstractions for hybrid systems. In *Proceedings of the Eleventh ACM International Conference on Embedded Software (EMSOFT '13)* (pp. 14:1–14:10). Piscataway: IEEE Press.
110. National Transportation Safety Board (NTSB) (2016). Collision between a car operating with automated vehicle control systems and a tractor-semitrailer truck. https://www.ntsb.gov/news/events/Documents/2017-HWY16FH018-BMG-abstract.pdf
111. Nghiem, T., Sankaranarayanan, S., Fainekos, G.E., Ivancic, F., Gupta, A., & Pappas, G.J. (2010). Monte-Carlo techniques for falsification of temporal properties of non-linear hybrid systems. In *Proceedings of Hybrid Systems: Computation and Control* (pp. 211–220). New York: ACM.
112. Nguyen, L. V., Kapinski, J., Jin, X., Deshmukh, J. V., & Johnson, T. T. (2017). Hyperproperties of real-valued signals. In *Proceedings of the 15th ACM-IEEE International Conference on Formal Methods and Models for System Design* (pp. 104–113). New York: ACM.
113. Nicolescu, G., & Mosterman, P. J. (2009). *Model-based design for embedded systems* (1st ed.). Boca Raton: CRC Press.
114. Nilsson, P., Hussien, O., Chen, Y., Balkan, A., Rungger, M., Ames, A., et al. (2014). Preliminary results on correct-by-construction control software synthesis for adaptive cruise control. In *2014 IEEE 53rd Annual Conference on Decision and Control (CDC)* (pp. 816–823). Piscataway: IEEE.
115. Norris, J. (1998). *Markov chains*. Cambridge: Cambridge University Press.
116. Øksendal, B. K. (2000). *Stochastic differential equations: An introduction*. Berlin: Springer.
117. Pajic, M., Mangharam, R., Sokolsky, O., Arney, D., Goldman, J., & Lee, I. (2014). Model-driven safety analysis of closed-loop medical systems. *IEEE Transactions on Industrial Informatics, 10*(1), 3–16.
118. Papachristodoulou, A., & Prajna, S. (2005). Analysis of non-polynomial systems using the sum of squares decomposition. In *Positive Polynomials in Control* (pp. 23–43). Berlin: Springer.

119. Pei, Y., Entcheva, E., Grosu, R., & Smolka, S. (2005) Efficient modeling of excitable cells using hybrid automata. In *Proceedings of the Computational Methods in Systems Biology* (pp. 216–227).
120. Platzer, A. (2008). Differential dynamic logic for hybrid systems. *Journal of Automated Reasoning, 41*(2), 143–189.
121. Platzer, A. (2010). Logical analysis of hybrid systems: Proving theorems for complex dynamics. Heidelberg: Springer. https://doi.org/10.1007/978-3-642-14509-4
122. Platzer, A., & Clarke, E. M. (2008). Computing differential invariants of hybrid systems as fixedpoints. In A. Gupta & S. Malik (Eds.), *Proceedings of computer aided verification. Lecture Notes in Computer Science* (Vol. 5123, pp. 176–189). Berlin: Springer.
123. Pnueli, A. (1977). The temporal logic of programs. In Proceedings of Symposium on Foundations of Computer Science (pp. 46–57). Piscataway: IEEE.
124. Podelski, A., & Wagner, S. (2007). *Region stability proofs for hybrid systems* (pp. 320–335). Berlin: Springer.
125. Prabhakar, P., Duggirala, P. S., Mitra, S., & Viswanathan, M. (2013). Hybrid automata-based CEGAR for rectangular hybrid systems. In R. Giacobazzi, J. Berdine, I. Mastroeni (Eds.), *Verification, model checking, and abstract interpretation* (pp. 48–67). Berlin: Springer.
126. Prajna, S. (2005). *Optimization-based methods for nonlinear and hybrid systems verification.* Ph.D. thesis, California Institute of Technology, Caltech, Pasadena, CA, USA.
127. Prajna, S., & Jadbabaie, A. (2004). Safety verification of hybrid systems using barrier certificates. In *Hybrid Systems: Computation and Control* (pp. 477–492). Berlin: Springer.
128. Pulina, L., & Tacchella, A. (2012). Challenging smt solvers to verify neural networks. *AI Communications, 25*(2), 117–135.
129. Ratschan, S., & She, Z. (2005). Safety verification of hybrid systems by constraint propagation based abstraction refinement. In *International Workshop on Hybrid Systems: Computation and Control. Lecture Notes in Computer Science* (Vol. 3414, pp. 573–589). Berlin: Springer.
130. Ratschan, S., She, Z.: Safety verification of hybrid systems by constraint propagation-based abstraction refinement. *ACM Transactions on Embedded Computing Systems, 6*(1), 8. http://doi.acm.org/10.1145/1210268.1210276
131. Reactive Systems Inc. (2003). *Model-based testing and validation of control software with reactis.* http://www.reactive-systems.com/papers/bcsf.pdf
132. Roohi, N., Prabhakar, P., & Viswanathan, M. (2016). Hybridization based CEGAR for hybrid automata with affine dynamics. In M. Chechik, & J. F. Raskin (Eds.), *Tools and algorithms for the construction and analysis of systems* (pp. 752–769). Berlin: Springer.
133. Ruan, W., Wu, M., Sun, Y., Huang, X., Kroening, D., & Kwiatkowska, M. (2018). *Global robustness evaluation of deep neural networks with provable guarantees for L0 norm.* http://arxiv.org/abs/1804.05805
134. Sankaranarayanan, S., & Fainekos, G. E. (2012). Falsification of temporal properties of hybrid systems using the cross-entropy method. In *ACM International Conference on Hybrid Systems: Computation and Control* (pp. 125–134). New York: ACM.
135. Sankaranarayanan, S., Kumar, S. A., Cameron, F., Bequette, B. W., Fainekos, G., & Maahs, D. M. (2017). Model-based falsification of an artificial pancreas control system. *ACM SIGBED Review, 14*(2), 24–33.
136. Sankaranarayanan, S., & Tiwari, A. (2011). Relational abstractions for continuous and hybrid systems. In *International Conference on Computer Aided Verification. Lecture Notes in Computer Science* (Vol. 6806, pp. 686–702). Berlin: Springer.
137. Siper, M. J. (2005). *An Introduction to mathematical theory of computation* (2nd ed.). Toronto: Thompson Publishing (Course Technology)
138. Skyler, J. S. (Ed.). (2012). *Atlas of diabetes* (4th ed.). Berlin: Springer.
139. Sontag, E. D. (1981). Nonlinear regulation: The piecewise linear approach. *IEEE Transactions on Automatic Control, 26*(2), 346–358.
140. Steil, G., Panteleon, A., & Rebrin, K. (2004). Closed-sloop insulin delivery—the path to physiological glucose control. *Advanced Drug Delivery Reviews, 56*(2), 125–144.

141. Steil, G. M. (2013). Algorithms for a closed-loop artificial pancreas: The case for proportional-integral-derivative control. *Journal of Diabetes Science and Technology, 7,* 1621–1631.
142. Sutton, R. S., & Barto, A. G. (1998). *Reinforcement learning: An introduction* (Vol. 1). Cambridge: MIT Press.
143. Teixeira, R. E., & Malin, S. (2008). The next generation of artificial pancreas control algorithms. *Journal of Diabetes Science and Technology, 2,* 105–112.
144. Tjeng, V., & Tedrake, R. (2017). Verifying neural networks with mixed integer programming. http://arxiv.org/abs/1711.07356
145. Topcu, U., & Packard, A. (2009). Stability region analysis for uncertain nonlinear systems. *IEEE Transactions on Automatic Control, 54,* 1042–1047.
146. Topcu, U., Seiler, P., & Packard, A. (2008). Local stability analysis using simulations and sum-of-squares programming. *Automatica, 44,* 2669–2675.
147. Tuncali, C. E., Fainekos, G., Ito, H., & Kapinski, J. (2018). Simulation-based adversarial test generation for autonomous vehicles with machine learning components. In *Proceedings of IEEE Intelligent Vehicles Symposium (IV)*
148. Tuncali, C. E., Kapinski, J., Ito, H., & Deshmukh, J. V. (2018). Reasoning about safety of learning-enabled components in autonomous cyber-physical systems. In *Proceedings of the 55th Annual Design Automation Conference, DAC 2018* (pp. 30:1–30:6). New York: ACM.
149. Ulus, D. (2017). Montre: A tool for monitoring timed regular expressions. In *Proceedings of the International Conference on Computer Aided Verification* (pp. 329–335). Berlin: Springer.
150. Ulus, D., Ferrère, T., Asarin, E., & Maler, O. (2014). Timed pattern matching. In *Proceedings of the International Conference on Formal Modeling and Analysis of Timed Systems* (pp. 222–236). Berlin: Springer.
151. Zutshi, A., Sankaranarayanan, S., Deshmukh, J., & Jin, X. (2016). Symbolic-numeric reachability analysis of closed-loop control software. In *Hybrid Systems: Computation and Control (HSCC)* (pp. 135–144). New York: ACM Press.
152. Zutshi, A., Sankaranarayanan, S., Deshmukh, J., & Kapinski, J. (2013). A trajectory splicing approach to concretizing counterexamples for hybrid systems. In *IEEE Conference on Decision and Control (CDC)* (pp. 3918–3925). New York: IEEE Press.
153. Zutshi, A., Sankaranarayanan, S., Deshmukh, J., & Kapinski, J. (2014). Multiple-shooting CEGAR-based falsification for hybrid systems. In *International Conference on Embedded Software (EMSOFT)* (pp. 5:1–5:10). New York: ACM Press.
154. Zutshi A., Sankaranarayanan S., & Tiwari A. (2012). Timed relational abstractions for sampled data control systems. In P. Madhusudan & S. A. Seshia (Eds.), *Computer Aided Verification. Lecture Notes in Computer Science* (Vol. 7358). Berlin: Springer.

Chapter 5
Data-Driven Safety Verification of Complex Cyber-Physical Systems

Chuchu Fan and Sayan Mitra

5.1 Introduction

Cyber-physical systems (CPS) are often safety critical and are expected to work in uncertain environments. Ensuring design correctness and safety of CPS has significant financial and legal implications. Existing design and test methodologies are inadequate for providing the needed level of safety assurances. For example, Koopman [55] argues how naïve test driving for reasonable catastrophic failure rates for a fleet of vehicles can grow to hundreds of billions of miles—a figure that is beyond the capabilities of even for large corporations. Formal verification, designed and deployed properly, can be the first line of defense against design bugs making their way into unsafe products [16].

A formal verification algorithm takes as input a cyber-physical system's (CPS) model and a requirement, and decides whether or not all the behaviors of the system meet the requirement. If the decision is "yes," the algorithm provides a supporting proof of this fact, which can then be used for certification, documentation, and for future testing, and maintenance. If the decision is "no," the algorithm produces a supporting counterexample or a "bug trace." This is a particular behavior of the systems resulting from specific initial states and inputs, which violates the requirement. For cyber-physical systems (CPS), the mathematical model may be a dynamical, switched, or a hybrid system, and the requirement may be a safety property, a stability property, or a temporal logic property.

Most instances of this model-based formulation of the verification problem for CPS are known to be undecidable [39, 67]. Significant progress has been made in the last decade and many powerful tools have been developed to solve approximate

C. Fan · S. Mitra (✉)
University of Illinois at Urbana-Champaign, ECE Department, Champaign, IL, USA
e-mail: cfan10@illinois.edu; mitras@illinois.edu

© Springer Nature Switzerland AG 2019
M. A. Al Faruque, A. Canedo (eds.), *Design Automation of Cyber-Physical Systems*, https://doi.org/10.1007/978-3-030-13050-3_5

versions of these problems for specific model classes [7, 15, 35, 36, 54]. Yet, these purely model-based techniques do not handle nonlinear and hybrid models that arise in practice. Real-world systems are often described by a heterogeneous mix of simulation code, differential equations, block diagrams, look-up tables, and machine learning modules, and it is unreasonable to even expect complete and precise models in the first place.

In the last 5 years, data-driven verification algorithms have gained momentum. Data-driven algorithms use executions (or numerical simulations) of the model in addition to statically analyzing the model itself. Thus, the verification algorithm can use powerful numerical simulators as a subroutine, which is particularly relevant for nonlinear models that do not permit a closed-form analytical solution. This opens the door to also verifying autonomous systems without complete and precise models.[1]

The basic principle of data-driven verification combines model-based *reachability analysis* with *sensitivity analysis* of the complex or unknown parts of the system. Sensitivity analysis algorithms give (probabilistic or worst-case) bounds on how much the states or outputs of a module change, with small changes in the input parameters. Under certain assumptions about the underlying system, we show that data-driven verification can indeed provide rigorous guarantees about system safety. An earlier sequence of papers culminating in [24] developed sensitivity analysis algorithms for nonlinear and hybrid systems with known models. These techniques are implemented in the C2E2 tool, which has been effectively used to verify an engine control system [46], a NASA-developed collision alerting protocol [63], and satellite controllers [24, 29]. For systems with unknown models, the deterministic sensitivity analysis algorithms have to be replaced with methods that only rely on execution data. In [32], we have shown how this problem can be cast as the well-known problem of learning a linear separator, and therefore, can be solved with probabilistic correctness guarantees. The resulting tool DRYVR was used to analyze several autonomous and ADAS-based[2] maneuvers [31, 32]. Other successful applications range across medical devices [40, 44], automotive [6, 22, 28, 47], air-traffic management [25], and energy systems [26]. A noteworthy related approach is *simulation-driven falsification*, which addresses the problem of finding bugs, but does not aim to prove their absence [1]. The search for bugs is formulated as an optimization problem, and since this typically works out to be a nonlinear and non-convex problem, stochastic optimization tools are employed to guide the search. The preeminent tool implementing this approach is S-taliro [5]; it has been effectively used to search for bugs in several practical applications [27, 65].

We present a broad and unified overview of data-driven verification with several case studies using both C2E2 and DRYVR . We classify the verification problems

---

[1] Autonomous systems sometimes also have incomplete requirements. The black-box approach described here does not address that problem.

[2] ADAS stands for advanced driving assistance systems such as adaptive cruise control, automatic emergency braking, etc.

regarding both the nature of the model and the requirement. First, in Sect. 5.2 we provide the necessary mathematical preliminaries; experienced readers can skip this. In Sect. 5.3, we set up the bounded verification problem and the related subproblem of sensitivity analysis. The existing techniques are described in the context of dynamical systems in Sect. 5.4, and extended for hybrid systems in Sect. 5.5. In Sect. 5.6, we discuss the black-box verification as in DRYVR. Two recent applications of data-driven verification are discussed in some detail, including a spacecraft rendezvous maneuver in Sect. 5.7.2 and an engine control challenge in Sect. 5.7.3. In Sect. 5.8, we conclude with a short summary of open problems and future research directions. Finally, in Sect. 5.9, we present pointers to additional works for further reading.

5.2 Mathematical Preliminaries

We will begin by defining the concepts and notations used throughout the chapter.

Matrix Norms For any matrix $A \in \mathbb{R}^{n \times n}$, A^T is its transpose; $\lambda_{\max}(A)$ and $\lambda_{\min}(A)$ are the maximum and minimum eigenvalues; a_{ij} denotes the element in the ith row and jth column. $\|A\|_1, \|A\|_2, \|A\|_\infty, \|A\|_F$ denote, respectively, the $1, 2,$ infinity, and the Frobenius norms of A. $|A|$ is the matrix obtained by taking the element-wise absolute value of matrix A.

Given a positive definite $n \times n$ real-valued matrix M, the M-*norm* of a vector $x \in \mathbb{R}^n$, $\|x\|_M = \sqrt{x^T M x}$ is the norm of x under the transformation M. Such M-norm will be used to represent reach sets of the system as ellipsoids. For any $M \succ 0$, there exists a nonsingular matrix $C \in \mathbb{R}^{n \times n}$, such that $M = C^T C$ and we write C as $M^{\frac{1}{2}}$. So, $\|x\|_M = \sqrt{x^T C^T C x} = \|Cx\|$. That is, $\|x\|_M$ is the 2-norm of the linearly transformed vector Cx. When $M = I$ is the identity matrix, $\|x\|_I$ coincidences with the 2-norm.

For sets $S_1, S_2 \subseteq \mathbb{R}^n$, $\text{hull}(S_1, S_2)$ is their convex hull. The hull of a set of $n \times n$ matrices is defined in the usual way, by considering each matrix as a vector in $\mathbb{R}^{n^2}$. The diameter of a compact set S is defined as $\text{Dia}(S) = \sup_{x,y \in S} \|x - y\|$. $E_{M,\delta}(x_0) = \{x \mid \|x - x_0\|_M \le \delta\}$ represents an ellipsoid centered at $x_0 \in \mathbb{R}^n$, with *shape* M and *size* δ. The δ ball around x_0: $B_\delta(x) = \{x \mid \|x - x_0\| \le \delta\}$ is a special case of $E_{M,\delta}(x_0)$ where M is the identity matrix I. A predicate over $\mathbb{R}^n$ is a computable function $\phi : \mathbb{R}^n \to \mathbb{B}$ that maps each state in $\mathbb{R}^n$ to either True or False.

Interval Matrices For a pair of matrices $B, C \in \mathbb{R}^{n \times n}$ with the property that: $b_{ij} \le c_{ij}$ for all $1 \le i, j \le n$, we define the set of matrices $\text{Interval}([B, C]) \triangleq \{A \in \mathbb{R}^{n \times n} | b_{qij} \le a_{ij} \le c_{ij}, 1 \le i, j \le n\}$. Any such set of matrices is called an *interval matrix*. Interval matrices will be used to linearly over-approximate behaviors of nonlinear models. Two useful notions are the *center matrix* and the *range matrix*, defined, respectively, as $\text{CT}([B, C]) = (B + C)/2$ and $\text{RG}([B, C]) = (C - B)/2$. Then, $\text{Interval}([B, C])$ can also be written as $\text{Interval}([A_c -$

$A_r, A_c + A_r]$), where $A_c = \text{CT}([B, C])$, $A_r = \text{RG}([B, C])$. A *vertex matrix* of an interval matrix $\text{Interval}([B, C])$ is a matrix V whose every element is either b_{ij} or c_{ij}. Let $\text{VT}(\text{Interval}([B, C]))$ be the set of all the vertex matrices of the interval matrix $\text{Interval}([B, C])$. The cardinality of $\text{VT}(\text{Interval}([B, C]))$ with $B, C \in \mathbb{R}^{n \times n}$ is 2^{n^2}.

Dynamical Systems Let us denote the set of all the real-valued variables in the model as the set X. For this set of variables, the set of all values the variables can take, denoted as $val(X)$, is isomorphic to $\mathbb{R}^n$.

A continuous behavior of the system is modeled as a trajectory. A *trajectory* ξ is defined as a function $\xi : dom \rightarrow val(X)$ where dom is the time domain of evolution, and it is either $[0, T]$ for some $T > 0$, or it is $[0, \infty)$. The domain of ξ is referred as $\xi.dom$. The state of the system along the trajectory at time $t \in \tau.dom$ is $\xi(t)$. For a bounded trajectory with $\xi.dom = [0, T]$, the *duration* $\xi.dur = T$. For unbounded trajectories, $\xi.dur$ is defined as ∞. The *first state* $\xi(0)$ is denoted by $\tau.fstate$, and for a bounded trajectory the *last state* $\xi.lstate = \xi(T)$ and $\xi.ltime = T$.

A T_1-*prefix* of ξ, for any $T_1 \in \xi.dom$, is the trajectory $\xi_1 : [0, T_1] \rightarrow \mathbb{R}^n$, such that for all $t \in [0, T_1]$, $\xi_1(t) = \xi(t)$. A set of trajectories $\mathscr{T}$ is *prefix-closed* if for any $\xi \in \mathscr{T}$, any of its prefix of ξ is also in $\mathscr{T}$. A set $\mathscr{T}$ is *deterministic* if for any pair $\xi_1(t), \xi_2(t) \in \mathscr{T}$, if $\xi_1(0) = \xi_2(0)$ then one is a prefix of the other. See, for example, [52] for detailed explanation of trajectories closed under prefix, suffix, and concatenation.

The continuous evolution of an n-dimensional *dynamical system* is given by an ordinary differential equation (ODE):

$$\dot{x} = f(x), \qquad\qquad (5.1)$$

where $f : \mathbb{R}^n \rightarrow \mathbb{R}^n$ is a locally Lipschitz and continuously differentiable function. A trajectory ξ is a *solution* of Eq. (5.1) if $\forall t \in \xi.dom$, $d\frac{\xi(t)}{dt} = f(\xi(t))$. The existence and uniqueness of solutions are guaranteed by the Lipschitz continuity of f. With an initial states and a time bound, an ODE defines a unique trajectory. Therefore, we abuse the notation and let $\xi(x_0, t)$ denote the solution $\xi(t)$ starting from $\xi(0) = x_0$. The *Jacobian* of f, $J_f : \mathbb{R}^n \rightarrow \mathbb{R}^{n \times n}$, is a matrix-valued function of all the first-order partial derivatives of f with respect to x, that is:

$$\left[J_f(x) \right]_{ij} = \frac{\partial f_i(x)}{\partial x_j}.$$

Example 5.1 The Moore–Greitzer model of a jet engine compression system is studied in [56] to understand and prevent two types of instabilities: rotating stall and surge. With a stabilizing feedback controller operating in the no-stall mode, it has the following dynamics:

$$\begin{cases} \dot{u} = -v - \frac{3}{2}u^2 - \frac{1}{2}u^3 \\ \dot{v} = 3u - v \end{cases}. \qquad\qquad (5.2)$$

The Jacobian of the system is

$$J_f(x) = \begin{bmatrix} -3u - \frac{3}{2}u^2 - 1 & -1 \\ 3 & -1 \end{bmatrix}. \tag{5.3}$$

5.3 Overview of Data-Driven Verification

5.3.1 Simulations and Reachable States

Obtaining closed-form or analytical solutions for nonlinear ordinary differential equations (ODEs) is generally impossible; however, libraries such as VNODE-LP [62] and CAPD [11] use validated numerical integration to generate a sequence of evaluations of ξ with guaranteed error bounds. We define a *simulation* as a sequence of time-stamped hyper-rectangles that contain a solution of the system.

Definition 5.1 (Simulation) For any $x_0 \in \mathbb{R}, \tau > 0, \epsilon > 0, T > 0$, a (x_0, τ, ϵ, T)-*simulation* of the system described in Eq. (5.1) is a sequence of time-stamped sets $\{(R_i, t_i)_{i=0}^k\}$ satisfying the following:

1. $0 < t_i - t_{i-1} \leq \tau$, for each $i = 1, \ldots, k$, and $t_0 = 0$ and $t_k = T$; τ is called the *maximum sampling period*.
2. Each R_i is a hyper-rectangle in $\mathbb{R}^n$ with a diameter smaller than ϵ.
3. $\xi(x_0, t_i) \in R_i$, for each $i = 0, 1, \ldots, k$, and $\forall t \in (t_{i-1}, t_i), \xi(x_0, t) \in \text{hull}(R_{i-1}, R_i)$, for $i = 1, \ldots, k$.

That is, at each time point t_i, the trajectory of the system $\xi(x_0, t_i)$ is contained in the hyper-rectangle R_i, and during the time intervals $t \in (t_{i-1}, t_i)$, the trajectory $\xi(x_0, t)$ is contained in the convex hull of R_{i-1} and R_i.

For a given initial set $\Theta \subseteq \mathbb{R}^n$, a state $x \in \mathbb{R}^n$ is said to be *reachable* if there exist a state $\theta \in \Theta$ and a time $t \geq 0$ such that $\xi(\theta, t) = x$. We denote by $\xi(\Theta, [t_1, t_2])$ the set of states that are reachable from Θ at any time $t \in [t_1, t_2]$. The set of reachable states at time t from initial set Θ is denoted by $\xi(\Theta, t)$. Given an n-dimensional dynamical system as in Eq. (5.1), a compact initial set $\Theta \subset \mathbb{R}^n$, an unsafe set $U \subseteq \mathbb{R}^n$, and a time bound $T > 0$, the *safety verification* problem (also called the bounded invariant verification) is to decide whether $\xi(\Theta, [0, T]) \cap U = \emptyset$. This problem is of fundamental importance as it captures many practical requirements.

Next, we define *reachtubes*, which are also sequences of time-stamped hyper-rectangles, but unlike simulations, they contain $\xi(\Theta, [0, T])$.

Definition 5.2 (Reachtube) For any $\Theta \subset \mathbb{R}^n, T > 0$, a (Θ, T)-*reachtube* is a sequence of time-stamped compact sets $\{(O_i, t_i)_{i=0}^k\}$, such that for each i in the sequence, $\xi(\Theta, [t_{i-1}, t_i]) \subseteq O_i$.

As we shall see in Sect. 5.3.3, computing precise reachtubes is sufficient for safety verification. Data-driven verification algorithms compute reachtubes from simulations using sensitivity analysis that we will discuss next.

5.3.2 Discrepancy Functions

Sensitivity of the solutions to changes in the initial states is formalized by discrepancy functions. Specifically, a discrepancy function bounds the distance between two neighboring trajectories as a function of the distance between their initial states and time [23, 30].

Definition 5.3 (Discrepancy Function) A continuous function $\beta : \mathbb{R}^{\geq 0} \times \mathbb{R}^{\geq 0} \to \mathbb{R}^{\geq 0}$ is a discrepancy function of (5.1) with initial set Θ if:

(1) for any pair of states $x_1, x_2 \in \Theta$, and any time $t \geq 0$,

$$\|\xi(x_1, t) - \xi(x_2, t)\| \leq \beta(\|x_1 - x_2\|, t), \, and$$

(2) for any t,

$$\lim_{\|x_1 - x_2\| \to 0^+} \beta(\|x_1 - x_2\|, t) = 0.$$

In Definition 5.3, the norm can be any norm. We will make specific choices for designing algorithms. Consider the system (5.1), and suppose with $L > 0$ is the Lipschitz constant for $f(x)$. Then, it can be shown that $\beta(\|x_1 - x_2\|_2, t) = e^{Lt}\|x_1 - x_2\|_2$ is a discrepancy function (Proposition 1 in [21]). For Example 5.1, $L = 2$ is a Lipschitz constant, and therefore, $e^{2t}\|x_1 - x_2\|_2$ can be used as a discrepancy function for the jet engine system.

According to the definition of discrepancy function, for system (5.1), at any time t, the ball centered at $\xi(x_0, t)$ with radius $\beta(\delta, t)$ contains every solution of (5.1) starting from $B_\delta(x_0)$. Therefore, by bloating the simulation trajectories using the corresponding discrepancy function, we can obtain an over-approximation of the reachtube. We remark that this definition of discrepancy function is similar to the incremental lyapunov functions [4]; however, here we do not require that trajectories converge to each other.

5.3.3 Verification Algorithm

We are now ready to present the verification algorithm (Algorithm 1). The basic idea is simple and appeared in [18, 23] at different levels of generality. Recall, the goal is to have an algorithm that answers bounded safety queries correctly: given

---

Algorithm 1: Simulation-driven verification algorithm

---

input: $\Theta, T, U, \epsilon_0, \tau_0$

1 $\delta \leftarrow \text{Dia}(\Theta); \epsilon \leftarrow \epsilon_0; \tau \leftarrow \tau_0; \text{RT}_{all} \leftarrow \emptyset;$
2 $C \leftarrow \text{Cover}(\Theta, \delta, \epsilon);$
3 **while** $C \neq \emptyset$ **do**
4   **for** $\langle \theta, \delta, \epsilon \rangle \in C$ **do**
5          $\psi = \{(R_i, t_i)_{i=0}^k\} \leftarrow \text{Simulate}(\theta, T, \epsilon, \tau);$
6          $\text{RT} \leftarrow \text{Bloat}(\psi, \delta, \epsilon);$
7   **if** $RT \cap U = \emptyset$ **then**
8              $C \leftarrow C \backslash \{\langle \theta, \delta, \epsilon \rangle\}; \quad \text{RT}_{all} \leftarrow \text{RT}_{all} \cup \text{RT};$
9   **else if** $\exists j, R_j \subseteq U$ **then**
10   **return** (U, ψ)
11   **else**
12             $C \leftarrow C \cup \text{Cover}(B_\delta(\theta), \frac{\delta}{2}, \frac{\epsilon}{2}) \backslash \{\langle \theta, \delta, \epsilon \rangle\};$
13             $\tau \leftarrow \frac{\tau}{2};$
14 **return** $(\text{SAFE}, \text{RT}_{all});$

---

system (5.1), a compact initial set $\Theta \subset \mathbb{R}^n$, an unsafe set $U \subseteq \mathbb{R}^n$, and a time bound $T > 0$, it answers whether $\xi(\Theta, [0, T]) \cap U = \emptyset$. A verification algorithm is said to be *sound* if it answers the safety question correctly and it is said to be *complete* if it is guaranteed to terminate on any input. We know that for general nonlinear and hybrid models, the unbounded time verification problem is undecidable, that is, no algorithm exists that is both sound and complete. Even for the bounded time, version of this problem is known to be undecidable. Algorithm 1 is sound and is guaranteed to terminate under a mild assumption on the inputs.

If there exists some $\epsilon > 0$ such that $B_\epsilon(\xi(\Theta, [0, T])) \cap U = \emptyset$, we say the system is *robustly safe*. That is, all states in some envelope around the system behaviors are safe. If there exist some $\epsilon, x \in \Theta$, such that $B_\epsilon(\xi(x, t)) \subseteq U$ over some interval $[t_1, t_2], 0 \leq t_1 < t_2 \leq T$, we say the system is *robustly unsafe*. An algorithm is said to be *relatively complete* if it is guaranteed to terminate when the system is either robustly safe or robustly unsafe. Algorithm 1 is relatively complete. Another way of saying this is that Algorithm 1 is a semidecision procedure for robust safety verification.

The algorithm consists of the following three main steps: (1) Simulate the system from a finite set of states (θ) that are chosen from the compact initial set Θ. The union of a set of balls of diameter δ centered at each of the states should contain Θ. (2) Bloat the $\{(R_i, t_i)_{i=0}^k\}$ simulations using a discrepancy function such that the bloated sets are reachtubes from the initial covers. (3) Check each of these over-approximations, and decide if the system is safe or not. If such a decision cannot be made, then we should start from the beginning with balls with smaller diameter δ.

There are several functions referred to in Algorithm 1. Functions $\text{Dia}()$ and $\text{Simulate}()$ are defined to return the diameter of a set and a simulation result, respectively. The $\text{Bloat}()$ function takes as the inputs the simulation ψ starting from θ, the size of the initial cover δ, and the simulation precision ϵ, and returns a reachtube that contains all the trajectories starting from the initial cover $B_\delta(\theta)$. This can be done by bloating the simulation using a discrepancy function as described in Sect. 5.4, which is an over-approximation of the distance between any neighboring trajectories starting from $B_\delta(\theta)$. Function Cover() returns a set of triples $\{\langle \theta, \delta, \epsilon \rangle\}$, where θs are sample states, the union of $B_\delta(\theta)$ covers Θ, and ϵ is the precision of simulation.

Initially, C contains a singleton $\langle \theta_0, \delta_0 = \text{Dia}(\Theta), \epsilon_0 \rangle$, where $\Theta \subseteq B_{\delta_0}(\theta_0)$ and ϵ_0 is a small positive constant. For each triple $\langle \theta, \delta, \epsilon \rangle \in$ C, the **while**-loop from Line 3 checks the safety of the reachtube from $B_\delta(\theta)$, which is computed in Lines 5–6. ψ is a $(\theta, T, \epsilon, \tau)$-simulation, which is a sequence of time-stamped rectangles $\{(R_i, t_i)\}$ and is guaranteed to contain the trajectory $\xi(\theta, T)$ by Definition 5.1. Bloating the simulation result ψ by the discrepancy function to get RT, a $(B_\delta(\theta), T)$-reachtube, we have an over-approximation of $\xi(B_\delta(\theta), [0, T])$. The core function Bloat() will be discussed in detail next. If RT is disjoint from U, then the reachtube from $B_\delta(\theta)$ is safe and the corresponding triple can be safely removed from C. If for some j, R_j (one rectangle of the simulation) is completely contained in the unsafe set, then we can obtain a counterexample in the form of a trajectory that violates the safety property. Otherwise, the safety of $\xi(B_\delta(\theta), [0, T])$ is not determined, and a refinement of $B_\delta(\theta)$ needs to be made with smaller δ and smaller ϵ, τ.

Figure 5.1 gives a conceptual demonstration of Algorithm 1 running on the jet engine example (Example 5.1).

Theorem 5.1 *Algorithm 1 is sound. That is, if it returns SAFE, then indeed $\xi(\Theta, [0, T]) \cap U = \emptyset$; if it returns UNSAFE, then it also finds a counterexample, the simulation ψ which enters U. Algorithm 1 is also relatively complete. That is, for any robustly safe or unsafe system, it will terminate and decide either SAFE or UNSAFE.*

A crucial and challenging aspect of Algorithm 1 is choosing an appropriate discrepancy function with which to implement the Bloat() function. In the next section, we introduce algorithms that implement this function.

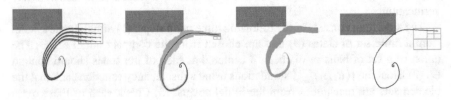

Fig. 5.1 Conceptual demonstration of verification algorithm. Red rectangle: unsafe set, cyan rectangle: cover of initial set K. Simulations (blue lines) cannot guarantee safety, but together with sensitivity analysis give reachsets (gray region) to prove safety (green region) or identify bug traces

5.4 Computing Discrepancy

In this section, we discuss several approaches for computing discrepancy functions of dynamical systems. We start with the simplest case of stable linear systems where Lyapunov equations can be used for computing discrepancy. Then, we move on to discuss nonlinear models and contraction metrics, and finally, we discuss locally optimal methods for general nonlinear systems.

5.4.1 Linear Models

For a linear time invariant (LTI) system $\dot{x} = Ax$, if the system is asymptotically stable we can find a discrepancy function by solving the Lyapunov equation:

Theorem 5.2 *For asymptotically stable linear system $\dot{x} = Ax$, given any positive definite matrix $Q \in \mathbb{R}^{n \times n}$, $\beta(\|x_1 - x_2\|_M, t) = e^{-\gamma t}\|x_1 - x_2\|_M$ is a discrepancy function, where $M \succ 0$ can be found by solving the Lyapunov equation $A^T M + MA + Q = 0$ and $\gamma = \frac{\lambda_{\min}(Q)}{2\lambda_{\max}(M)}$.*

Proof Fix any $x_1, x_2 \in \mathbb{R}^n$, and let $y(t) = \xi(x_1, t) - \xi(x_2, t)$, we have

$$d\frac{\|y(t)\|_M^2}{dt} = \dot{y}^T(t)My(t) + y(t)M\dot{y}(t) = y^T(t)(A^T M + MA)y(t)$$
$$= -y^T(t)Qy(t) \leq -\lambda_{\min}(Q)y^T(t)y(t)$$
$$\leq -\frac{\lambda_{\min}(Q)}{\lambda_{\max}(M)}y^T(t)My(t) = -\frac{\lambda_{\min}(Q)}{\lambda_{\max}(M)}\|y(t)\|_M^2$$

By applying Grönwall's inequality, we obtain

$$\|y(t)\|_M \leq e^{-\frac{\lambda_{\min}(Q)}{2\lambda_{\max}(M)}}\|y(0)\|_M. \tag{5.4}$$

5.4.2 Nonlinear Models: Optimization-Based Approaches

For nonlinear systems with trajectories that exponentially converge to each other, contraction metrics can be used as a certificate for this convergence [58]. Discrepancy functions can be computed from contraction metrics.

Definition 5.4 (From [58]) A uniform metric $M : \mathbb{R}^n \times \mathbb{R}^{\geq 0} \to \mathbb{R}^{n \times n}$ is called a contraction metric for (5.1) if $\exists \gamma \in \mathbb{R}^{\geq 0}$ such that:

$$J_f^T(x)M(x, t) + M(x, t)J_f(x) + \dot{M}(x, t) + \gamma M(x, t) \preceq 0.$$

Theorem 5.3 (Theorem 2 from [58]) *For system given by (5.1) that admits a contraction metric* M, *the trajectories converge exponentially with time, i.e.,* $\exists k \geq 1, \gamma > 0$ *such that,* $\forall x_1, x_2 \in \mathbb{R}^n$, $y^T(t)y(t) \leq k y^T(0)y(0)e^{-\gamma t}$, *where* $y(t) = \xi(x_1, t) - \xi(x_2, t)$.

Proposition 5.1 (Proposition 5 from [21]) *For system given by (5.1) that admits a contraction metric* M, $\beta(\|x_1 - x_2\|_2, t) = \sqrt{k}e^{-\frac{\gamma}{2}t}\|x_1 - x_2\|_2$ *is a discrepancy function, where* k, γ *are from Theorem 5.3.*

In [8], a technique for establishing exponential convergence among trajectories using sum of squares (SOS) optimization is proposed. Informally, it searches for a contraction metric that satisfies conditions given in Definition 5.4 as follows:

1. Select the degree of the polynomial d for contraction metric $M(x)$. That is, all the terms in the contraction metric are fixed degree polynomial terms in the n real variables. For example, the general form of $M(x)$ for a two-dimensional system with variables u and v is given as $\begin{bmatrix} \sum a_{ij}v^i u^j & \sum b_{ij}v^i u^j \\ \sum c_{ij}v^i u^j & \sum d_{ij}v^i u^j \end{bmatrix}$.

2. Calculate $R(x) = J_f^T(x)M(x) + M(x)J_f(x) + \dot{M}(x)$ and enforce constraints on a_{ij}, b_{ij}, c_{ij} and d_{ij} such that $R(x)$ is symmetric.

3. Impose the restrictions that polynomials $y^T M(x)y$ and $-y^T R(x)y$ are sum of squares polynomials and solve for the feasibility using SOS tools. If the solution exists, then the SOS solver will find values of coefficients of polynomials.

4. If the solution is feasible, compute the exponential rate of convergence by computing the value of γ such that:

$$J_f^T(x)M(x) + M(x)J_f(x) + \dot{M}(x) + \gamma M(x) \prec 0.$$

5. If SOS solver returns infeasible, then increase the degree of the polynomial terms in M and repeat.

For a given nonlinear ordinary differential equation (ODE), a contraction metric that is a sum of squares polynomial is not guaranteed to exist, and hence, the above procedure is not guaranteed to terminate.

5.4.3 Nonlinear Models: Local Discrepancy

The main obstacle to finding a (global) discrepancy function for general nonlinear systems is the difficulty to globally bound the convergence (or divergence) rates across all trajectories. By restricting the definition of discrepancy functions over carefully computed parts of the state space, we will gain two benefits. First, such local discrepancy functions will still be adequate to compute Bloat needed in Algorithm 1. Second, it will become possible to compute a *local* discrepancy function automatically from simulation traces.

We begin by observing that, over a compact set $S \subseteq \mathbb{R}^n$, the Jacobian J_f of the system described by Eq. (5.1) can be over-approximated by an interval matrix. Then, we establish that the distance between two trajectories in S satisfies a differential equation from a set of differential equations described using the interval matrix. By bounding the matrix measure of the interval matrix, we can get a discrepancy function.

Since we assume that the system is continuously differentiable, the Jacobian matrix is continuous, and therefore, over a compact set S, the elements of $J_f(x)$ are bounded. That is, there exists an interval matrix $\mathscr{A}$ such that $\forall x \in S, J_f(x) \in \mathscr{A}$. For interval matrix $\mathscr{A} = \texttt{Interval}(B, C)$, the bounds B and C can be obtained using interval arithmetic or an optimization toolbox by maximizing and minimizing the terms of J_f over S. (The set S can be chosen to be a coarse over-approximation of the reach set, obtained using the Lipschitz constant as in Sect. 5.4.) Once the bounds are obtained, we use the interval matrix that over-approximates the behavior of $J_f(x)$ over S to analyze the rate of convergence or divergence between trajectories:

Lemma 5.1 (Lemma 3.4 from [29]) *For system* (5.1) *with initial set* Θ *starting from time* t_1, *suppose* $S \subseteq \mathbb{R}^n$ *is a compact convex set, and* $[t_1, t_2]$ *is a time interval such that for any* $\xi(\Theta, [t_1, t_2]) \subseteq S$. *If there exists an interval matrix* $\mathscr{A}$ *such that* $\forall x \in S, J_f(x) \in \mathscr{A}$, *then for any* $x_1, x_2 \in \Theta$, *and for any* $t \in [t_1, t_2]$, *the distance* $y(t) = \xi(x_2, t) - \xi(x_1, t)$ *satisfies* $\dot{y}(t) = A(t)y(t)$, *for some* $A(t) \in \mathscr{A}$.

$\dot{y}(t) = A(t)y(t)$ used in Lemma 5.1 can be used to define a discrepancy function. Given any matrix $M \succ 0$, $\|y(t)\|_M^2 = y^T(t)My(t)$, and by differentiating $\|y(t)\|_M^2$, we have that for any fixed $t \in [t_1, t_2]$:

$$\frac{d\|y(t)\|_M^2}{dt} = \dot{y}^T(t)y(t) + y^T(t)\dot{y}(t) = y^T(t)(A(t)^T M + M A(t))y(t), \quad (5.5)$$

for some $A(t) \in \mathscr{A}$. We write $A(t)$ as A in the following for brevity. If there exists a $\hat{\gamma}$ such that $A^T M + M A \preceq \hat{\gamma}M, \forall A \in \mathscr{A}$, then (5.5) becomes $\dfrac{d\|y(t)\|_M^2}{dt} \leq \hat{\gamma}\|y(t)\|_M^2$. After applying Grönwall's inequality, we have

$$\|y(t)\|_M \leq \|y(t_1)\|_M e^{\frac{\hat{\gamma}}{2}(t-t_1)}, \forall t \in [t_1, t_2].$$

The above provides a discrepancy function: $\beta(\|x_1 - x_2\|_M, t) = \|x_1 - x_2\|_M e^{\frac{\hat{\gamma}}{2}(t-t_1)}$. This discrepancy function could result in more or less conservative reachtubes, depending on the selection of M and $\hat{\gamma}$. Ideally, we would like to identify the optimal M such that we can obtain the tightest bound $\hat{\gamma}$. This problem is formulated as follows:

$$\min_{\hat{\gamma} \in \mathbb{R}, M \succ 0} \hat{\gamma} \quad (5.6)$$

$$\text{s.t } A^T M + M A \preceq \hat{\gamma}M, \quad \forall A \in \mathscr{A}.$$

Solving (5.6) to obtain the optimal $\hat{\gamma}$ for each time interval involves solving optimization problems with infinite numbers of constraints (imposed by the infinite set of matrices in $\mathscr{A}$). To overcome this problem, we introduce a strategy to transform (5.6) to an equivalent problem with finitely many constraints based on the vertex matrices.

Lemma 5.2 (Lemma 4.1 from [29]) *For system* (5.1) *with initial set* Θ *starting from time* t_1, *suppose* $S \subseteq \mathbb{R}^n$ *is a compact convex set, and* $[t_1, t_2]$ *is a time interval such that for any* $x \in \Theta$, $t \in [t_1, t_2]$, $\xi(x, t) \in S$. *Let* M *be a positive definite* $n \times n$ *matrix. If there exists an interval matrix* $\mathscr{A}$ *such that:*

(a) $\forall x \in S$, $J_f(x) \in \mathscr{A}$, *and*
(b) $\exists \hat{\gamma} \in \mathbb{R}$, $\forall A_i \in \text{VT}(\mathscr{A})$, $A_i^T M + M A_i \preceq \hat{\gamma} M$,

then for any $x_1, x_2 \in \Theta$ *and* $t \in [t_1, t_2]$:

$$\|\xi(x_1, t) - \xi(x_2, t)\|_M \le e^{\frac{\hat{\gamma}}{2}(t - t_1)} \|x_1 - x_2\|_M.$$

Lemma 5.2 suggests the following bilinear optimization problem for finding discrepancy over compact subsets of the state space:

$$\min_{\hat{\gamma} \in \mathbb{R}, M > 0} \quad \hat{\gamma} \tag{5.7}$$

$$\text{s.t. for each } A_i \in \text{VT}(\mathscr{A}), A_i^T M + M A_i \preceq \hat{\gamma} M.$$

Letting $\hat{\gamma}_{\max}$ be the maximum of the eigenvalues of $A_i^T + A_i$ for all i, then $A_i^T + A_i \preceq \hat{\gamma}_{\max} I$ (i.e., $M = I$) holds for every A_i, so a feasible solution exists for (5.7). To obtain a minimal feasible solution for $\hat{\gamma}$, we choose a range of $\gamma \in [\gamma_{\min}, \gamma_{\max}]$, where $\gamma_{\min} < \gamma_{\max}$ and perform a line search of $\hat{\gamma}$ over $[\gamma_{\min}, \gamma_{\max}]$. Note that if $\hat{\gamma}$ is fixed, then (5.7) is a semidefinite program (SDP), and a feasible solution can be obtained by an SDP solver. As a result, we can solve (5.7) using a line search strategy, where an SDP is solved at each step.

This approach is computationally intensive due to the potential $O(2^{n^2})$ matrices in $\text{VT}(\mathscr{A})$ that appear in the SDP (5.7). In [29], a second method is shown to avoid the exponential increase in the number of constraints in (5.7), at the expense of lower accuracy (i.e., increasing the conservativeness).

5.4.4   Algorithm to Compute Local Optimal Reach Set

Given an initial set $B_\delta(x)$ and time bound T, Lemma 5.2 provides discrepancy functions over compact subsets of the state space, and over a bounded time horizon. To compute the reach set of a nonlinear model from a set of initial states over a long time horizon $[0, T]$, we will divide the time interval $[0, T]$ into smaller intervals

$[0, t_1], \ldots, [t_{k-1}, t_k = T]$, and compute a piece-wise discrepancy function, where each piece is relevant for a smaller portion of the state space and time.

Consider two adjacent subintervals of $[0, T]$, $a = [t_1, t_2]$ and $b = [t_2, t_3]$. Let $E_{M_a, c_a(t_2)}(\xi(x_0, t_2))$ be an ellipsoid that contains $\xi(B_\delta(x), t_2)$, and suppose we are given a matrix M_b and we want to select a $c_b(t)$ such that $\xi(B_\delta(x), t_2) \subseteq E_{M_b, c_b(t_2)}(\xi(x_0, t_2))$. To over-approximate the reach set for the interval b, we require that $c_b(t_2)$ is chosen so that at the transition time t_2:

$$E_{M_a, c_a(t_2)}(\xi(x_0, t_2)) \subseteq E_{M_b, c_b(t_2)}(\xi(x_0, t_2)). \tag{5.8}$$

This is a standard SDP problem to compute the minimum value for $c_b(t_2)$ that ensures (5.8) (see, for example, [10]). This minimum value is used as $c_b(t_2)$ for computing the reachtube for time interval b.

Let Ea denote the ellipsoid $E_{M_a, c_a(t_2)}(\xi(x_0, t_2))$ and Eb denote the ellipsoid $E_{M_b, c}(\xi(x_0, t_2))$. The problem of minimizing $c_b(t_2)$, given $M_a, M_b, c_a(t_2)$, such that Eq. (5.8) holds, is the following optimization problem:

$$\begin{aligned} \min \quad & c \\ s.t. \quad & Eb \supseteq Ea. \end{aligned} \tag{5.9}$$

In what follows, let $c_b(t_2)$ be equal to a solution of the above. We can transfer problem (5.9) to the following *sum-of-squares* problem as the "S procedure" [57] to make it solvable by SDP solvers:

$$\begin{aligned} \min \quad & c \\ s.t. \quad & c - \|x - \xi(x_0, t_2)\|_{M_b}^2 - \lambda \left(c_a(t_2) - \|x - \xi(x_0, t_2)\|_{M_a}^2 \right) \geq 0, \lambda \geq 0. \end{aligned} \tag{5.10}$$

We present an algorithm to compute a $(B_\delta(x), T)$-reachtube for system (5.1) using the results from Lemmas 5.2. The inputs to Algorithm Bloat are as follows: (1) A simulation ψ of the trajectory $\xi(x, t)$, where $x = \xi(x, t_0)$ and $t_0 = 0$, represented as a sequence of points $\xi(x, t_0), \ldots, \xi(x, t_k)$ and a sequence of hyper-rectangles $Rec(t_{i-1}, t_i) \subseteq \mathbb{R}^n$. That is, for any $t \in [t_{i-1}, t_i]$, $\xi(x, t) \in Rec(t_{i-1}, t_i)$. (2) The Jacobian matrix $J_f(\cdot)$. (3) A Lipschitz constant L for the vector field (this can be replaced by a local Lipschitz constant for each time interval). (4) A matrix M_0 and constant c_0 such that $B_\delta(x) \subseteq E_{M_0, c_0}(x)$. The output is a $(B_\delta(x), T)$-Reachtube. We assume that the exact simulation of the solution $\xi(x, t)$ exists and can be represented as a sequence of points and hyper-rectangles for ease of exposition.

Algorithm Bloat uses Lemma 5.2 to update the coordinate transformation matrix M_i to ensure an optimal exponential rate γ_i of the discrepancy function in each time interval $[t_{i-1}, t_i]$. It will solve the optimization problem (5.7) in each time interval to get the local optimal rate, and solve the optimization problem (5.8) when it moves forward to the next time interval.

Algorithm 2: Algorithm `Bloat`

---

input : ψ, $J_f(\cdot)$, L, M_0, c_0
initially: RT $\leftarrow \emptyset$, $\gamma_0 \leftarrow -100$
1   $\delta_0 = \mathrm{Dia}\left(E_{M_0,c_0}(x)\right)$;
2   **for** $i = 1{:}k$ **do**
3      $\Delta t \leftarrow t_i - t_{i-1}$;
4      $S \leftarrow B_{\delta_{i-1}e^{L\Delta t}}(Rec(t_{i-1}, t_i))$;
5      $\mathscr{A} \leftarrow \mathrm{Interval}[B, C]$ such that $J_f(x) \in \mathrm{Interval}[B, C], \forall x \in S$;
6   **if** $\forall V \in \mathrm{VT}(\mathscr{A}) : V^T M_{i-1} + M_{i-1}V \le \gamma_{i-1}M_{i-1}$ **then**
7         $M_i \leftarrow M_{i-1}$; ;
8         $\gamma_i \leftarrow \underset{\gamma\in\mathbb{R}}{\arg\min} \ \forall V \in \mathrm{VT}(\mathscr{A}) : V^T M_i + M_i V \le \gamma M_i$;
9         $c_{tmp} \leftarrow c_{i-1}$
10   **else**
11   compute M_i, γ_i from Eq. (5.7) ;
12   compute minimum c_{tmp} such that $E_{M_{i-1},c_{i-1}}(\xi(x, t_{i-1})) \subseteq E_{M_i,c_{tmp}}(\xi(x, t_{i-1}))$;
13      $c_i \leftarrow c_{tmp}e^{\gamma_i \Delta t}$;
14      $\delta_i \leftarrow \mathrm{Dia}(E_{M_i,c_i}(\xi(x, t_i)))$;
15      $O_i \leftarrow B_{\delta'/2}(Rec(t_{i-1}, t_i))$ where $\delta' = \max\{dia\left(E_{M_i,c_{tmp}}(\xi(x, t_{i-1}))\right), \delta_i\}$;
16   RT $\leftarrow$ RT $\cup [O_i, t_i]$;
17 **return** RT ;

The algorithm proceeds as follows. The diameter of the ellipsoid containing the initial set $B_\delta(x)$ is computed as the initial set size (Line 1). At Line 4, $Rec(t_{i-1}, t_i)$, which contains the trajectory between $[t_{i-1}, t_i]$ is bloated by the factor $\delta_{i-1}e^{L\Delta t}$ which gives the set S that is guaranteed to contain $\xi(B_\delta(x), t)$ for every $t \in [t_{i-1}, t_i]$. Next, at Line 5, an interval matrix $\mathscr{A}$ containing $J_f(x)$, for each $x \in S$, is computed. The "if" condition in Line 6 determines whether the M_{i-1}, γ_{i-1} used in the previous iteration satisfy the conditions of Lemma 5.2 (γ_0 when $i = 1$, where γ_0 is an initial guess). This condition will avoid performing updates of the discrepancy function if it is unnecessary. If the condition is satisfied, then M_{i-1} is used again for the current iteration i (Lines 7–9) and γ_i will be computed as the smallest possible value such that Lemma 5.2 holds (Line 8) without updating the shape of the ellipsoid (i.e., $M_i = M_{i-1}$). In this case, the γ_i computed using M_{i-1} in the previous iteration $(i - 1)$ may not be ideal (minimum) for the current iteration (i), but we assume that it is acceptable. If M_{i-1} and γ_{i-1} do not satisfy the conditions of Lemma 5.2, that means the previous coordinate transformation can no longer ensure an accurate exponential converging or diverging rate between trajectories. Then, M_i and γ_i are recomputed at Line 11. For the vertex matrix constraints case, (5.7) is solved to update M_i and γ_i.

At Line 12, an SDP is solved to identify the smallest constant c_{tmp} for discrepancy function updating such that $E_{M_{i-1},c_{i-1}}(\xi(x, t_{i-1})) \subseteq E_{M_i,c_{tmp}}(\xi(x, t_{i-1}))$. At Line 13, we compute the updated ellipsoid size c_i such that $E_{M_i,c_i}(\xi(x, t_i))$ contains $\xi(B_\delta(x), t_i)$. At Line 14, the diameter of $E_{M_i,c_i}(\xi(x, t_i))$ is assigned to δ_i for next iteration. At Line 15, the set O_i is computed such that it contains the reach set during time interval $[t_{i-1}, t_i]$. Finally, at Line 16 RT is returned as an over-approximation of the reach set.

The next lemma states that the γ produced by Line 11 is a local optimal exponential converging or diverging rate between trajectories.

Lemma 5.3 (Lemma 5.1 from [29]) *In the ith iteration of Algorithm* Bloat, *suppose $\mathscr{A}$ is the approximation of the Jacobian over $[t_{i-1}, t_i]$ computed in Line 5. If E_{i-1} is the reach set at t_{i-1}, then for all M' and γ' such that $\xi(E_{i-1}, t_i) \subseteq E_{M',c'}(\xi(x, t_i))$ where c' is computed from γ' (Line 13), we have that the γ produced by Line 11 satisfies $\gamma \leq \gamma'$.*

Theorem 5.4 ensures soundness of the verification algorithm.

Theorem 5.4 (Theorem 5.2 from [29]) *For any (x, T)-simulation $\psi = \xi(x, t_0), \ldots, \xi(x, t_k)$ and any constant $\delta \geq 0$, a call to* Bloat(ψ, δ) *returns a $(B_\delta(x), T)$-reachtube.*

Proof By Lemma 5.2, at any time $t \in [t_{i-1}, t_i]$, any other trajectory $\xi(x', t)$ starting from $x' \in E_{M_{i-1}, c_{i-1}}(\xi(x, t_{i-1}))$ is guaranteed to satisfy

$$\|\xi(x, t) - \xi(x', t)\|_{M_i} \leq \|\xi(x, t_{i-1}) - x'\|_{M_i} e^{\frac{\gamma_i}{2}(t - t_{i-1})}. \tag{5.11}$$

Then, at time t_i, the reach set is guaranteed to be contained in the ellipsoid $E_{M_i, c_i}(\xi(x, t_i))$.

At Line 15, we want to compute the set O_i such that it contains the reach set during time interval $[t_{i-1}, t_i]$. According to Eq. (5.11), at any time $t \in [t_{i-1}, t_i]$, the reach set is guaranteed to be contained in the ellipsoid $E_{M_i, c(t)}(\xi(x, t))$, where $c(t) = c_{tmp} e^{\gamma_i(t - t_{i-1})}$. O_i should contain all the ellipsoids during time $[t_{i-1}, t_i]$. Therefore, it can be obtained by bloating the rectangle $Rec(t_{i-1}, t_i)$ using the largest ellipsoid's radius (half of the diameter). Since $e^{\gamma_i(t - t_{i-1})}$ is monotonic (increasing when $\gamma_i > 0$ or decreasing when $\gamma_i < 0$) with time, the largest ellipsoid during $[t_{i-1}, t_i]$ is either at t_{i-1} or at t_i. So, the largest diameter of the ellipsoids is $\max\{dia\left(E_{M_i, c_{tmp}}(\xi(x, t_{i-1}))\right), \delta_i\}$. Thus, at Line 15, O_i computed at Line 15 is an over-approximation of the reach set during time interval $[t_{i-1}, t_i]$.

When $i = 1$, because the initial ellipsoid $E_{M_0, c_0}(x)$ contains the initial set $B_\delta(x)$, we have that $E_{M_1, c_1}(\xi(x, t_1))$ defined at Line 14 contains $\xi(B_\delta(x), t_1)$. Also at Line 15, O_1 contains $\xi(B_\delta(x), [t_0, t_1])$. Repeating this reasoning for subsequent iterations, we have that $E_{M_i, c_i}(\xi(x, t_i))$ contains $\xi(B_\delta(x), t_i)$, and O_i contains $\xi(B_\delta(x), [t_{i-1}, t_i])$. Therefore, RT returned at Line 16 is a $(B_\delta(x), T)$-Reachtube.

Remark 5.1 It is straightforward to modify Algorithm 2 to accept validated simulations and the error bounds introduced. At Line 4 and Line 15, instead of bloating $Rec(t_{i-1}, t_i)$, we need to bloat $\text{hull}(\{R_{i-1}, R_i\})$, which is guaranteed to contain the solution $\xi(x, t), \forall t \in [t_{i-1}, t_i]$. Also, at Line 12 and Line 14, when using the ellipsoid $E_{M_i, c_i}(\xi(x, t_i))$, we use $E_{M_i, c_i}(0) \oplus R_i$.

5.5 Hybrid System Verification

Hybrid systems are a natural and popular model for representing cyber-physical systems [3, 38, 51, 61]. One can view a hybrid system as a collection of ODEs—one for each *mode*—and a set of discrete transition rules for switching between the ODEs or modes. Thus, the continuous behavior of a hybrid system is described by differential equations, and discrete behavior is described by a set of transition rules that can be defined in terms of a labeled control graph, a program, or an automaton. In this section, we present extensions of the data-driven verification approach to fit hybrid models.

5.5.1 Hybrid Model

We will use L to denote a finite set of *modes, locations,* or discrete states. We will use a Euclidean space $X \subseteq \mathbb{R}^n$ for the continuous state. The combined *hybrid* state space is $L \times X$. The discrete behavior or mode transitions will be specified a control graph over L with labels defining the guards and resets on X. A guard on X is predicate $G : X \rightarrow \mathbb{B}$, and reset function is a mapping $R : X \rightarrow X$.

Definition 5.5 Given a hybrid state space $L \times X$, a control graph on $L \times X$ is a labeled directed graph $G = \langle V, E, elab \rangle$, where:

1. $V \subseteq L$ is the set of vertices,
2. $E \subseteq V \times V$ is the set of edges, and
3. *elab* labels each edge $e \in E$ with finitely many guards and reset maps on X.

The evolution of the system's continuous state variables is formally described by the continuous functions of initial states and time called trajectories (see Sect. 5.2). For a hybrid system with L modes, each trajectory is labeled by a mode in L. A *trajectory labeled by* L is a pair $\langle \xi(x_0, t), \ell \rangle$ where $\xi(x_0, t)$ is a trajectory starting from x_0, and $\ell \in L$. A deterministic, prefix-closed set of labeled trajectories TL describes the behavior of the continuous variables in modes L.

In this section, we consider hybrid system with explicit continuous dynamics expressions. That is, the dynamical evolution of the hybrid system's continuous state variables in each mode is expressed by ODEs. Therefore, a hybrid system is formally defined as follows:

Definition 5.6 A hybrid system $\mathcal{H}$ is a tuple $\langle X, L, \Theta, L_{init}, G, TL \rangle$, where:

1. $X \times L$ is the hybrid state space,
2. $\Theta \times L_{init} \subseteq X \times L$ is a compact set of initial states,
3. $G = \langle V, E, elab \rangle$ is a control graph on $X \times L$, and
4. TL is a set of deterministic, prefix-closed labeled trajectories. For each $\ell \in L$, a set of trajectories TL_ℓ is specified by differential equations $f_\ell : \mathbb{R}^n \rightarrow \mathbb{R}^n$ and an invariant $I_\ell \subseteq \mathbb{R}^n$, such that over any trajectory $\langle \xi, \ell \rangle \in TL_\ell$, ξ evolves according to $d\frac{\xi}{dt} = f_\ell(\xi)$ at each time in the domain of ξ, and ξ satisfies the invariant I_ℓ.

Semantics of $\mathcal{H}$ is given in terms of executions which are sequences of trajectories consistent with the modes defined by the control graph. An execution of $\mathcal{H}$ starting from $x_0 \in \Theta$ and $\ell_{init} \in \mathsf{L}_{init}$ is a sequence of labeled trajectories $exec(x_0, \ell_{init}) = \langle \xi_{\ell_1}, \ell_1 \rangle, \cdots, \langle \xi_{\ell_k}, \ell_k \rangle$ such that:

1. $\xi_{\ell_1}.fstate = x_0 \in \Theta$ and $\ell_1 = \ell_{init} \in \mathsf{L}_{init}$,
2. $\sum_{j=1}^{k} \xi_{\ell_j}.dur = T$,
3. $\ell_1, \cdots, \ell_k$ follow the control graph G. That is, for each $i > 1$, there is an edge $e \in \mathsf{E} : v_{i-1} \rightarrow v_i$ with the edge label $elab = [Guard_e]\{Reset_e\}$, such that v_{i-1} corresponds to the mode ℓ_{i-1} and v_i corresponds to the mode ℓ_i, $\xi_{\ell_{i-1}}.lstate$ satisfies the guard: $Guard_e(\xi_{\ell_{i-1}}.lstate) = \text{True}$, and $\xi_{\ell_i}.fstate$ satisfies the reset map: $Reset_e(\xi_{\ell_i}.fstate) = \text{True}$.

The set of all executions of $\mathcal{H}$ is denoted by $\mathsf{Execs}_{\mathcal{H}}$. A state $\langle x, \ell \rangle$ is *reachable* at vertex ℓ (of graph G) if there exists an execution $\langle \xi_{\ell_1}, \ell_1 \rangle, \ldots, \langle \xi_{\ell_k}, \ell_k \rangle \in \mathsf{Execs}_{\mathcal{H}}, i \in \{1, \ldots k\}$, and $t' \in \xi_i.dom$ such that $\ell = \ell_i, x = \xi_{\ell_i}(t')$. The set of reachable states is defined as:

$$\xi(\mathcal{H}, T) = \{\langle x, \ell \rangle \mid \text{for some } \ell, \ \langle x, \ell \rangle \text{ is reachable at vertex } \ell\}.$$

Given a set of (unsafe) states $U \subseteq X \times L$, the bounded safety verification problem is to decide whether $\xi(\mathcal{H}, T) \cap U = \emptyset$.

Example 5.2 A hybrid system that models the behavior of a cardiac pacemaker system is given in Fig. 5.2a. The hybrid system has two modes, namely, Stim_on and Stim_off. The continuous variables u and v model the voltage and the current on the tissue membrane and the timer t measures the time spent in each location.

a

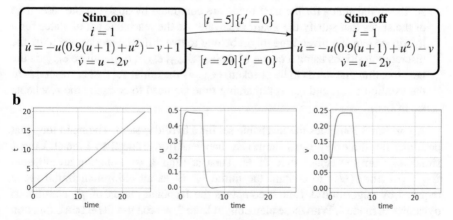

b

Fig. 5.2 (a) Hybrid system model of a cardiac cell with a pacemaker. (b) Sample execution of the cardiac cell-pacemaker system from the initial state $[0, 0.1, 0.1]$. Blue and green trajectories correspond to the Stim_on and Sim_off modes, respectively

The system stays in Stim_on location when the pacemaker gives a stimulus to the cell and is in Stim_off when the stimulus is absent. The discrete transition from Stim_on to Stim_off is enabled when $t = 5$; and t is reset to 0 after a transition; u and v are left unchanged. Transition from Stim_off to Stim_on is enabled when $t = 20$; and both these transitions are urgent. Thus, the pacemaker gives a stimulus every 25 time units for a duration of 5 time units. The behavior of the continuous variables t, u, v within a time period is given in Fig. 5.2b.

5.5.2 Hybrid System Verification Algorithm

We outline the hybrid extension of Algorithm 2 now presented as Algorithm 3. Algorithm 2 computes the set of reachable states for a given continuous system as described in Eq. (5.1) for a given time interval. Therefore, one can essentially apply this algorithm for each of the relevant modes of a hybrid system. For simplicity, let us assume that all the mode invariants and transition guards to be convex polyhedra, and that all the reset mappings are linear functions. Without loss of generality, we assume that there is only one mode ℓ_{init} in the set of initial locations L_{init}. Algorithm 3 performs the following three steps iteratively until the time horizon for verification:

1. For the given mode ℓ and a given initial set Θ, the algorithm first simulates from the center of Θ, computes the Jacobian of the continuous dynamics in mode ℓ, and then computes the reachable set RT_ℓ for that mode from Θ for the bounded remaining time specified using Algorithm 2.
2. The reachable set is pruned by removing all the states that violate the mode invariant.
3. The reachable set is checked to satisfy any guards for discrete transitions, and if so, the initial states for the next mode are computed by applying the reset map of the states that satisfy the guard predicate. As the reachable set of states for a hybrid system at a given time might belong to two different modes, we track the discrete transitions using a queue of tuples $\langle \Theta_{next}, \ell_{next}, t_{left} \rangle$, where ℓ_{next} is the next location that needs to be checked, Θ_{next} is the initial set that corresponds to the location ℓ_{next}, and t_{left} is remaining time we need to compute the reachable set in ℓ_{next}.

Algorithm 3 computes the reachable set for a hybrid system. The main loop that performs the three key steps iteratively happens from Line 2 to Line 9. Line 2 simulates from the center state of Θ. Then at Line 3, we compute an ellipsoid $E_{M_0,c_0}(\text{center}(\Theta))$ to contain the initial set Θ as an ellipsoidal initial set is required by Algorithm 2. Line 4 computes the Jacobian matrix of f_ℓ, continuous dynamics in mode ℓ. With these elements, at Line 5, we can use the Bloat function as Algorithm 2 to get the reachable set of states from Θ for the corresponding

---

Algorithm 3: Algorithm `HybridReachtube`

---

    input   : Hybrid System $\mathcal{H} = \langle X \cup \{\ell\}, \Theta, \ell_{init}, T, G, \mathsf{TL}\rangle$, Time bound T, Lipschitz
              constants $\{L_\ell\}_{\ell \in \mathsf{L}}$, Parameters for validated simulation ϵ, τ.
    initially: $\mathsf{Q} \leftarrow \langle \Theta, \ell_{init}, T\rangle$, $RT_{hybrid} \leftarrow \emptyset$
1 **for** *each* $\langle \Theta, \ell, t_{left}\rangle \in Q$ **do**
2 $\psi = \{(R_i, t_i)_{i=0}^k\} \leftarrow \mathtt{Simulate}(\mathtt{center}(\Theta), t_{left}, \epsilon, \tau)$;
3   Compute M_0, c_0 such that $\Theta \subseteq E_{M_0, c_0}(\mathtt{center}(\Theta))$;
4 $J_{f_\ell}(x) \leftarrow$ Jacobian matrix of f_ℓ in mode ℓ;
5 $\mathsf{RT}_\ell \leftarrow \mathtt{Bloat}(\psi, J_{f_\ell}(x), L_\ell, M_0, c_0)$;
6 $\mathsf{RT}_\ell \leftarrow \mathsf{RT}_\ell \cap I_\ell$;
7 $\{\langle \Theta_{next}, \ell_{next}, t_{left}\rangle\} \leftarrow \mathtt{discreteTransitions}(\mathsf{RT}_\ell)$;
8 $RT_{hybrid} \leftarrow RT_{hybrid} \cup \mathsf{RT}_\ell$;
9 $\mathsf{Q}.append(\{\langle \Theta_{next}, \ell_{next}, t_{left}\rangle\})$;
10 **return** RT_{hybrid} ;

---

mode ℓ. Line 6 checks the invariant for the reachable set and line 7 computes the states reached Θ_{next} and the remained time t_{left} to be checked after discrete transitions.

C2E2

Algorithms 1–3 are the core procedures implemented in the verification tool *Compute Execute Check Engine*(C2E2) developed at University of Illinois [24, 33]. C2E2 is a software tool for simulating and verifying hybrid automata models. Hybrid models and the requirements have to be specified in an xml format. The tool parses the xml model to generate C++ libraries for numerical simulations and computes other relevant quantities like the Jacobians of the different modes. Using the data-driven verification algorithms, C2E2 can automatically check bounded time invariant properties of nonlinear hybrid automata. The tool also supports compositional modeling, a graphical user interface for model editing, and plotting. C2E2 has been used for modeling and analyzing robots, autonomous cars, and medical devices. Some of these applications are discussed in Sect. 5.7.

Example 5.3 (Example 5.2 Continued) Figure 5.3 shows the reachtubes of the continuous variables u and v of the cardiac cell-pacemaker system computed using the verification tool C2E2.

5.6   Verification of Models with Black-Box Components

In hybrid system models, we have discussed thus far the evolution of the continuous state variables that is explicitly described by differential equations and trajectories. In real-world control systems, "models" are typically a heterogeneous mix of simulation code, differential equations, block diagrams, and hand-crafted look-up tables. Extracting clean mathematical models (e.g., ODEs) from these descriptions

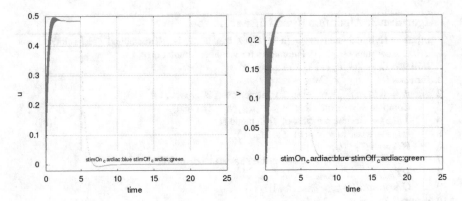

Fig. 5.3 Reachtubes of the cardiac cell-pacemaker system produced by C2E2 with initial set $t \in$ $[0, 0]$, $u, v \in [0, 0.2]$. Left: u vs time. Right: v vs time. Blue and green regions correspond to the Stim_on and Stim_off modes, respectively

is usually infeasible. The high-level logic deciding the transitions of when and for how long the system stays in each mode is usually implemented in a relatively clean piece of code and this logical module can be seen as the control graph as in Definition 5.5. In contrast, the dynamics of physical plant, with hundreds of parameters, is more naturally viewed as a "black-box." That is, it can be simulated or tested with different initial conditions and inputs, but it is nearly impossible to write down a nice mathematical model. This unavailability of explicit "white-box" models is a major roadblock for formal techniques becoming practical for CPS. In this section, we address this problem in the context of data-driven verification. We will view hybrid systems as a combination of a "white-box" control graph that specifies the mode switches and a "black-box" that can simulate the continuous evolution in each mode.

5.6.1 A Hybrid Formalism Accommodating Black-Boxes

Suppose the hybrid system has a set of modes L and continuous state space X as in Definition 5.6. The mode switches are defined by a control graph over L and X, as in Definition 5.5. The black-box generates a set of trajectories TL in X for each mode in L. We denote by $\mathsf{TL}_{init,\ell} = \{\xi.fstate \mid \langle \xi, \ell \rangle \in \mathsf{TL}\}$, the set of initial states of trajectories in mode ℓ. Without loss of generality, we assume that $\mathsf{TL}_{init,\ell}$ is a connected, compact subset of X.

Instead of a closed-form description of TL as in Definition 5.6, we have a *simulator* that can generate sampled data points on individual trajectories. We will develop techniques that avoid over-reliance on the models generating the trajectories and instead, work with sampled data of $\xi(\cdot)$ generated from the simulators. Of course, in order to obtain safety guarantees we will need to make assumptions about the underlying system generating the data.

Definition 5.7 A simulator for a (deterministic and prefix-closed) set TL of trajectories labeled by L is a function (or a program) SIM that takes as input a mode label $\ell \in$ L, an initial state $x_0 \in$ TL$_{init,\ell}$, and a finite sequence of time points $t_1, \ldots, t_k$, and returns a sequence of states SIM$(x_0, \ell, t_1), \ldots,$ SIM(x_0, ℓ, t_k) such that there exists $\langle \xi, \ell \rangle \in \mathscr{T}$ with $\xi.fstate = x_0$ and for each $i \in \{1, \ldots, k\}$, SIM$(x_0, \ell, t_i) = \xi(t_i)$.

For simplicity, we assume that the simulations are perfect (as in the last equality of Definition 5.7). Formal guarantees of soundness are not compromised if we use validated simulations instead. Our new definition of a hybrid system, therefore, is analogous to Definition 5.6 except that TL is a set of deterministic trajectories labeled by L that can be simulated but does not necessarily come from any known differential equations. Executions and reachable states are defined analogously to those in Sect. 5.5.1.

5.6.2 Learning Discrepancy from Simulations

The key subroutine needed for computing the reachable states with Algorithm 1 has to compute a discrepancy function which upper bounds the distance between trajectories. Owing to the absence of ODE models, the `Bloat` function of Algorithm 2 is useless. We will use a probabilistic algorithm for estimating the discrepancy from the data generated by black-box simulators [32].

Recall that a discrepancy function is a continuous function $\beta : \mathbb{R}^n \times \mathbb{R}^{\geq 0} \to \mathbb{R}^{\geq 0}$, such that for any pair of identically labeled trajectories $\langle \xi_1, \ell \rangle, \langle \xi_2, \ell \rangle \in$ TL, and any $t \in \xi_1.dom \cap \xi_2.dom$: (a) β upper bounds the distance between the trajectories, that is:

$$\|\xi_1(t) - \xi_2(t)\| \leq \beta(\|\xi_1.fstate - \xi_2.fstate\|, t), \tag{5.12}$$

and (b) β converges to 0 as the initial states converge, i.e., for any trajectory ξ and $t \in \xi.dom$, if a sequence of trajectories $\xi_1, \ldots, \xi_k, \ldots$ has $\xi_k.fstate \to \xi.fstate$, then $\beta(\|\xi_k.fstate - \xi.fstate\|, t) \to 0$. We present a simple method for discovering discrepancy functions that only uses simulations. Our method is based on a classical result in PAC learning theory [53]. We revisit this result before applying it to finding discrepancy functions.

Learning Linear Separators
For $\Gamma \subseteq \mathbb{R} \times \mathbb{R}$, a *linear separator* is a pair $(a, b) \in \mathbb{R}^2$ such that:

$$\forall (x, y) \in \Gamma. \, x \leq ay + b. \tag{5.13}$$

Let us fix a subset Γ that has a (unknown) linear separator (a_*, b_*). Our goal is to discover some (a, b) that is a linear separator for Γ by sampling points in Γ.[3] The assumption is that elements of Γ can be drawn according to some (unknown) distribution $\mathcal{D}$. With respect to $\mathcal{D}$, the *error* of a pair (a, b) from satisfying Eq. (5.13) is defined to be $\text{err}_{\mathcal{D}}(a, b) = \mathcal{D}(\{(x, y) \in \Gamma \mid x > ay + b\})$ where $\mathcal{D}(X)$ is the measure of set X under distribution $\mathcal{D}$. Thus, the error is the measure of points (w.r.t. $\mathcal{D}$) that (a, b) is not a linear separator for. There is a very simple (probabilistic) algorithm that finds a pair (a, b) that is a linear separator for a large fraction of points in Γ, as follows.

1. Draw k pairs $(x_1, y_1), \ldots (x_k, y_k)$ from Γ according to $\mathcal{D}$; the value of k will be fixed later.
2. Find $(a, b) \in \mathbb{R}^2$ such that $x_i \leq ay_i + b$ for all $i \in \{1, \ldots k\}$.

Step 2 involves checking feasibility of a linear program, and so can be done quickly. This algorithm, with high probability, finds a linear separator for a large fraction of points.

Proposition 5.2 (Proposition 4 from [32]) *Let $\epsilon, \delta \in \mathbb{R}^{\geq 0}$. If $k \geq \frac{1}{\epsilon} \ln \frac{1}{\delta}$, then, with probability $\geq 1 - \delta$, the above algorithm finds (a, b) such that $\text{err}_{\mathcal{D}}(a, b) < \epsilon$.*

Proof The result follows from the PAC learnability of concepts with low VC dimension [53]. However, since the proof is very simple in this case, we reproduce it here for completeness. Let k be as in the statement of the proposition, and suppose the pair (a, b) identified by the algorithm has error $> \epsilon$. We will bound the probability of this happening.

Let $B = \{(x, y) \mid x > ay + b\}$. We know that $\mathcal{D}(B) > \epsilon$. The algorithm chose (a, b) only because no element from B was sampled in Step 1. The probability that this happens is $\leq (1 - \epsilon)^k$. Observing that $(1 - s) \leq e^{-s}$ for any s, we get $(1 - \epsilon)^k \leq e^{-\epsilon k} \leq e^{-\ln \frac{1}{\delta}} = \delta$. This gives us the desired result. $\qquad \blacksquare$

5.6.3 Discrepancy Functions as Linear Separators

Using the above result, we will compute discrepancy functions from simulation data, independently for each mode. Let us fix a mode $\ell \in L$, and a domain $[0, T]$ for each trajectory. The special type of discrepancy functions that we will learn from simulation data are called *global exponential discrepancy (GED)* and have the special form:

$$\beta(\|x_1 - x_2\|, t) = \|x_1 - x_2\| K e^{\gamma t}.$$

---

[3] We prefer to present the learning question in this form as opposed to the one where we learn a Boolean concept because it is closer to the task at hand.

Here, K and γ are constants. Thus, for any pair of trajectories ξ_1 and ξ_2 (for mode ℓ), we have

$$\forall t \in [0, T].\ \|\xi_1(t) - \xi_2(t)\| \leq \|\xi_1.fstate - \xi_2.fstate\| K e^{\gamma t}.$$

Taking logs on both sides and rearranging terms, we have

$$\forall t.\ \ln \frac{\|\xi_1(t) - \xi_2(t)\|}{\|\xi_1.fstate - \xi_2.fstate\|} \leq \gamma t + \ln K.$$

It is easy to see that a global exponential discrepancy is nothing but a linear separator for the set Γ consisting of pairs $\left(\ln \frac{\|\xi_1(t) - \xi_2(t)\|}{\|\xi_1.fstate - \xi_2.fstate\|}, t \right)$ for all pairs of trajectories ξ_1, ξ_2 and time t. Using the sampling-based algorithm described before, we could construct a GED for a mode $\ell \in \mathsf{L}$, where sampling from Γ reduces to using the simulator to generate traces from different states in $\mathsf{TL}_{init,\ell}$. Proposition 5.2 guarantees the correctness, with high probability, for any separator discovered by the algorithm. However, for our reachability algorithm to not be too conservative, we need K and γ to be small. Thus, when solving the linear program in Step 2 of the algorithm, we search for a solution minimizing $\gamma T + \ln K$.

Learned Discrepancy and Guarantees in Practice

In theory, there is some probability that the learned discrepancy function β is incorrect. That is, some pair of executions $\xi, \xi' \in \mathsf{TL}$ of the system, starting from the same initial state Θ, diverges more than the bound given by the computed β. However, experiments in [32] on dozens of modes with complex, nonlinear trajectories suggest that this almost never happens. In the reported experiments, for each mode a set S_{train} of simulation traces that start from independently drawn random initial states in $\mathsf{TL}_{init,\ell}$ are used to learn a discrepancy function. Each trace has between 100–$10,000$ data points, depending on the relevant time horizon and sample times. Then, another set S_{test} of $1,000$ simulations traces are drawn for validating the computed discrepancy. For every pair of trace in S_{test} and for every time point, it is checked whether the computed discrepancy satisfies Eq. (5.12). It is observed that for $|S_{\mathsf{train}}| > 10$ the computed discrepancy function is correct for 96% of the points S_{test} in and for $|S_{\mathsf{train}}| > 20$ it is correct for more than 99.9%, across all experiments.

DryVR

Replacing the `Bloat` function in Algorithm 3 with a subroutine for learning discrepancy, we can obtain a complete verification algorithm for black-box hybrid models. This is the core of the approach implemented in the open-source DRYVR verification tool [32]. The tool supports other forms of discrepancy functions (for example, piece-wise exponential and polynomial) that can also be learned from simulation data with the same type of guarantees. DryVR has been effectively employed to analyze space-craft control systems and maneuvers involving multiple autonomous and semiautonomous vehicles (see Sect. 5.7 for some examples).

5.7 Verification Case Studies

Data-driven verification algorithms have been implemented in a number of software tools such as Breach [19], C2E2[4] [33], and DryVR[5] [32]. These tools have been effective in verifying challenging benchmark applications from the automotive, aerospace, energy, and medical devices domain. In the following, we discuss three applications that were beyond the capabilities of automatic verification tools until recently, and help paint a picture of the rapid developments in this area over the last 5 years.

5.7.1 Automatic Braking and Forward Collision Avoidance System

Growth of autonomy and advanced driver assist (ADAS) features in cars has led to significant pressures for assuring system-level safety at design time. The broad topic of safety certification for such systems is currently a big open problem. While this topic touches multiple technical challenges in several disciplines that are beyond the scope of our discussion (for example, human-autonomy interactions, traffic modeling, and testing for different weather conditions), formal verification, and in particular data-driven verification can play an effective role for creating safety assurance cases needed for certification with standards like the ISO2626 [64]. Here, we summarize a comprehensive case study from [31] which looks at the most common type of rear-end crashes involving automatic emergency braking (AEB) and forward collision avoidance systems.

Each scenario for safety verification is constructed by composing several hybrid automaton models—one for each vehicle or road agent. Each vehicle has several continuous variables including the x, y-coordinates of the vehicle on the road, its velocity, heading, and steering angle. The detailed dynamics of each vehicle comes from a black-box simulator (for example, written in Python or MatLab). The higher-level decisions about the modes (for example, for "cruising," "speeding," "merging left," etc.) followed by the vehicles are captured by control graphs. In more detail, a vehicle can be controlled by two input signals, namely the throttle (acceleration or brake) and the steering. By choosing appropriate values for these input signals, the modes are defined— cruise: move forward at constant speed, speedup: constant acceleration, brake: constant (slow) deceleration, and em_brake: constant (hard) deceleration. The switching rules (guards) between the modes is defined by "driver models." For example, one such rule might state that if the distance between the ego vehicle and its leading car drops below a threshold S_{safe}, then the ego vehicle

---

[4]C2E2 available from:http://publish.illinois.edu/c2e2-tool/.

[5]DryVR available from:https://gitlab.engr.illinois.edu/dryvrgroup/dryvrtool.

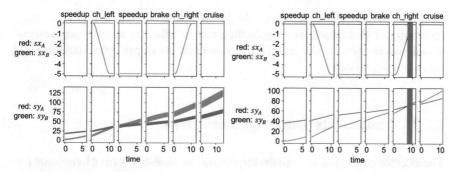

Fig. 5.4 Verification of the vehicles overtake scenario. Left: safe reachtube. Right: unsafe execution. Vehicle A's (red) modes are shown above each subplot. Vehicle B (green) is in cruise. Top: sx_A, sx_B vs time. Bottom: sy_A, sy_B vs time

switches to **brake** after a delay of T_{react}, where T_{react} is a parameter corresponding to driver's reactions time. Typical values of these parameters were obtained from previously available driving data. The composed hybrid automaton graph is then presented to DRYVR as the input model.

Consider a scenario with Vehicle A behind B in the same lane starting with the same speed, and A wanting to overtake B. A will switch to the left lane after it approaches B, and then switch back to the right lane once it is ahead of B. In some cases, A may fail to get ahead of B, in which case it times out and returns back in the right lane behind B. The safety requirement is that the vehicles maintain safe separation. Figure 5.4 (*left*) shows a version of this scenario that is verified to be safe by DRYVR. The plots show the reachtube over-approximations computed by DRYVR. Vehicle B stays in the **cruise** always but Vehicle A goes through a sequence of modes **speedup, change_left, speedup, brake,** and **change_right, cruise** to overtake B. Figure 5.4 left top shows the projection of reachtubes on lateral positions (sx_A in red and sx_B in green) subplot, and the bottom plot shows the positions along the lane (sy_A in red and sy_B in green, in the bottom plot). Initially, for both $i \in \{A, B\}$, $sx_i = vx_i = 0$ and $vy_i = 1$, i.e., both are cruising at constant speed at the center of the right lane, initial positions along the lane are $sy_A \in [0, 2]$, $sy_B \in [15, 17]$. As time advances, Vehicle A moves to left lane (sx decreases) and then back to the right, while B remains in the right lane, as A overtakes B (bottom plot). With a different initial set, $sy_B \in [30, 40]$, DRYVR finds counterexample demonstrating unsafe behavior of the system (Fig. 5.4 (right)). In both of these instances, the running time for verification is of the order of minutes.

In [31], hundreds of scenarios are analyzed for 2 and 3 vehicles, with different ranges of initial velocities of the cars, different reaction times (T_{react}), and different braking profiles. DRYVR proves certain scenarios to be safe and for others it computes the severity of accidents based on the worst-case relative velocity of collisions. In [31], it is shown how these verification results can be aggregated with information about the distribution of model parameters (T_{react}, S_{safe}, etc.), to assess the system-level risk, which in turn is essential for determining automotive safety integrity levels (ASIL) for standards like the ISO26262 . In summary, this case study demonstrated that data-driven verification can be effective in analyzing relevant vehicle autonomy scenarios involving complex composition of hybrid automata and black-box simulators.

5.7.2   *Autonomous Spacecraft Rendezvous*

The extreme cost of failures and the infeasibility of terrestrial testing have made formal methods singularly attractive for space systems. Reachability-based automatic safety verification for satellite control systems was first studied in [48]. At the time of that study, hybrid verification tools were available only for linear hybrid systems, which have restricted applicability because many satellite control problems involve nonlinear orbital dynamics and nonlinear constraints. Here, we present a case study based on the ARPOD problem introduced in [43]. ARPOD stands for autonomous rendezvous proximity operations and docking. It captures an overarching mission needed to assemble a new space station that has been launched in separate modules. Our discussion here is based on the results presented in [12, 14].

A generic ARPOD scenario involves a passive module or a *target* (launched separately into orbit) and a *chaser* spacecraft that must transport the passive module to an on-orbit assembly location. The chaser maintains a relative bearing measurement to the target, but initially it is too far to use its range sensors. Once range measurements become available, the chaser gets more accurate relative positioning data and it can stage itself to dock with the target. Docking must happen with a specific angle of approach and closing velocity, in order to avoid collision and to ensure that the docking mechanisms on each spacecraft will mate.

For simplicity, here we discuss the planar (or 2-dimensional) version of the model. The variables of the hybrid model include position (relative to the target) x, y (in meters), time t (in minutes), and horizontal and vertical velocity v_x, v_y. The modes of the hybrid automaton capture four phases of the docking maneuver. Each phase is defined by a separation distance $\rho = \sqrt{x^2 + y^2}$ between the chaser and target spacecraft, closing this distance from up to 10 km down to 0, and then performing a maneuver once the satellites are docked. As seen in Fig. 5.5 (left), the chaser spacecraft begins in **Phase 1** while the separation distance ρ is not available but only has angular of approach $\theta = \mathtt{atan}(\frac{y}{x})$ available, and the system is unobservable. While ρ gets small enough, the mission moves into **Phase 2**, where the chaser spacecraft now has a ranging measurement to the chaser spacecraft and must position itself for the **Phase** 3 docking. After the chaser moves such that $\rho \leq 100$, the docking phase, **Phase 3** is initiated and additional docking port constraints are active. Once the spacecraft dock (i.e., $\rho = 0$), both spacecraft move into **Phase 4**, where the joint assembly must move to the relocation position.

The chaser must adhere to different sets of constraints in each discrete mode. In [13], a switched linear quadratic regulator (LQR) is designed to meet these constraints while maintaining liveness in navigating toward the target spacecraft.

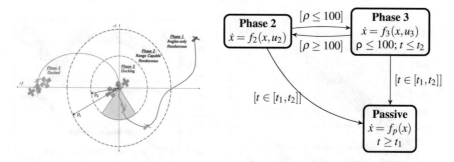

Fig. 5.5 Left: description of the overall mission phases (not to scale). Right: hybrid system model of the autonomous spacecraft rendezvous mission

Figure 5.5 (right) gives the hybrid system model of interest. In addition to the existing mode, the model also has a **Passive** mode in which the chaser has the thrusters shut down. The system may nondeterministically transition to the **Passive** mode as a result of a failure or loss of power. The nonlinear dynamic equations describing the motion of the chaser spacecraft relative to the target is given by:

$$\begin{cases} \dot{x} = v_x \\ \dot{y} = v_y \\ \dot{v}_x = n^2 x + 2n v_y + \frac{\mu}{r^2} - \frac{\mu}{r_c^3}(r + x) + \frac{u_x}{m_c} \\ \dot{v}_y = n^2 y - 2n v_x - \frac{\mu}{r_c^3} y + \frac{u_y}{m_c}. \end{cases}$$

The parameters are $\mu = 3.986 \times 10^{14} \times 60^2$ [m^3 / min^2], $r = 42164 \times 10^3$ [m], $m_c = 500$ [kg], $n = \sqrt{\frac{\mu}{r^3}}$, and $r_c = \sqrt{(r + x)^2 + y^2}$. The linear feedback controllers for the different modes are defined as $[u_x, u_y]^T = K_1 \underline{x}$ for mode **Phase 2**, and $[u_x, u_y]^T = K_2 \underline{x}$ for mode **Phase 2**, where $\underline{x} = [x, y, v_x, v_y]^T$ is the vector of system states. The feedback matrices K_i were determined with an LQR approach applied to the linearized system dynamics, where the detailed number can be found at [13]. In mode **Passive**, the system is uncontrolled $[u_x, u_y]^T = [0, 0]^T$. The spacecraft starts from the initial set $x \in [-925, -875]$ [m], $y \in [-425, -375]$ [m], $v_x = 0$ [m/min] and $v_y = 0$ [m/min]. For the considered time horizon of $t \in [0, 200]$ [min], the following specifications have to be satisfied:

- **Line-of-sight:** In mode **Phase 3**, the spacecraft has to stay inside line-of-sight cone:

$$\{[x, y]^T \mid (x \geq -100) \wedge (y \geq x \tan(30°)) \wedge (-y \geq x \tan(30°))\}.$$

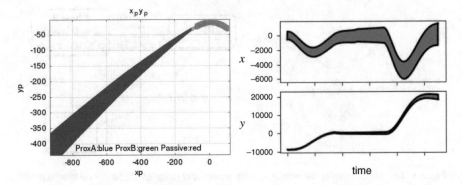

Fig. 5.6 Left: reachtube of x (x-axis) vs y (y-axis) produced by C2E2. Right: reachtube of x vs time (above) and y vs time (below) produced by DRYVR

- **Collision avoidance:** In mode Passive, the spacecraft has to avoid a collision with the target, which is modeled as a box B with 0.2 m edge length and the center placed at the origin.
- **Velocity constraint:** In mode Phase 3, the absolute velocity has to stay below 3.3 [m/min]:

$$\sqrt{v_x^2 + v_y^2} \leq 3.3 \text{ [m/min]}.$$

C2E2 was used to prove that the autonomous rendezvous system with the LQR controller satisfying the above requirements. Figure 5.6 (left) shows the reachtube of x (x-axis) vs y (y-axis) produced by C2E2. A different control strategy for ARPOD was proposed in [60] which characterizes the family of individual controllers and the required properties they should induce for the closed-loop system to solve the problem within each phase, then use a supervisor that robustly coordinates the individual controllers. Using these controlled subsystems as a black-box, we have been able to check the safety of the overall system using DRYVR. Figure 5.6 (right) shows the reachtube of x and y produced by DRYVR.

5.7.3 Powertrain Control System

The demand of greater fuel efficiency and lower emissions constantly challenges automotive companies to improve control software in the powertrain systems. Recently, a suite of benchmarks were published in [45] to introduce realistic, industrial scale models to the formal verification community. The suite consists of three Simulink® models with increasing levels of complexity and sophistication. These models capture the behavior of chemical reactions in internal combustion engines, and hybrid models are deemed suitable for capturing the discrete transitions of control software and the continuous parameters in these models. At a high level, the models take inputs from a driver (throttle angle) and the environment (sensor failures), and define the dynamics of the engine. The key controlled quantity is the air-to-fuel ratio which in turn influences the emissions, the fuel efficiency, and torque generated.

The most complicated model (Model 1) in the suite captures all the interactions taking place in a physical process and faithfully models the control software. It contains several hierarchical components in Simulink® with look-up tables, and delay differential equations. Model 1 is simplified to a model with periodic inputs to ordinary differential equations using several heuristics (Model 2), which as per the authors, exhibit similar behavior of Model 1. Then, Model 2 is further simplified to a hybrid system with only polynomial ODEs (Model 3). At the time of publication of [45], these models were beyond the reach of the then available verification tools, but within a year the simplified models were verified using C2E2 [22], and subsequently, the more complex models were handled by DryVR in [32].

In more detail, Model 2 and 3 have four variables: intake manifold pressure p, air-fuel ratio λ, intake manifold pressure estimate p_e, and integrator state i, and four modes: **Start_up**, **Normal**, **Power**, and **Sensor_fail**. The hybrid model also receives an input signal θ_{in} (throttle angle) as the user input. The required safety specification of powertrain control systems was given in [45] as a number of Signal Temporal Logic properties. Here, we only illustrate one primary result for each model, with the simple unsafe set U: in **Power** mode, $t > 4 \vee \lambda \notin [12.4, 12.6]$, in **Normal** mode, $t > 4 \vee \lambda \notin [14.6, 14.8]$. We refer readers to [22, 32] for more comprehensive studies involving other scenarios and requirements.

Figure 5.7 (left) shows the hybrid model of the powertrain control system Model 2. The physical plant dynamics are modeled using continuous variables $x_p = [p, \lambda]$,

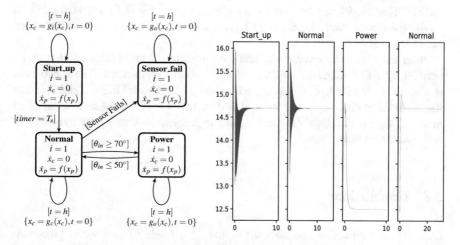

Fig. 5.7 Left: hybrid system model of the powertrain control system Model 2. Right: reachtube for λ vs time of Model 2 produced by DRYVR

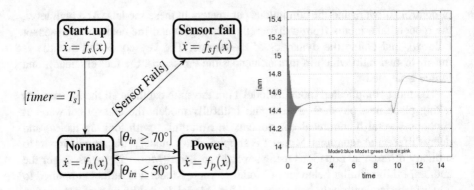

Fig. 5.8 Left: hybrid system model of the powertrain control system Model 3. Right: reachtube for λ vs time of Model 3 produced by C2E2; blue and green regions correspond to the Start_up and Normal modes, respectively

which evolve according to a nonlinear ODE $\dot{x}_p = f(x_p)$. The controller variables $x_c = [p_e, i]$ are, instead, updated periodically every h time units by the reset functions $g_i(x_c), g_o(x_c), g_c(x_c)$ in different modes. We treat the entire system as a black-box simulator with the four given variables and four modes. With the initial set $p \in [0.6115, 0.6315], \lambda \in [14.6, 14.8], p_e \in [0.5555, 0.5755], i \in [0, 0.01]$, DRYVR is able to prove that the system satisfies the safety requirements as stated above. Figure 5.7 (right) shows a safe reachtube of the Air/Fuel variable λ computed using DRYVR going through the sequence of modes Start_up, Normal, Power, and Normal.

Model 2 got further simplified such that all four variables are continuous and follow a set of polynomial differential equations in Model 3 (see [45] for detailed ODEs). This model can be handled by C2E2. Figure 5.8 (left) shows the hybrid model (Model 3), and Fig. 5.8 (right) gives a safe reachtube of λ from the same initial set as above.

Both the spacecraft rendezvous and the powertrain control applications can be verified by either C2E2 or DRYVR within a couple of minutes. These two case studies show that for hybrid systems with complex nonlinear ODEs, C2E2 can take the verification challenge, and when it is difficult to get a complete mathematical model of the system, DRYVR can address the problem by treating the dynamics in each mode as a black-box.

5.8 Conclusions

Data-driven verification has shown promise in a range of real-world problems. The key to its success is the powerful amalgamation of the speed of numerical simulations with the guarantees coming from sensitivity analysis.

Nevertheless, these are the early days of exploration of these ideas; and the current approaches have several limitations: First, we observe that the tools produce better results when the system is stable. This is because the proposed methods can usually find a tighter discrepancy function for stable systems, which in turn decreases the number of refinements needed to conclude safety or find a counterexample. For unstable systems, the over-approximation of reachable sets can get very conservative, and therefore, the algorithm may not terminate in a reasonable amount of time. Second, our proposed algorithm mainly looks at safety requirements, although the computed reachtubes can be used to check for much more general specifications such as linear temporal logic. Usability of the tools remains to be improved if they are to be adopted commercially. Modeling, inter-operation with simulators, editing properties, and analyzing verification results—all of this has to become user-friendly. Finally, as usual, scalability remains a challenge. The dimension of the state space of the biggest examples the current tools have handled within a reasonable amount of time (around 2 h) is 12 for nonlinear systems and 350 for linear systems. High dimensionality will not only increase the difficulty of computing discrepancy functions but also introduce a huge number of refinements as the number of initial covers needed to cover the initial set in data-driven verification will increase exponentially.

Other important directions that call for further investigation are broadly *compositional techniques* for handling networked and distributed CPS. Examples of such systems are abundant in automotive control systems, power networks, and embedded medical devices. The naïve approach to consider such systems is to compute the cross-product of all components. However, in this way, the resulting hybrid system will become inevitably complicated with huge dimensionality and a tremendous amount of mode switches. Methods to make the analysis scalable for networked CPS with large-scale components will become a necessity. As an early step towards this direction, the notion of input-to-state discrepancy was introduced in [41, 42], and has been used to conduct a compositional sensitivity analysis of closed networked dynamical and hybrid systems [40]. The learning-based discrepancy function approach can be seen as learning an envelope which safely contains the possibly trajectories of the system. It is worth to explore more interesting learning models for identifying the dynamics of the black-box systems. There has been a methodology with a long history for building mathematical models of dynamic systems using the system's input and output behaviors called system identification. However, methods for identifying and verifying systems with guarantees remain to be developed.

5.9   Further Reading

Many new works on verification of CPS got published every year. The major conferences in this area include but not limit to International Conference on Hybrid Systems: Computation and Control (HSCC), International Conference on Computer

Aided Verification (CAV), and Applied Verification for Continuous and Hybrid Systems (ARCH).

Recently, verification tools such as Flow* [15], NLTOOLBOX [17], iSAT [34], dReach [54], and CORA [2] have demonstrated the feasibility of verifying nonlinear dynamic and hybrid models. These tools are still limited in terms of the complexity of the models and the type of external inputs they can handle, and they require quite often manual tuning of algorithmic parameters. Some of these tools' approaches for reach set estimation operate directly on the vector field involving higher-order Taylor expansions [15, 54]. However, this method suffers from complexity that increases exponentially with both the dimension of the system and the order of the model.

Several approaches have been proposed to obtain proofs about (bounded time) invariant or safety properties from simulations [20, 37]. A technique that is very close to discrepancy functions is called *sensitivity matrix*, a matrix that captures the sensitivity of the system to its initial condition x_0. This is then used to give an upper bound on the distance between two system trajectories. In [49], the authors provided sound simulation-driven methods to over-approximate the distance between trajectories, but these methods are mainly limited to affine and polynomial systems. For general nonlinear models, this approach may not be sound, as higher-order error terms are ignored when computing this upper bound.

The idea of computing the reach sets from trajectories is similar to the notions of incremental Lyapunov function [4]. In this work, we do not require systems to be incrementally stable. Similar ideas have also been considered for control synthesis in [68]. The work closest to this paper involves reachability analysis using matrix measures [59], where the authors use the fact that the matrix measure of the Jacobian matrix can bound the distance between neighboring trajectories [9, 66]. Unlike the approach in this paper which automatically computes the bounds on matrix measures, the technique there relies on user-provided closed-form matrix measure functions, which are in general difficult to compute.

Although data-driven verification is a young field, the literature in this area is growing and interesting results are published every year. For an alternative view of this topic from the modeling, testing, and verification of embedded control system perspective, we refer the interested readers to [50].

References

1. Abbas, H., & Fainekos, G. E. (2011). Linear hybrid system falsification through local search. In *Proceedings of the 9th International Symposium on Automated Technology for Verification and Analysis (ATVA 2011)*, Taipei, Taiwan, October 11–14, 2011 (pp. 503–510). https://doi.org/10.1007/978-3-642-24372-1_39.
2. Althoff, M., & Grebenyuk, D. (2016). Implementation of interval arithmetic in CORA 2016. In *ARCH Workshop* (pp. 91–105). Manchester: EasyChair.

3. Alur, R., Courcoubetis, C., Henzinger, T. A., & Ho, P. H. (1993). Hybrid automata: an algorithmic approach to the specification and verification of hybrid systems. In R. L. Grossman, A. Nerode, A. P. Ravn, & H. Rischel (Eds.), *Hybrid systems. Lecture notes in computer science* (Vol. 736, pp. 209–229). Berlin: Springer.

4. Angeli, D. (2002). A Lyapunov approach to incremental stability properties. *IEEE Transactions on Automatic Control, 47*(3), 410–421.

5. Annapureddy, Y., Liu, C., Fainekos, G., & Sankaranarayanan, S. (2011). S-TaLiRo: a tool for temporal logic falsification for hybrid systems. In *TACAS*. Berlin: Springer.

6. Aréchiga, N., Kapinski, J., Deshmukh, J. V., Platzer, A., & Krogh, B. (2015). Numerically-aided deductive safety proof for a powertrain control system. *Electronic Notes in Theoretical Computer Science, 317*, 19–25.

7. Asarin, E., Bournez, O., Dang, T., & Maler, O. (2000). Approximate reachability analysis of piecewise-linear dynamical systems. In B. Krogh & N. Lynch (Eds.), *Hybrid systems: computation and control. Lecture notes in computer science* (Vol. 1790, pp. 20–31). Berlin: Springer.

8. Aylward, E.M., Parrilo, P.A., & Slotine, J. -J. E. (2008). Stability and robustness analysis of nonlinear systems via contraction metrics and SOS programming. *Automatica, 44*(8), 2163–2170.

9. Boichenko, V.A., & Leonov, G.A. (1998). Lyapunov's direct method in estimates of topological entropy. *Journal of Mathematical Sciences, 91*(6), 3370–3379.

10. Boyd, S., El Ghaoui, L., Feron, E., & Balakrishnan, V. (1994). *Linear matrix inequalities in system and control theory. Studies in applied mathematics* (Vol. 15). Philadelphia, PA: SIAM.

11. CAPD. (2002). Computer assisted proofs in dynamics.

12. Chan, N., & Mitra, S. (2017). Verified hybrid LQ control for autonomous spacecraft rendezvous. In *56th IEEE Annual Conference on Decision and Control, CDC 2017, Melbourne, December 12–15, 2017* (pp. 1427–1432). Piscataway: IEEE.

13. Chan, N., & Mitra, S. (2017) Verified hybrid LQ control for autonomous spacecraft rendezvous. In *2017 IEEE 56th Annual Conference on Decision and Control (CDC)* (pp. 1427–1432). Piscataway: IEEE.

14. Chan, N., & Mitra, S. (2017). Verifying safety of an autonomous spacecraft rendezvous mission. In *ARCH17. 4th International Workshop on Applied Verification of Continuous and Hybrid Systems, Collocated with Cyber-Physical Systems Week (CPSWeek), Pittsburgh, PA, April 17, 2017* (pp. 20–32).

15. Chen, X., Ábrahám, E., & Sankaranarayanan, S. (2013). Flow*: an analyzer for non-linear hybrid systems. In *CAV* (pp. 258–263). Berlin: Springer.

16. Cook, B. (2018). Formal reasoning about the security of amazon web services. In *Computer Aided Verification—30th International Conference, CAV 2018, held as part of the Federated Logic Conference, FloC 2018, Oxford, July 14–17, 2018, Proceedings, Part I* (pp. 38–47). New York: Springer International Publishing.

17. Dang, T., Le Guernic, C., & Maler, O. (2009). Computing reachable states for nonlinear biological models. In *CMSB. Lecture notes in computer science* (Vol. 5688, pp. 126–141). Berlin: Springer.

18. Donzé, A. (2010). Breach, a toolbox for verification and parameter synthesis of hybrid systems. In *CAV* (pp. 167–170). Berlin: Springer.

19. Donzé, A. (2010). Breach, a toolbox for verification and parameter synthesis of hybrid systems. In *Computer Aided Verification. CAV 2010. Lecture Notes in Computer Science* (Vol. 6174). Berlin: Springer.

20. Donzé, A., & Maler, O. (2007). Systematic simulation using sensitivity analysis. In *HSCC* (pp. 174–189). Berlin: Springer.

21. Duggirala, P. S. (2015). *Dynamic Analysis of Cyber-Physical Systems*. PhD thesis. Champaign: University of Illinois at Urbana-Champaign.

22. Duggirala, P. S., Fan, C., Mitra, S., & Viswanathan, M. (2015). Meeting a powertrain verification challenge. In *Computer Aided Verification* (pp. 536–543). Berlin: Springer.

23. Duggirala, P. S., Mitra, S., & Viswanathan, M. (2013). Verification of annotated models from executions. In *EMSOFT* (pp. 26:1–26:10). Piscataway: IEEE Press.
24. Duggirala, P. S., Mitra, S., Viswanathan, M., & Potok, M. (2015). C2E2: A verification tool for stateflow models. In *TACAS* (pp. 68–82). Berlin: Springer.
25. Duggirala, P. S., Wang, L., Mitra, S., Viswanathan, M., & Muñoz, C. (2014). Temporal precedence checking for switched models and its application to a parallel landing protocol. In *Formal methods* (pp. 215–229). Cham: Springer.
26. El-Guindy, A., Han, D., & Althoff, M. (2016) Formal analysis of drum-boiler units to maximize the load-following capabilities of power plants. *IEEE Transactions on Power Systems* (99), 1–12.
27. Fainekos, G. E. (2015). Automotive control design bug-finding with the s-taliro tool. In *American Control Conference, ACC 2015, Chicago, IL, July 1–3, 2015* (p. 4096). Piscataway: IEEE.
28. Fainekos, G. E., Sankaranarayanan, S., Ueda, K., & Yazarel, H. (2012) Verification of automotive control applications using S-TaLiRo. In *American Control Conference (ACC), 2012* (pp. 3567–3572). Citeseer. Piscataway: IEEE.
29. Fan, C., Kapinski, J., Jin, X., & Mitra, S. (2016). Locally optimal reach set over-approximation for nonlinear systems. In *EMSOFT* (pp. 6:1–6:10). New York: ACM.
30. Fan, C., & Mitra, S. (2015). Bounded verification with on-the-fly discrepancy computation. In *ATVA* (pp. 446–463). Berlin: Springer.
31. Fan, C., Qi, B., & Mitra, S. (2018). Data-driven formal reasoning and their applications in safety analysis of vehicle autonomy features. *IEEE Design & Test, 35*(3), 31–38.
32. Fan, C., Qi, B., Mitra, S., Viswanathan, M. (2017). Dryvr: data-driven verification and compositional reasoning for automotive systems. In *Computer Aided Verification, CAV 2017* (pp. 441–461). Heidelberg: Springer International Publishing
33. Fan, C., Qi, B., Mitra, S., Viswanathan, M., & Duggirala, P. S. (2016). Automatic reachability analysis for nonlinear hybrid models with C2E2. In *Computer Aided Verification–28th International Conference, CAV 2016, Toronto, ON, July 17–23, 2016, Proceedings, Part I* (pp. 531–538). Cham: Springer.
34. Fränzle, M., Herde, C., Teige, T., Ratschan, S., & Schubert, T. (2007). Efficient solving of large non-linear arithmetic constraint systems with complex boolean structure. *JSAT, 1*(3–4), 209–236.
35. Frehse, G. (2005). Phaver: algorithmic verification of hybrid systems past hytech. In M. Morari & L.Thiele (Eds.), *HSCC* (Vol. 3414, pp. 258–273) *Lecture notes in computer science* . Berlin: Springer.
36. Frehse, G., Guernic, C. L., Donzé, A., Cotton, S., Ray, R., Lebeltel, O., Ripado, R., Girard, A., Dang, T, & Maler, O. (2011). SpaceEx: scalable verification of hybrid systems. In S. Qadeer & G. Gopalakrishnan (Eds.), *CAV. Lecture Notes in Computer Science*. Berlin: Springer.
37. Girard, A., Pola, G., & Tabuada, P. (2010). Approximately bisimilar symbolic models for incrementally stable switched systems. *IEEE Transactions on Automatic Control, 55*(1), 116–126.
38. Henzinger, T. A. (1996). The theory of hybrid automata. In *11th Annual IEEE Symposium on Logic in Computer Science* (pp. 278–292). Washington: IEEE Computer Society.
39. Henzinger, T. A., Kopke, P. W., Puri, A., & Varaiya, P. (1998). What's decidable about hybrid automata? *Journal of Computer and System Sciences, 57*, 94–124.
40. Huang, Z., Fan, C., Mereacre, A., Mitra, S., & Kwiatkowska, M. Z. (2014). Invariant verification of nonlinear hybrid automata networks of cardiac cells. In *CAV* (pp. 373–390). Berlin: Springer.
41. Huang, Z., Fan, C., & Mitra, S. (2017). Bounded invariant verification for time-delayed nonlinear networked dynamical systems. *Nonlinear Analysis: Hybrid Systems, 23*, 211–229.
42. Huang, Z., & Mitra, S. (2014). Proofs from simulations and modular annotations. In *HSCC, Berlin, Germany*. New York: ACM press.

43. Jewison, C., & Erwin, R. S. (2016). A spacecraft benchmark problem for hybrid control and estimation. In *2016 IEEE 55th Conference on Decision and Control (CDC)* (pp. 3300–3305). Piscataway: IEEE.
44. Jiang, Z., Pajic, M., Moarref, S., Alur, R., & Mangharam, R. (2012). Modeling and verification of a dual chamber implantable pacemaker. In *TACAS* (pp. 188–203). Berlin: Springer.
45. Jin, X., Deshmukh, J. V., Kapinski, J., Ueda, K., & Butts, K. (2014). Powertrain control verification benchmark. In *Proceedings of the 17th International Conference on Hybrid Systems: Computation and Control, HSCC '14* (pp. 253–262). New York, NY: ACM.
46. Jin, X., Deshmukh, J. V., Kapinski, J., Ueda, K., & Butts, K. R. (2014). Powertrain control verification benchmark. In *17th International Conference on Hybrid Systems: Computation and Control (Part of CPS Week), HSCC'14, Berlin, April 15–17, 2014* (pp. 253–262). New York: ACM.
47. Jin, X., Donzé, A., Deshmukh, J. V., & Seshia, S. A. (2015). Mining requirements from closed-loop control models. *IEEE Transactions on Computer-Aided Design of Integrated Circuits and Systems, 34*(11), 1704–1717.
48. Johnson, T. T., Green, J., Mitra, S., Dudley, R., & Erwin, R. S. (2012). Satellite rendezvous and conjunction avoidance: case studies in verification of nonlinear hybrid systems. In *FM 2012: Formal Methods—18th International Symposium, Paris, France, August 27–31, 2012. Proceedings* (pp. 252–266). Berlin: Springer.
49. Julius, A. A., & Pappas, G. J. (2009). Trajectory based verification using local finite-time invariance. In *HSCC* (pp. 223–236). Berlin: Springer.
50. Kapinski, J., Deshmukh, J. V., Jin, X., Ito, H., & Butts, K. (2016). Simulation-based approaches for verification of embedded control systems: an overview of traditional and advanced modeling, testing, and verification techniques. *IEEE Control Systems, 36*(6), 45–64.
51. Kaynar, D. K., Lynch, N., Segala, R., & Vaandrager, F. (2005). *The theory of timed I/O automata*. Synthesis Lectures on Computer Science. Morgan Claypool, November. Also available as Technical Report MIT-LCS-TR-917.
52. Kaynar, D. K., Lynch, N., Segala, R., & Vaandrager, F. (2010). The theory of timed I/O automata. *Synthesis Lectures on Distributed Computing Theory, 1*(1), 1–137.
53. Kearns, M. J., & Vazirani, U. V. (1994) *An introduction to computational learning theory*. Cambridge: MIT press.
54. Kong, S., Gao, S., Chen, W., & Clarke, E. (2015) dReach: δ-reachability analysis for hybrid systems. In *TACAS* (pp. 200–205). Berlin: Springer.
55. Koopman, P., & Wagner, M. (2016) Challenges in autonomous vehicle testing and validation. *SAE International Journal of Transportation Safety, 4*(2016-01-0128), 15–24.
56. Krstic, M., Kokotovic, P. V., & Kanellakopoulos, I. (1995). *Nonlinear and adaptive control design* (1st ed.). New York, NY: Wiley.
57. Liberzon, D. (2012). *Switching in systems and control*. Berlin: Springer Science & Business Media.
58. Lohmiller, W., & Slotine, J. -J. E. (1998) On contraction analysis for non-linear systems. *Automatica, 34*(6), 683–696.
59. Maidens, J., & Arcak, M. (2015). Reachability analysis of nonlinear systems using matrix measures. *IEEE Transactions on Automatic Control, 60*(1), 265–270.
60. Malladi, B. P., Sanfelice, R. G., Butcher, E., & Wang, J. (2016). Robust hybrid supervisory control for rendezvous and docking of a spacecraft. In *2016 IEEE 55th Conference on Decision and Control (CDC)* (pp. 3325–3330). Piscataway: IEEE.
61. Mitra, S. (September 2007). *A Verification Framework for Hybrid Systems*. PhD thesis. Cambridge, MA: Massachusetts Institute of Technology, 02139.
62. Nedialkov, N. (2006). VNODE-LP: validated solutions for initial value problem for ODEs. Technical report. Hamilton: McMaster University.
63. Perry, R. B., Madden, M. M., Torres-Pomales, W., & Butler, R. W. (2013). *The simplified aircraft-based paired approach with the ALAS alerting algorithm*. Technical Report NASA/TM-2013-217804. Hampton: NASA, Langley Research Center.

64. Road vehicles—Functional safety. (November 2011). Standard, International Organization for Standardization (ISO), Geneva, Switzerland.
65. Sankaranarayanan, S., Kumar, S. A., Cameron, F., Bequette, B. W., Fainekos, G., & Maahs, D. M. (March 2017) Model-based falsification of an artificial pancreas control system. *SIGBED Review, 14*(2), 24–33.
66. Sontag, E. D. (2010). Contractive systems with inputs. In *Perspectives in mathematical system theory, control, and signal processing* (pp. 217–228). Berlin: Springer.
67. Vladimerou, V., Prabhakar, P., Viswanathan, M., & Dullerud, G. E. (2008). Stormed hybrid systems. In *ICALP (2)*. *Lecture Notes in Computer Science* (Vol. 5126, pp. 136–147). Berlin: Springer.
68. Zamani, M., Pola, G., Mazo, M., & Tabuada, P. (2012). Symbolic models for nonlinear control systems without stability assumptions. *IEEE Transactions on Automatic Control, 57*(7), 1804–1809.

Chapter 6
System Assurance in the Design of Resilient Cyber-Physical Systems

Thomas A. McDermott, Arquimedes Canedo, Megan M. Clifford, Gustavo Quirós, and Valerie B. Sitterle

6.1 Background on Dependable and Secure Computing and the Cyber-Physical System Context

Cyber-physical systems (CPS) are "engineered systems that are built from, and depend upon, the seamless integration of computational algorithms and physical components" [1]. A CPS has computers and networks which control physical processes, often characterized by feedback loops that affect computations and the physical outcomes of those computations. Figure 6.1 provides a general layered depiction of a CPS framework [2]. Insertion or disruption of CPS control activities is a unique concern for security. As shown in the figure, the design of CPS must address these control activities in a device of interest but also with respect to the interconnected human and machine systems that interact with it. In current times, that interaction may be at long ranges, which introduces benefits in efficiency but also increases security risk [3].

Traditional approaches to security, privacy, reliability, resilience, and safety may be insufficient to address the risks to CPS. In the cybersecurity domain, CPS have exposed cyber and physical world interfaces that are vulnerable to new types of intrusions from both local and remote adversaries. CPS are frequently systems of

T. A. McDermott (✉) · M. M. Clifford
Stevens Institute of Technology, Hoboken, NJ, USA
e-mail: tmcdermo@stevens.edu; mcliffor@stevens.edu

A. Canedo · G. Quirós
Siemens Corporate Technology, Princeton, NJ, USA
e-mail: arquimedes.canedo@siemens.com; gustavo.quiros@siemens.com

V. B. Sitterle
Georgia Tech Research Institute, Atlanta, GA, USA
e-mail: valerie.sitterle@gtri.gatech.edu

© Springer Nature Switzerland AG 2019
M. A. Al Faruque, A. Canedo (eds.), *Design Automation of Cyber-Physical Systems*, https://doi.org/10.1007/978-3-030-13050-3_6

Fig. 6.1 CPS conceptual
model [2]

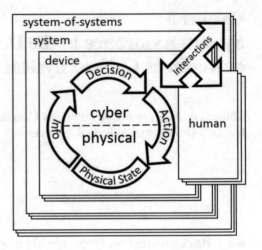

systems (SoS), which increases attack surfaces, system diversity, and complexity, and difficulty in identifying system boundaries. The architectural constructs used in the design of CPS should be able to be applied recursively or iteratively to support the nested nature of CPS in the world, where the sensing, control, and computational nature of CPS generally lead to emergent higher levels of behavior and system intelligence [2]. This construct can be viewed as a set of hierarchical layers of control. In the CPS hierarchy, resilience is a primary architectural attribute. Characteristics of resilience in CPS include:

- Supervisory control methods that support graceful degradation of the system in the presence of failures or malicious attacks.
- Maintenance of system availability using dynamic and potentially distributed control system elements.
- Mechanisms to ensure the integrity of control functions in the presence of failures or malicious attacks.
- Ability to address a range of human interactions across differing system performance levels.
- Ability to recognize and respond to failures that disrupt CPS elements or control functions to maintain levels of performance or to maintain safety and security.
- Ability to recognize and respond to intentional disruptions of CPS elements or control functions in order to maintain levels of performance or to maintain safety and security.
- Ability to protect and maintain the privacy of CPS system data or control states.
- Ability to evaluate CPS system resilience in different environments.
- Need to determine trade-offs in system performance, safety, and security in conditions of complexity, changing threat environments, and emerging or unplanned use cases.

These characteristics have been studied and employed in systems for several years. What is changing in the context of CPS is the complexity of the system control methods and the evolution of the external domain that constitutes a threat to the system. The changes are a product of the computer/network interactions and increased use of digital data in the control functions, along with the connected nature of the cyber threat. In a CPS, cybersecurity meets physical security. Even if the cyber domain is completely secure and the physical domain is completely secure, the system still may not be secure because of the domain interactions and interdependencies.

6.1.1   General Concepts of Resilience with Respect to CPS

General concepts of CPS resilience are found in engineering concepts of dependable and secure computing systems. Resilience is the evaluated capability of the CPS to resist threats using sets of design attributes and resulting patterns that produce resilient effects. The general framework is shown in Fig. 6.2. CPS resilience can be categorized into a set of dependability and security attributes. Dependability and security are the ability of a system to avoid service failures and cover the interrelated foundational attributes of availability, reliability, safety, integrity, confidentiality, and maintainability [4]. These attributes work together to ensure the system's successful application. A vital aspect of these attributes is the overlap between the characteristics of dependability (including availability, reliability, safety, integrity, and maintainability) and security (including confidentiality, integrity, and availability). The resilience definition of dependability is the ability of a system to avoid service failures that are more frequent or more severe than is acceptable [4]. These include security failures. Specific characteristics of security also include the availability of authorized access to the system, authentication of access, confidentiality of both system functions and data on the system's design, and integrity in terms of unauthorized functionality [4].

In the CPS hierarchy, there is also a defined relationship between dependability and trust. This relationship is defined by the dependence of one system on another, and the acceptance that the other system is also dependable [4]. In CPS, this dependence can be either human/machine or machine/machine. The NIST framework highlights the concern of resilience in terms of trustworthiness related to the ability of the CPS to withstand instability, unexpected conditions, and gracefully return to predictable, but possibly degraded, performance. This is a system-of-systems (SoS) concern, and CPS resilience must be considered as both a characteristic of an individual CPS and as relationships between CPS and other systems.

CPS designs must consider this holistic framework, as there will be trade-offs between the interrelated attributes. Designs that improve safety may reduce the effectiveness of the actions for cybersecurity, resilience, or reliability [2]. Trustworthy CPS architectures must be based on a detailed understanding of the physical properties and constraints of the system in the context of threats, attributes, and effects. Analyses in support of design activities must include the creation and

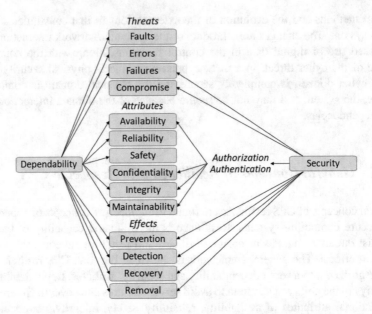

Fig. 6.2 Related concerns for dependability and security in CPS (adapted from [4])

simulation of up-to-date adversary models as well as physical and functional models of the CPS [2].

The outcome of CPS resilience is expressed in the operational characteristics of the total system-of-systems. Using a U.S. Department of Defense (DoD) definition, hardware and software components of systems "have the ability to reconfigure, optimize, self-defend, and recover with little or no human intervention." The DoD further defines three aspects of resilience in an operational context: the systems are trustworthy, missions of these systems can tolerate degradation or loss of resources, and the systems have designs that provide means to prevail in the presence of adverse events [5].

With respect to knowledge and skills, a CPS systems engineer must master all knowledge areas related to resilient CPS: the threat, the system operation, and human interactions, system vulnerabilities, approaches for resilient design, and validation of the system. Design principles of resilient CPS [6] include the following with related disciplinary domains:

1. Concepts of secure access control to and use of the system and system resources (domain of system security engineering).
2. Understanding of design attributes that minimize exposure of vulnerabilities to external threats (systems security engineering and dependable computing).
3. Understanding of design patterns to produce effects that protect and preserve system functions or resources (dependable computing).
4. Approaches to monitor, detect, and respond to threats and security anomalies (cybersecurity).

5. Understanding of network operations and external security services (information systems).
6. Approaches to maintain system availability under adverse conditions (all of the above).

6.1.2 Basis for a Functional Approach to Analysis of Dependability and Security

This chapter focuses on the need for an interdisciplinary approach, founded on rigorous system modeling, to CPS design and evaluation. Cybersecurity in the broader concept of resilience in that security concentrates on protecting defense systems from sentient adversaries. CPS are generally designed by initially specifying critical and other necessary functionality. The high-level functionality is decomposed into specific functional capabilities, and system requirements derive from these functional needs. Boehm and Kukreja [7] distinguish between functional and non-functional requirements as what the system does and how well it does those things, respectively. Non-functional requirements relate to those attributes shown in Fig. 6.1. Security, another non-functional requirement, is assessed on how well a given security design pattern protects the system as intended—without adversely impacting the intended functional operational and performance capabilities.

Traditionally, systems theory has distinguished how a system is constituted internally (i.e., its structure) from how the system manifests itself externally (i.e., its behavior) and deduced the latter from the former. This notion is the foundation underlying hierarchical construction of systems with defined input and output interfaces across multiple modular components. It has resulted in the extensive focus on structural system representations to date in current cybersecurity protection methods. CPS, however, are heterarchical in nature. They are comprised of numerous, heterogeneous elements acting both independently and interdependently. Because of this complexity, traditional decomposition and predictive methods are insufficient. In all complex systems, form (structure) and function (behavior) are intrinsically linked. A system's structural characteristics and what processes and behaviors are possible within and as produced by that system are not separable.

Even so, many current methods focus extensively on structural system representations. The structural bias results in an overt conflation of resilience (in the broader sense) with robustness or reliability, resulting in a focus on perimeter protection methods to prevent threat intrusion or viewing threat attacks the same way one would view failure modes. Cyber threats attack both structural components of a system and its functions and create their effects in ways that do not exhibit characteristics system failures. Even graph-based methods, a core approach highlighted in this chapter, commonly evaluate the resilience of complex systems based on structural properties. For example, approaches like random node or edge removal are used to determine the impact on structural properties such as connectedness

[8]. There has been a significant emphasis on how structural statistical properties of a network topology will be affected by additions (growth or augmentation) and removals (failures or attacks) of nodes and links, and particularly how networks can be more robust against the latter [9]. Those additions and removals are typically assumed as perturbations coming from external sources, not incorporated into the dynamics of the network itself. The literature is also replete with studies on the *dynamics of networks*, meaning how a network changes its structure over time. *Dynamics of networks* is distinct from studies of *dynamical processes on networks,* which is necessary to capture functional preservation of CPS effectively.

6.2   Model-Based Assurance of CPS

The disciplinary aspects of engineering design for resilient CPS are further complicated by the need for system assurance—"the justified confidence that the system functions as intended and is free of exploitable vulnerabilities, either intentionally or unintentionally designed or inserted as part of the system at any time during the life cycle. This confidence is achieved by system assurance activities, which include a planned, systematic set of multi-disciplinary activities to achieve the acceptable measures of system assurance and manage the risk of exploitable vulnerabilities" [10]. Assurance is a set of engineering practices that serve to evaluate the design of the CPS for a reasonable set of dependability and security requirements based on the operational use of the system. In the CPS design context, this is the selection of technologies and processes that provide confidence the system operates as intended in the presence of both accidental and intentional vulnerabilities [11]. The concept of "designing in" CPS dependability and security requires a multidisciplinary set of methods and tools. These are described in the remainder of the chapter with respect to threats, tools, modeling methods, and assurance methods.

Model-based assurance (MBA) is a process that supports system verification and validation requirements using both conceptual and analytical modeling and simulation techniques. As digital engineering and model-based systems engineering (MBSE) become more prevalent, there is potential to transform traditional system assurance processes to more holistic and more evidence-based forms using models. The current state of practice in system assurance is in need of a paradigm change. Verification and validation of mission-critical systems through test and evaluation has historically been the gold standard for assurance but is significantly expensive and increasingly fraught with difficulty as systems become more complex, more expansive, and more inter-dependent on other systems to realize their intended capabilities. Bringing models into this process aims to relieve some of the expense and make the entire process more flexible and amenable to changes that can occur across a system's lifecycle. The use of models to support analysis of assurance, while promising and necessary, still faces an uphill challenge to establish the best practices and systems engineering foundations required to produce what can be counted as evidence to support assurance judgments. This is especially true for

CPS, often employed in system-of-systems operational configurations, increasingly connected and increasingly complex when considered in the context of their higher order dynamics with other systems in the environment and facing increasingly diverse and sophisticated threats.

MBA is the use of a model or group of models as a basis of understanding and to produce evidence that a given system will perform as intended in various potential environments, operational conditions, arrangements with other systems, etc. Part of MBA includes ensuring that a model allows determination of whether a system design meets functional and non-functional requirements. In turn, to do so requires ensuring that the model, with all of its levels of abstraction, represents accurate system functional performance and characteristics. Any final system implementations using model-based design approaches in an MBA context should lead to a system design that is safe and secure, or analyzable with respect to its safety and security. While most models may not exist at a level of specification that a system may be completely built from their template, many of these models can specify diverse categories of functional or other performance requirements that a system will need to be safe and secure within a reasonable level of uncertainty.

CPS create new challenges to the concept of MBA. Specifically, what model-based approaches capture relevant and representative levels of abstraction sufficient to help validate the integrity of the system requirements and the integrity of the design? CPS are generally designed by initially specifying critical and other necessary functionality. The high-level functionality is decomposed into specific functional capabilities, and system requirements derive from these functional needs. From this perspective, MBA seeks to build a model-based process in concert with existing MBSE practices to produce an evidentiary case that a system is trustworthy with respect to the properties its stakeholders legitimately rely upon within acceptable levels of risk.

The focus on analytic decomposition (i.e., identification of component failures) in many current model-based approaches needs to be augmented by approaches that enforce safe behavior (e.g., state-based evaluation of dynamic control). Assurance must cover any undesired or unplanned event that results in a loss, and address hazards and vulnerabilities as a system state or set of conditions that, together with worst-case environmental conditions, will lead to a loss. To satisfy analysis for the purpose of assurance, a model that is an abstraction of a system's functional behavior should represent failure effects in the system, how failures propagate through the system, and observable conditions those failures manifest. A mission-critical cyber-physical system must consider of all classes of system failures, whether inherent or malicious, in rapidly changing external system-of systems contexts. Future MBA methods, processes, and tools must go beyond traditional quality assurance scope to include emergent dimensionality of the design space through the evolving quantification of concepts such as flexibility and resilience. They must also create a framework for describing these concepts as patterns of design that can be captured into standard practices and tool libraries. The next sections discuss these generalized patterns.

6.2.1   CPS Threat Patterns

The threats of concern to CPS are different than IT systems—they are constrained by the nature of CPS and also functionally different. Griffor summarizes the CPS specific concerns as (1) tolerating intrusion and disruption of signal/information flow, and (2) ensuring there is no insertion, fabrication, or replay of legitimate control commands [3]. Table 6.1 enumerates a number of CPS-attack types. The attacks are categorized according to what is being exploited: software vulnerabilities or flawed network implementations (e.g., leading to cyber attacks), physical vulnerabilities, or cyber-physical combinations [12].

Wan et al. further defined a set of generalized control system attack models of specific concern to CPS shown in Table 6.2 [13].

Attack trees and attack-graph-based visualization offer an intuitive, well-understood approach to capture the complexity of coordinated attack models and to view the effects of the attacks and combined countermeasures. Attack graph analysis is the most widely accepted method to assess how an attacker can gain access to cyber assets in a CPS and how that can be used to exploit the system functions. Up-front human-centered workshops capture the definition of cyber assets in terms of system functions as well as the attack graph analysis that explores the dependent functions that can be exploited by the attack. These provide a tool basis to explore these basic patterns in multiple combinations. CPS functional modeling must consider threat parameters as additional functions to be captured in the control system design decomposition process, producing a graph of both the control function and the attack.

6.2.2   CPS Countermeasure Patterns

Wan et al. also defined a set of threat countermeasure patterns of specific concern to CPS shown in Table 6.3 [13].

These also can be captured into a functional model of the system, effectively combining intended control functions, threat functions, and countermeasure functions into a single model using selected patterns. This allows analysis of "reasonable assurance," which is effectively a cost trade. While the effects of the attack–countermeasure interactions cannot be directly costed due to the lack of details, CPS designers can assess the effectiveness or "residual risk" associated with selected countermeasure patterns as shown in Table 6.3. For example, the control parameter attack model requires an adversary with moderate knowledge of the system functional design, low attack specific technical ability, and generally low resources, while the countermeasure approach would have a low to medium implementation cost, and low collateral impact to the system. On the other hand, coordinated attack models that would be effective in identical binary copies of the critical assets would have higher attacker specific technical ability and medium to

Table 6.1 Taxonomy of CPS-attack types

	Description
Cyber attacks	
Network denial of service	Disrupt network operations to stop control signal/flow during the attack period
Malware	Installation of malicious software, or viruses, into the device
Software vulnerability exploitation	Exploitation of software vulnerabilities such as buffer overruns, numerical overflows, or backdoor applications; can be defects or intentionally created
Physical attacks	*Description*
Sensor spoofing	Providing false sensor data to the CPS, either by injecting false information into the sensor or into the communication paths
Signal jamming	Disrupting control/signal flow by preventing or changing the signal, primarily using external signals
Hardware vulnerability exploitation	Exploitation of hardware vulnerabilities such as network protocol errors, or side channel attacks on power, cooling, or other flows
Physical damage	Physically disrupting the CPS to disrupt control signal/flow, such as damaging interconnects or sensing surfaces
Cyber-physical attacks	*Description*
Insider threat	Stealing and/or modifying design or operational data for exploitation; emphasizes the importance of the CPS design environment
Identity spoofing	Providing signals or information to the CPS that appears to be legitimate, such as false network packets; used in man-in-the-middle attacks
Supply chain compromise	Introducing a flawed or vulnerable hardware or software component into a CPS when it is being manufactured or configured
Information disclosure	Monitoring CPS information to gather information needed for additional attacks or to steal private information
Social engineering	Deception or influence on the CPS engineer or operator in order to gain information for other attacks, or possible to induce poor safety or security decisions into the design
Replay attacks	Recording the control signal/flow over a period of time, in order to replace the actual signal/flow with the recorded data to confuse the system
Control system instability	Disrupting control/signal flow in order to produce control system instability

Table 6.2 Generalized control system attack models

Attack model	Description
Interruption attack	Also called denial-of-service attack, stops the control signal/flow during the attack period
Man-in-the middle attack	Mimics the human attack behavior. When the attack happens, the control signal/flow is changed to a different manipulated signal/flow controlled by the attacker
Replay attack	Records the control signal/flow over a period of time in a vector, and when the attack starts, it replaces the actual signal/flow with the recorded data to confuse the system
Control parameter attack	Modifies the vulnerable control parameters of the system to the attacker's defined parameters. This changes the quality of control of the system
Coordinated attack	Combines two or more basic attack models, for example, combining a man-in-the-middle attack with a control parameter attack

high resources to implement but would also require more intrusive design impacts to the CPS. Attack graph tools already have support for such cost/risk analyses.

The basic countermeasure patterns can be extended to more complex attack strategies by combining patterns. For example, diverse redundancy is a pattern that employs redundant control systems using differing hardware and/or algorithms whose outputs are voted or averaged [13, 14]. This mitigates the ability for a cyber threat to compromise all redundant elements in the system. The diversity and redundancy can be provided by components and algorithms added to the functional design of the control system. Horowitz and colleagues collected a number of these redundancy strategies as summarized in Table 6.4.

The outcome of this process is a "cyber-protected system model." The result is a set of security design patterns described as to (a) their functional capabilities, (b) the cyber assets they require to achieve their functional capabilities, (c) the critical cyber assets and/or functions they will protect, and potentially if applicable (d) the specific threat functional capabilities and/or threat cyber assets they are designed to detect or counter through direct connective action. These patterns (actually a set of new functions) can be designed into the CPS or a monitor device that tracks the CPS functional behaviors externally. Using a general model-based approach, different security patterns and countermeasure approaches can be implemented as updates to the CPS as threats evolve. There are a number of processes and tools that support this analysis. The most used are highlighted in the next section.

6.3 Tools to Evaluate Threat and Countermeasure Patterns

New methods and tools are needed for design and analysis of CPS dependability and security. Tools and techniques exist for design and analysis of fault-tolerant systems, and these tools may be adapted for application in the CPS domain. However,

Table 6.3 Generalized control system countermeasure patterns [13]

Countermeasure	Description	Attack model countered
Isolation	Creates an isolated runtime environment (sandbox) for the critical asset that is resistant against attacks	Escalation, interruption attacks
Redundancy	Replicates the functionality of the critical asset in order to create multiple paths for high availability and fault tolerance in the case of individual function failures	Attacks that disable individual instances of critical assets and functionality
Diversification	Produces functionally equivalent variations of binaries running in software critical assets. This is an enhancement of the redundancy countermeasure	Coordinated attacks, zero-day attacks effective in identical binary copies of the critical assets
Physically unclonable function	Secures the integrity and privacy of the messages in the system using a physical unclonable function (PUF) that is hard to predict and duplicate	Attacks that hijack the communication channels such as man-in-the-middle attacks
Obfuscation	Obscures the real meaning of data/signals/flows by making them difficult for an attacker to understand. It can use random sources of noise from the environment of the critical assets to increase the entropy	Attacks that require knowledge of the inner workings of the system, its functions, and its mission
Parameter assurance	Compares input data to a table of values in the system to check for large, unexpected deviations	Attacks that manipulate data files or messages that are sent to the system
Data consistency checking	Verifies the source of a parameter change	Attacks that use operator specific data entry

Table 6.4 Specific countermeasure patterns for CPS

Countermeasure	Description
Redundancy	Checks for errors by comparing outputs from multiple components at different levels of the system. This can range from logic levels (parity or cyclic codes) to structural component levels to system levels. Checking accuracy and latency of detection is improved by lower level techniques, while higher-level techniques may be more cost effective
Forward error recovery	Allows continued system operation in the presence of errors by compensating for the error with correction strategies. This provides redundancy beyond just error detection. It can also be implemented at multiple levels. From cyclic correction codes and lower levels to voting strategies at component or system levels
Triple modular redundancy	Provides forward error recovery with masking redundancy—the system has three identical modules that perform redundant operations. The output of these modules is checked by a voting module, which forwards the output based on majority vote. An error in one of the modules is masked by the majority
Backward error recovery	Tracks known good process states as recovery points. When an error is encountered the process returns to the last known good state. This is effective for temporary errors but can recur continuously (known as livelock) in the presence of permanent errors
Dynamic redundancy	Detects errors and reconfigures the system in response. This is often referred to as "hot standby" when the reconfiguration selects a continuously running system and "cold standby" when the standby system has to be started
Diverse redundancy	Schemes like triple modular redundancy can use identical or diverse modules. Use of diverse, or non-identical, modules counters errors that may occur in common designs but also complicates the voting design
Data consistency checking	Verifies the source of a parameter change

traditional tools are insufficient for resilient CPS due to the complexity of these systems and the potential for cyber attack.

Most failure analysis tools used today represent the process as a set of linear causal models that sequence a series of events over time. Analysis of the system anomaly is traced backwards in time from the system failure event (if it has occurred) or forward in time from the analyzed potential event causes to their effects (in a preventive design process). These analyses are essentially event chains with branching logic [15]. They trace preconditions and effects of errors in the system as a chain of events that progresses sequentially in time, then apply a countermeasure pattern that changes the chain of events to a more desired effect.

These are a useful and logical way for the system designer to think about natural and intentional causes of system disruption but ignore non-linear relationships such as feedback. They also tend to ignore long timespan chains of events such as a control attack that is placed in the system long before the compromise is planned or intended. This leads to great subjectivity in the analysis. For example, a design feature intentionally placed into a counterfeit programmable logic part can exist for years before a separate piece of malicious software is inserted into a CPS by a human attacker who gains control of the systems data through a third vulnerability in a communications protocol. Together, none of this manifests themselves as errors, and the effect might create a new system behavior that takes control of the CPS.

The design of the attack creates an additional set of control and feedback systems in the CPS, which must be countered not by tracing linear chains of events and preventing each, but by creating new design patterns that serve to limit these new control behaviors from being exploited by a determined threat. The primary shortfall of causal event models is that they limit the analysis of countermeasure designs to similar linear cause and effect relationships [16].

A resilience analysis will convert the concept of a failure in a CPS to one of availability, or continuity of service. A resilience analysis will ask the questions:

- What will disrupt the control operation, what are the vulnerable aspects?
- How will that affect the availability of the control function to other parts of the system or to its human user?
- What other system components is that control function dependent upon, and to what extent?
- Can those components also be trusted?
- What is the risk (and associated cost) to ensure trust?

In a linear causal analysis, these dependencies are often omitted because they cannot be quantified, or an assumption is made subjectively to omit that dependency for convenience. Although failure analysis remains important, it must be embedded in a larger process that considers the functions that result from the CPS and its external interactions and dependencies. Due to the complexity of the systems involved, the analysis should not attempt to purely identify all interactions and dependencies but should define hierarchies of control over essential CPS functions. Resilience is a process of reducing or eliminating dysfunctional interactions. These are interactions that can disrupt the availability of critical control functions or lead to potentially hazardous states in the controlled process [16].

6.3.1 Mission/Operational Resilience Analysis

The resilience analysis starts with human-centered workshops to capture the definition of cyber assets in terms of system functions as well as the threat attack analysis that explores the dependent functions that can be exploited by the attack. Besides being a manual technique, the main limitation of many resilience analysis approaches is that they are too high level and the threat parameters are not treated as additional functions during the functional decomposition process that creates the CPS design. Therefore, this does not allow the traditional systems engineering analyses to include functions and requirements that address cyber threat parameters as inherent to the system functional design.

The accepted processes for non-functional requirements derivation are generally human-driven and involve scenario analysis and modeling to derive lower level functions or requirements from higher-level architectural models. These are facilitated processes using subject matter experts from the operational context of the CPS, CPS designers, and the cyber experts with knowledge of potential attacks. The facilitator starts at a mission/operational level with questions about lost or changed functionality during operations. This creates shared understanding of mission objectives and operational tasks related to the CPS. The process also creates a set of information that can be used by CPS designers and cyber threat experts to conduct the analysis of attack and countermeasure patterns.

One approach to manually derive these patterns is describing "critical cyber assets" to link functional descriptions and early views of structural components. The MITRE Mission Assured Engineering (MAE) process framework [17] is often used in the context of defense assured systems, and is a starting point for the development of our approach. MITRE defines three different assessment processes which we can describe as mission analysis, cyber threat susceptibility assessment, and cyber risk remediation assessment. Established IT guidance links cyber requirements to "critical cyber assets" (crown jewels). DoD guidance links cyber requirements to critical mission threads. In the CPS design one needs to assess vulnerability of both critical information assets and critical control processes against criticality of the mission segment of interest. The three assessment processes are:

- Mission analysis or "mission-based critical asset identification." This process takes critical mission objectives or operational tasks and associates them with the critical system functions, creating an initial linking of functions to types of cyber-physical assets.
- Creation of a functional dependency graph that hierarchically links high-level mission objectives to operational tasks to information (or control) assets to sets of system assets as shown in Fig. 6.3. This is a traditional mission to

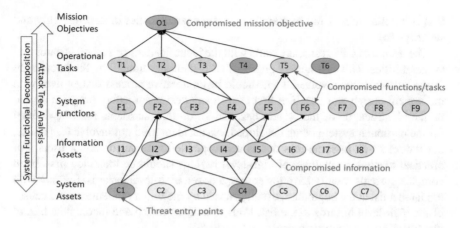

Fig. 6.3 Functional dependency graph linking mission objectives to system assets (adapted from [17])

function to structure decomposition approach, which can be captured in a model. System assets are those structural components that mighty be exploited by a cyberattacker.

• Creation of an equivalent failure model by tracing back up the graph. The definition model is effectively a causal event chain: "if 'asset' is compromised, then 'function' is compromised, then 'mission objective' is compromised."

Various analysis tools can be then used to assess the resulting functional and asset relationships. Three of particular interest are fault and attack tree analysis, goal-structure notation, and system-theoretic process analysis.

6.3.2   System-Theoretic Accident Model and Process (STAMP)

To address the shortcomings of linear failure models, Leveson developed the system-theoretic accident model and process (STAMP), and system-theoretic process analysis (STPA). STAMP has been applied to accident analysis and prevention, and also to general dependability and security design in CPS using a security specific form of STPA called STPA-Sec [18]. In the STAMP framework, understanding system disruptions requires the analysts to determine why the control structure was ineffective. STAMP replaces the concept of an event that results from a control failure with the concept of a constraint that enforces appropriate control. STAMP analyzes and imposes an equivalent structure of CPS information and control feedback. This feedback extends hierarchically from a central control structure to include larger feedback loops created by larger system dependencies. The potential interactions between dependent systems are changed from assumed trust to evidence of trust. This allows for the system to have adaptive feedback loops that either maintain or fail to maintain system characteristics of dependability and security [16].

The design of CPS resilience starts with the identifications of system hazards and vulnerabilities. This is a human process that is naturally limited by the knowledge of the human teams involved, so it should be an iterative process that evolves with the system design and use. The first step in any design for safety program should be the identification of the system hazards. To do this, accidents must be defined for the particular system being developed. An accident need not involve loss of life, but it does result in some loss that is unacceptable to the customers or users. For practical reasons, a small set of high-level hazards should be identified first. Even complex systems usually have fewer than a dozen high-level hazards. Starting with too large a list at the beginning, usually caused by including refinements and causes of the high-level hazards in the list, leads to a disorganized and incomplete hazard identification and analysis process.

A system is a recursive concept, that is, a system at one level may be viewed as a subsystem of a larger system. Unsafe behavior (hazards) at the system level can be translated into hazardous behaviors at the component or subsystem level. Note, however, that the reverse (bottom-up) process is not possible. Hazards can also be related to the interaction between components such as the interaction between attempts by air traffic control to prevent collisions and the activities of pilots to maintain safe control over the aircraft. Due to the potential vulnerabilities (i.e., threats to a system's intended safe function) stemming from an array of complex interactions and sequences of events, STAMP views accidents—by analogy to the CPS security case, threats—as a control problem. Vulnerabilities may consequently be prevented by enforcing certain constraints on system component behaviors and their interactions. In STAMP, a process model controls the actions to help expose what are deemed unsafe control actions: control commands required for safety are not given, unsafe control commands are given, commands are given too early or too late, or the control action stops too soon or is applied for too long [16].

This view implies a modeling approach that is able to capture the functional state space of a CPS and reveal whether that state space has been compromised or preserved in the face of threats and applied protection patterns. The focus on functional modeling espoused in this chapter is consequently synergistic with the goals of STAMP. Further, the frameworks and methods discussed in this chapter may serve as a direct compliment to existing model-based systems engineering processes and tools and, in turn, themselves be executable within a toolset that enables systems engineers to produce, navigate, and understand the complexity and scope of the problem.

STPA-Sec takes the mission objectives and operational tasks developed in the war-room setting and extends them down to dependability and security objectives. The first step is identification of unacceptable losses in the mission/operational context. This list is then used to derive a set of system hazards and associated system constraints. Take, for example, a simple digital engine control loop in an aircraft. Unacceptable losses would include loss of the aircraft, loss of human life or injury, unacceptable delays in travel, and loss of trust in air travel. Three (of many) specific hazards in an engine controller would include uncontrolled changes in engine thrust, incorrect engagement of engine thrust reversers, and incorrectly reported engine failures. An example system constraint would include a requirement such as "the system shall prevent engagement of thrust reversers while the aircraft is in flight."

This high-level descriptive model then can be used to create a functional model of the control structure, leading to a set of expected control actions and potentially hazardous control actions. The thrust reverser example is relatively simple. Two control actions likely define the function: engage and disengage reversers. This would be accompanied by several constraints defining functions such as "check for weight on wheels before engaging..." All of these would be typical system functions in a system functional model. The STPA process recommends evaluating the causes and effects associated with not providing the expected control action, providing it in a way that causes hazard, providing it too soon or too late, and providing it out of sequence. Detailed examples of STPA and STPA-Sec can be further explored in open literature.

STPA-Sec extends the analysis to identification of hazardous control actions and related security constraints, such as "thrust reversers shall not engage without direct physical indications of aircraft weight on wheels" or "weight on wheels indicators shall employ diverse redundancy." This process leads to a set of scenarios that relate hazardous CPS control actions to security related scenarios (and also dependability related scenarios as a set of control constraints). These can be further explored in the context of threat attacks and associated system loss of control (expressed as errors) using tree or graph models.

6.3.3   Fault and Attack Tree Analysis

A basic approach used by designers of fault-tolerant systems is fault tree analysis. A fault tree model is a graphical representation of the various parallel and sequential combinations of errors that could reasonably contribute to the occurrence of a top loss or hazard event. Because the fault tree focuses on its top event, the tree only includes faults that contribute to this top event. This fault list is not exhaustive. However, the STPA process helps to create a holistic model of these events and associated errors. As this chapter is focused on the cyber-attack domain, a detailed description of the system fault analysis is left to other references.

Fault tree analysis has been adapted for application in cyber physical security in the form of attack trees. Like fault trees they begin with top events but then analyze all steps an attacker might use to cause that event. A number of commercial tools support the development of attack trees or their graphical equivalents, attack graphs. In the attack tree analysis, the attacker's main goal is the root of the tree. In CPS this may be disruption of a control action but could be other effects such as stealing information. Although this chapter is focused on the CPS control actions, the analysis should cover all aspects of the system including its development environment. Attack trees are built from the attacker perspective.

Most attack tree tools branch from the root node into attacker subgoals using an AND/OR logic structure where the AND nodes imply the attacker must take multiple actions to accomplish the attack and OR nodes signify one of several alternative actions. As the tree is constructed it follows the structure of Fig. 6.3 where system control functions are decomposed into lower level system functions and information assets, then into system assets which serve as threat entry points.

After the trees are constructed, analysis prunes the tree to the most likely attack strategies. Most attack tree analysis tools support analysis capabilities that aid in calculating cost measures and risk measures.

The attack tree analysis serves to add detail to the higher-level STPA analysis so that more detailed models can be built to reflect countermeasure strategies. Actual attacks and countermeasures occur within the system and information assets in the accomplished design. It is important to gain this level of understanding in the CPS design, but also to understand that the analysis will be limited to the knowledge and experience of the subject matter experts used at the time of the analysis. Also, the resulting attack models are static and do not capture the actual dynamic control actions that are to be protected. For this, a functional modeling language and simulation environment should be used. Thus, the analysis should be used to inform the functional model of the protected system, not as a direct analysis of system assurance.

In the dependability and security analysis, the holistic analysis process of STPA is combined with more detailed fault and attack tree analyses to gain the detailed awareness of how the CPS will remain resilient to both internal failures and external threats. These come together in a functional modeling environment viewing system functions, functional models of threat attacks, and functional models of the countermeasure designs that become part of the overall CPS design. These are explained in a more detailed example to follow.

6.4 High-Level Functional Modeling with Critical Cyber Assets and Missions

This section provides a case study representing the modeling process for the case of a Navy ship chilled water system that provides cooling flow to two ship board radar systems [19, 20]. Figure 6.4 shows the functional diagram of the fluidic subsystem. The system's mission is to provide cold water from coolers (chillers) to heat loads (e.g., radar and generators), in order to regulate their temperature and maintain it within suitable operation conditions. There are four main component types: loads, chiller plants, valves, and pipes. The nomenclature is explained as follows: Loads: L01–L06, where L05 and L06 are vital loads and the remaining are non-vital loads; Flow meters: FM01–FM10, which can indicate if the chilled water service is in operation; Valves: V01–V26, which control the transport of water from chillers to loads for cooling purposes; and Chillers/pumps: P01–P04, which provide a pumping

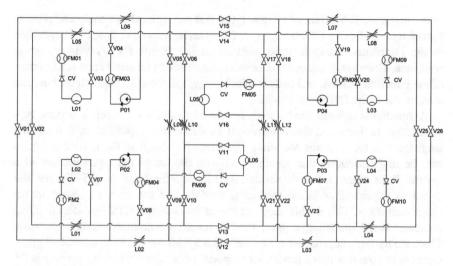

Fig. 6.4 Fluid subsystem of the navy ship chilled water distribution system

function as well as a cooling function. The plant has a total of four zones, with one chiller plant per zone.

The control logic for the model (not shown) must maximize the operability and the survivability of the system against kinetic attacks. The control logic must also maximize the efficiency of system's operation. Loads are a set of special components that directly interact with the environment outside of the system, in this case they are heat loads. One or more loads may combine to provide system functions. For example, the identification of airborne threats using shipboard radars requires effective functioning of the chilled water distribution system. It is through these functions that the system's purpose is realized. The primary goal for a controller is to ensure that functions currently demanded by the mission are satisfied. This use case will be used to explain concepts as relevant throughout the remainder of the chapter.

A functional model describes what the system does in a formal, high-level manner. This high level of abstraction allows CPS multidisciplinary engineering because it formalizes the functional requirements, makes design intentions explicit, and decouples intentions from implementation. Functional models can therefore be used as a formal early representation of the system to reason about and to measure the impact of cyber-resilience aspects at the system level. Functional modeling is one of the most widely used concept design formalisms in industry, including ship building, automotive, aerospace, robotics, and machine building.

A persistent challenge facing system engineers is the need to establish not only a consistent functional vocabulary across design types and extent but also a clear stopping point in the functional modeling process and a consistent level of functional detail. The Functional Basis articulated by NIST [21, 22] provides a vocabulary of eight function categories with a total of 32 elementary functions, three flow categories, and a total of 18 flow types that can be used as a multidisciplinary, formal model of early CPS design [23, 24]. The function and flow types specified in the Functional Basis may be used as anchor points for analysis of cyber-resilience. Concretely, attacks and countermeasures can be also expressed in terms of the Functional Basis, and pattern matching (graph isomorphism) algorithms can traverse the entire functional model.

A functional system model may be directly developed using a textual representation or imported from functional model views expressed in a modeling language like the Systems Modeling Language (SysML). In the latter case, there may be multiple views that together comprise the functional representation of a system. Prior research has shown that SysML activity diagrams [25–27] are best suited to express flow-based functional models like the ones expressed in the Functional Basis. This SysML diagram assists the design of CPS by enabling direct computer-based modeling support needed for semi-automated design synthesis. There are multiple approaches for converting a SysML model to an equivalent graph representing the functional model. An important consideration in this process is the proper definition of interrelationships used to create the graph based on various system functional and structural models, attributes, and performance metrics. A consistent approach appropriate to the level of abstraction is required either in terms of extracting a functional representation from the SysML model or in setting up front how the SysML functional representation should be described.

As per the MITRE MAE process, a functional model must be annotated with critical cyber assets that will be evaluated for vulnerabilities and potential attack surfaces. These assets can be logical or structural. For example, the signaling in a communication system, the parameters of a controller, and a GPS receiver device are examples of critical cyber assets. Assets are modeled as nodes in the functional model connected to specific functions or information objects. Also, system's missions must be linked to the key functions in the functional model that fulfill them. Missions are also modeled as nodes in the functional model connected to specific functions.

Figure 6.5 shows a notional functional model for a navy ship chilled water distribution system presented earlier where functions are represented by ovals, assets by rectangles, and missions by diamonds. Consider two electrically powered, water cooled "Radar 1" and "Radar 2" assets which provide a redundant capability for the "Identify Airborne Threats" mission. Both radars have associated attributes that represent their thermal state (current temperature) and the electrical power provided to the unit and are linked to the "Radar Operation" function. Notice that the "Identify Airborne Threats" mission can only be realized if the thermal state of the radar load is below a certain specified temperature to prevent overheating, and if sufficient electrical power is provided to the load. At this level of abstraction, the redundancy of the function is represented by the number of assets associated to a given function and how the function attributes combine their assets using AND, OR, and XOR operators. For example, the "Operate Radar" function has an OR attribute (not visible in the figure) and thus to operate requires the asset "Radar 1" OR "Radar 2" operating. Similarly, the "Navigate" function is fulfilled if and only

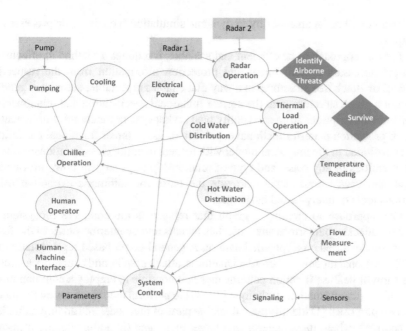

Fig. 6.5 Functional model with capability of identifying airborne threats using radars as loads

if the "GPS Receiver" OR "Inertial Navigation System" assets are available. By default, functions have AND-semantics.

6.4.1 Functional Modeling of Attacks and Countermeasures in Libraries

Threats to a CPS in the form of attacks and the design patterns intended to augment the original system to imbue it with secure, dependable ability to perform its function(s) are also abstractions. This means that attacks reduce or disable functionality, and countermeasures increase or enable functionality. Modeling the original system together with these abstractions as a directed graph supportive of simulation will reveal important behavioral dynamics across the CPS, threat, and protection elements. The original system components, functional representation, and assets critical to realizing those functions will derive from existing MBSE system component models, functional architectures, and activity and/or N^2 diagrams. However, graph theoretic measures based on aggregate statistics alone do not adequately characterize a system in terms of its throughput, performance, and vulnerabilities. Networks may have identical degree distributions, for example, and yet also have fundamentally different structures functional performance [28]. Key to the ability of a directed graph to elucidate CPS relationships and performance behaviors will be its amenability to dynamic simulation (i.e., dynamic processes on graphs).

For this approach to be efficient and scalable, it requires a method to effectively extract necessary system threat and protection information from the respective MBSE or attack tree structures in a way that enables repeatable, automated generation of a graph structure with the correct functional specification and relationships. Semantic descriptions from each entity type (cyber system functional component or asset, protection pattern functional capability or asset, threat functional capability) to be included in the graphical model with respect to entity class, relationships with other entities of any class, and type of relationship (e.g., logical flow, information flow, and causal dependency) provide the basis for automatic extraction using various tools or query-based languages.

One approach to creating a graph that represents the complete ecosystem of system-attacks-countermeasures attaches attacks and countermeasures to the functional model based on a specified pattern. A general graph-based representation for CPS functional models, cyber-attack patterns, attack graphs, and countermeasures in the form of defense function patterns makes it possible to employ a common graph rewriting (transformation) technique [29–32] to perform modifications on the model for two purposes: (1) identifying vulnerable parts of the model and adding individual or complex cyber-attack patterns that target them, and (2) adding countermeasures in the form of defense function patterns in order to neutralize these attacks.

The goal is to create a complete ecosystem model comprised of the original (unprotected) CPS, the threat functional capabilities and attack vectors revealing the critical cyber assets they will target, and the security functional architecture via one or more security design patterns showing how (functional capabilities) and where (cyber assets or threat assets) they will be applied. Such an ecosystem model is contextual with respect to the threat and protection spaces. In its final, directed graph form, nodes will represent the functional capabilities and their critical cyber assets, while edges (links between nodes) will represent a logical flow, information flow, or causal connectivity. Both direct and indirect connections can be consequential, together producing cascading effects in the event of a disruption.

6.4.2　Attack Models

There are various types of adversary attacks possible against the functional elements of a system that can interfere with the intended flow (and therefore capability) [13, 14]. In the static sense, these attacks can be either active or inactive, and can disable the corresponding function in the functional model to which they are attached. In the dynamic sense, these attacks are active or inactive over specific periods of time. The five primary types of attacks on system function are listed in Table 6.2 and their equations suitable for mathematical simulation are described in [14].

6.4.3   Countermeasure Models

Countermeasures are required to augment the initial designs of CPS to counteract the static and dynamic effects of attacks against system functions. These counter-measures may be thought of as protection patterns, also representable as functional nodes and edges that preferentially attach to the relevant original CPS functional graph nodes. Countermeasures are implementation dependent, that is, they require an understanding of likely or most significant attack vectors and intended system function as designed.

Scalability and generalization across different CPS require developing a library of countermeasures and their corresponding graph patterns from which a systems engineer can draw and augment the CPS-attack pattern graph to complete the ecosystem perspective. One approach to effective selection of the appropriate security protection patterns for a given attack pattern is to create a mapping between these pattern groups in the form of a bipartite graph that a user may modify as requirements evolve. Table 6.3 provided a listing of example countermeasure models that may be transformed into graph protection patterns at the appropriate level of abstraction for a given problem. While effects of attack–countermeasure interactions cannot be directly assessed with respect to cost at this level of abstraction, residual risk analyses already common in most attack graph tools can help evaluate relative cost and likelihood functions that assist with ranking attack and countermeasure patterns via multi-attribute or qualify function deployment (QFD).

6.4.4   Generation of Cyber-Physical Attack Environments

Initial requirements documentation and functional models are useful but incomplete. The requirements modeling starts with these artifacts, but one cannot trade maximal attack coverage and minimal requirements without baselining a cyber risk assess-ment process that relates risk of cyber "failures" to appropriate mission threads. The combination of STPA and tree analysis discussed earlier begins the process.

Attack graph analysis [33, 34] extends and partially automates the manually generated attack trees described earlier, and is the most widely accepted method to assess how an attacker can gain access to these cyber assets and how that can be used to exploit the system functions. In this section we use the attack graph formalism to represent plausible attacks in the form of functional graphs. The process captures the definition of cyber assets in terms of system functions as well as the attack graph analysis that explores the dependent functions that can be exploited by the attack. Future maturation of this process should include the consideration of threat parameters as additional functions to be captured in the decomposition process. This allows the traditional systems engineering capability/gap analyses to include

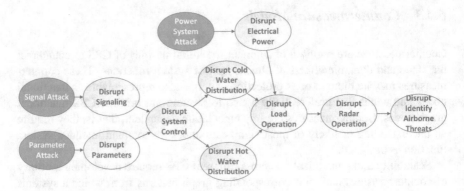

Fig. 6.6 Generated attack graph

functions and requirements that address the cyber threat parameters as inherent to the system functional design, not just as component of the mission design.

The use of functional models for representing a high-level system view allows automatic generation of attack graphs from functional system models. The generated attack graphs model coordinated attacks which target the system's mission. The attack surfaces that are exploited by these attacks are system functions and assets for which there are corresponding cyber-attack patterns in the library. This approach is extensible to new kinds of individual attacks and is able to generate complex attack graphs for coordinated attacks that specifically target any system mission in the model. In turn, this methodology allows the attack model to be accurate in pinpointing the mission-critical portions of the system.

Figure 6.6 shows a generated attack graph that targets the "Identify Airborne Threats" mission of the previously described navy use case. The attack graph is generated by traversing the functional model starting from the mission node and creating attack nodes and edges that correspond to the visited structure in the functional model. The process ends at functions or assets that are directly vulnerable to a CAP in the library, for which the specific individual attack is instantiated (filled nodes in the figure) and linked to the vulnerable node. In this example, the mission can be disrupted by affecting the system functions that are required for the proper operation of the loads: the cold and hot water distribution and the electrical power. The first two functions can be disrupted if the control system itself is compromised, which can be done by attacking the signals and parameters of the controller. The electrical power system can also be attacked directly. These cyber-attack patterns are selected from a library, and the method shows how the functional model may be traversed in order to generate attack nodes that target the capabilities of the system, represented as critical cyber assets.

Statically, these attack graphs employ OR-semantics by default, meaning that any successful child attack causes its parent attack to be successful. Regarding temporal semantics, the model of attack graphs is non-deterministic. Generalizing with respect to Boolean (AND/OR/XOR) and temporal (SEQUENCE, BEFORE,

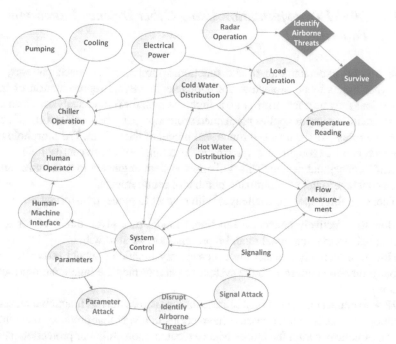

Fig. 6.7 Attacked functional model

AFTER) operators used in current variations of attack graphs and graphs [35], the attack graphs capture all possible interactions between the attack nodes, where the attack nodes can activate and deactivate non-deterministically over time. Thus, the attack graph semantics represent all possible behaviors of the coordinated attack over the targeted functions of the system, including any specifically modeled attack using temporal operators. This makes the approach more general and provides a more complete coverage of coordinated attacks, by using a dynamic model that tries all possible combinations of attacks.

The generated attack graph information can now be introduced into the functional model using the graph rewriting mechanism described earlier, producing the attacked functional model shown in Fig. 6.7. (For simplification, the intermediate attack nodes are omitted.) The figure illustrates that the new function aiming to disrupt the critical cyber asset is not affecting the asset directly, but rather indirectly through the functional dependency graph. The chosen points of attack correspond to those functions that are directly targetable through cyber-attack patterns found in the library. The attack functions have a set of possible attacker modes (not shown in the figure) as defined in the attack library. The attack graph model is linked to the functional model with the new attack function, and the attack graph analysis parameters can be maintained for multi-attribute tradeoff analyses.

6.4.5 Model Transformation Using Cyber Defense Functional Patterns

With a good understanding of the functional model, cyber-attack patterns, and the cyber assets to be protected, the complete functional representation of these dimensions may be transformed into a directed graph for further analysis. The model transformation process applies predefined protection patterns—in the form of graph rewriting rules—to the attacked functional model of the system. The morphological approach can be thought of as a systematic "patching" of cybersecurity holes in the system's conceptual design. This will allow system engineers to "maximize attack space coverage" while "minimizing number of requirements."

There are three specific challenges with respect to model transformation:

1. How to effectively connect a functional model of a given threat vector to the original system functional model representation we start with?
2. How to effectively connect the cybersecurity functional model associated with our protection pattern with the system-threat intermediate functional representation?
3. How to connect the functional representations across the full functional representation of attack–countermeasure interactions in a way that produces meaningful consequences within the model when executed upon—did our protection pattern implementation work or not?

A key aspect of (1) and (2) above will be to automatically extract semantic descriptions from each entity type (cyber system functional component or asset, protection pattern functional capability or asset, and threat functional capability) that will be included in the graphical model with respect to entity class, relationships with other entities of any class, and type of relationship (i.e., logical flow, information flow, and causal dependency). With these characteristics derived from the aforementioned functional model extracted from the system description, attack graphs (including temporal), and a library of cyber defense functional patterns, this will serve as the basis for a repeatable, directly executable model modification approach.

6.4.6 Countermeasure Analysis

The model transformation technique of graph rewriting can be applied to attacked functional models, thereby adding protection patterns that mitigate or neutralize the cyber-attack patterns. The outcome of this step is a "cyber-protected system model." A semantic mapping of security design patterns can be extracted with respect to (a) their functional capabilities, (b) the cyber assets they require to achieve their functional capabilities, (c) the critical cyber assets and/or functions they will protect, and potentially, if applicable, and (d) the specific threat functional capabilities

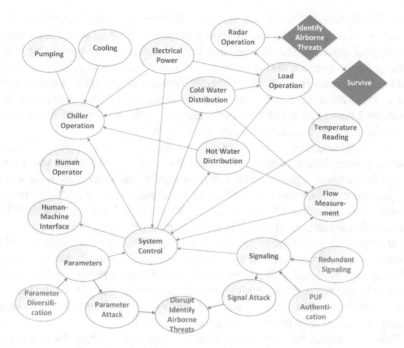

Fig. 6.8 Cyber-protected functional model

and/or threat cyber assets they are designed to detect or counter through direct connective action. An example of a protected functional model based on our navy chill water system use case is shown in Fig. 6.8. Here, protection patterns are linked to target design patterns in order to neutralize the threat functions associated with the attacked functional model.

Techniques relevant to simulating dynamic processes on graphs may then be used to assign various properties to the nodes and edges of the final graph that will demonstrate the propagation of a threat impact and security impact, the latter of which may compete directly with the former. For this to be scalable, this depends on generalizing a library of micropatterns used to assign node and edge attributes based on their extracted semantic description from the functional models. The result will be a single, scalable, executable methodology for dynamic simulation of cyber system graph models applicable to the cyber-physical system problem class.

Systems engineers can analyze the model and determine the effectiveness of the protection patterns added to the system based on the static and dynamic semantics of the modeling formalism. The functional models explicitly show dependencies between system functions, assets, and missions, and link these via AND-semantics, resulting in the assumption that all child functions are needed in order to realize the parent function. The addition of OR and XOR nodes is allowed in order to express different kinds of semantics for function dependencies, e.g., alternate or redundant functions. In contrast to attack graphs, functional models exhibit deterministic temporal semantics. All basic functions are active unless targeted by an active cyber attack, and higher-level functions activate or deactivate based on their logical dependencies. On the other hand, countermeasure models are defined in the protection pattern library to provide different methods to protect asset nodes and functions against attacks. Consequently, they employ static OR-semantics, such that a parent countermeasure function can be realized through any one of its child functions. Likewise, AND or XOR nodes can used in order to express different kinds of relationships between countermeasure functions.

The logical relations between nodes in the functional model (system functions, assets, missions, attacks, and countermeasures) enable analysis of the effectiveness of countermeasures against attacks both statically and dynamically. In a static analysis, the non-deterministic semantics of attack graphs can be used to compute all possible attack situations of coordinated attacks, even combining multiple individual and coordinated attacks at once. In all computed situations, effectiveness of the cyber protection functionality can be assessed by evaluating the realization of the mission objectives (represented by mission nodes). If situations can be identified where the system is not properly protected, then the cyber protection functionality needs to be improved. Also, multiple alternate protection strategies can be evaluated and compared in this manner, allowing the system to select those successful strategies with the least amount of required functionality.

This process has the potential to define a protection strategy for any class of threat, at any level of the system. The selection of a protection function, or cyber protection pattern, will in turn determine the requirements. The classes of threat covered will be based on both the understanding of the threat tactics, techniques, and procedures (TTPs) and the availability and cost function associated with various design patterns. "Protect against many classes of threats" would be a requirement evaluated through multi-attribute tradespace analyses methods.

In addition to the purely static analysis, it is possible to achieve greater accuracy by performing an analysis of the system's dynamic behavior. Relating to the notion of security for CPS, the dynamic problem is to determine if the functional state space of the original CPS system is preserved when attack patterns and protection patterns have combined to create a new, expanded functional structure. This requires node and edge functional/property motifs that make the ecosystem graph executable and meaningful with respect to evaluating the preservation of the functional state space. These motifs should be detailed enough to produce the needed overall system behavior but generalizable enough by type to be reusable and allow us to produce similarly executable graphs at much greater scale. Unlike the random or statistically generated graphs studies in other fields, we must develop techniques best suited and modifiable for simulating the impacts of highly correlated threats and protection implementations on functional capabilities of a system. This research task will develop ways to specify parameters in our graphical model that, when executed via discrete simulation, produce "consequences" most relevant and meaningful to the CPS security problem.

A key lesson from the science fields mentioned previously is that different graph topologies radically alter the dynamics possible. Different graph topologies may analogously alter the effectiveness of establishing security for a system under threat(s). The difficulty in modeling CPS is the number of different node types or classes more so than the number of nodes. To make executable models of CPS scalable, motifs, algorithmic representations that represent the behaviors and decision processes of each distinct class, will be required for reuse across different problems. Various graph properties may be assigned in ways that capture dynamic impact to best represent the problem, specifically node and/or edge state descriptions and weights. These techniques will, together with the graph's final structure, govern the spreading dynamics of the impact of the threat(s) and security implementation(s) and, consequently, the final state of the cyber system's functional capability. Rules governing node and edge states will derive from the node or edge type (i.e., node class) and relational nodes defined by a Markov blanket [36]. A key aspect of the work will be to simplify the relational state rules as much as possible to still reflect an abstraction of the contextual cyber system behavior over time.

6.5   Assurance Test and Evaluation

Assurance testing within the software community has been a long-standing discipline. In software, assurance cases have been developed to reason and communicate the trustworthiness of systems as well as the development of policy to assess the impact of security issues on safety regulation and the impact of cybersecurity.

Design assurance in engineering has sought to establish a sufficient, foundational set of design assurance requirements and processes that are analogous to product quality assurance. Aerospace in 2009 released a document that illustrated a cross-discipline, multi-company team named Design Assurance Topic Team that worked to formulate a risk-based design assurance process flow. The flow was to then serve as a roadmap for aerospace programs' design assurance activities [37]. In 1974, Caslake submitted a paper on the update to the Standards development in the quality assurance area by IEEE Nuclear Power Engineering Committee (IEEE/NPEC) [38].

With the onset of interconnectivity among devices, the boom of the IoT, and development of CPS, there was an increase in scrutiny of quality assurance in both hardware and software and the development of network assurance, which does differ from the software assurance community. In 1991, the proposed workstations with a shared computer/simulated environment were explored, and a key attribute was the delivery of better-quality assurance for hardware and software design with a more integrated, concurrent approach [39]. The networked reliability also had to be present through secure solutions. IT network assurance has grown into a discipline and is recognized for quantifying risk from an IT network perspective. It is the engineering process of formal verification, which specifically contrasts with design testing.

Embedded security and cyber-assurance also became clear disciplines and engineering processes. Embedded security is not a new concept, and has, in fact, been exhausted for engineered systems such as the IoT. Cyber-assurance is the justified confidence that networked systems are adequately secure to meet operational needs, even in the presence of attacks, failures, accidents, and unexpected events [40]. The internet-of-things (IoT) has the fundamental need and concept of embedded security. There are several tests and evaluations that have been developed to obtain embedded security and cyber-assurance. In fact, as discussed by Brooks, cyber-assurance concepts will have to include embedded security solutions, and information assurance for IoT networks to be resilient with confidence [41].

While all these are important to address in the components, the more holistic picture is still needed to address the challenges and opportunities in system assurance in the design for CPS, and therefore, the test and evaluation of such. System assurance should ensure that the system meets its functional and performance requirements, verify the ability of the system to perform at the limits of its design, verify that all configuration items/modules work correctly with each other, and design specifications for input, processing, and output are met. It should also ensure compatibility of the software, hardware, and documentation. Several areas of research attempt to argue system assurance through thorough model-based testing, architectures, and system synthesis of the implementation of the CPS [42].

Attack vectors use specific features or design elements of CPS to their advantages and are unparalleled in the traditional IT, and conventional computer security techniques are not able to address CPS. Due to the nature of complexity within these systems, a variety of guidelines and standards have been developed for the design of reliable CPS. For instance, the DO-178C was created by the authorities of FAA, EASA, and Transport Canada for commercial airborne software, which defined design assurance levels (DALs) as: catastrophic, hazardous, major, minor, and no safety effect. Each level defines a set of objectives to be satisfied and requires document traceability between the certification artifacts [43]. Specifically, during the last two decades, new standards have been defined to enable the development of safety-critical systems, including CPS, and have advocated for model-based development approaches.

6.5.1   Model-Based Assurance for Test and Evaluation

Hughues and Delange explore model-based design and automated validation of differing aircraft architectures using the Architecture Analysis and Design Language (AADL). In the research, model-based assurance is pursued through leveraging AADL language to capture the system architecture [44]. The system architecture captured is processed and validated against the system requirements; however, the team found it challenging to analyze due to the large quantity of inter-dependent results. Therefore, the team extended the analysis tool and auto-generated an assurance case from the validation results. This, in turn, shows the interdependencies of each requirement using a hierarchical notation and details which are not enforced.

In 2016, authors from Carnegie Mellon University proposed utilizing ModelPlex, a method ensuring that verification results about models, to apply to

CPS implementation, utilized as a method of formal verification and validation [45]. ModelPlex, according the authors, provides correctness guarantees for CPS executions at runtime: combining offline verification of CPS models with runtime validation of system executions for compliance with the model, and then initiates provably safe fallback actions. The team also developed a systematic technique to synthesize provably correct monitors automatically that form CPS proofs in differential dynamic logic by a correct-by-construction approach. The ModelPlex is a principle to build and verify high-assurance controllers for safety-critical computerized systems that interact physically to their environment, and also monitors relate states. The states are semantic objects and cannot be represented precisely in a program, but the research team developed a systematic logical characterization as syntactic expressions for the relations to monitor the conditions through computable programs.

Researchers designed and implemented a novel intrusion detection and response scheme to test, evaluate, and detect malicious anomalies that threaten unmanned aerial vehicle (UAV) networks [46]. They propose detection and response techniques to monitor the UAV behaviors and categorize them into the appropriate list (normal, abnormal, suspect, and malicious) according to the detected cyber attack. The team only focused on the most lethal cyber attacks that can target a UAV network, which are false information dissemination, CPS spoofing, jamming, and black hole and gray hole attacks. The hierarchical scheme relies on two mechanisms: an intrusion detection mechanism running at the UAV node level and an intrusion response mechanism running at the ground station level. The research team analyzed the performance of the scheme using NS-3, and the proposed hierarchical intrusion detection and response scheme demonstrated a high-level of security with high detection rate (more than 93%) and low false positive rate (less than 3%).

Another rising schema for identification of failures and attacks has been merging differing models that will assist in system assurance in the design of resilient CPS. For instance, the six-step model was created for CPS safety and security analysis, which incorporates six hierarchies of CPS. The six hierarchies include functions, structure, failures, safety countermeasures, cyber attacks, and security countermeasures. Relationship matrices are used to identify the interrelationships between hierarchies and determine the effect of failures and cyber attacks on CPSs. The SSM is based on two previously developed approaches: GTST-MLD and the three-step model, established in 1999 and 2009, respectively [47, 48]. Research on an approach for integrating six-step model (SSM) with informational flow diagrams (IFDs) was proposed in 2017 since SSM, while useful for modeling CPS safety and security, lacks guidance for identification of failures and attacks and selecting adequate countermeasures [49]. IFDs are behavioral diagrams used for information flow modeling and have been used for complex CPS safety analysis. While IFDs were used for CPS confidentiality analysis, for instance, Akella et al. proposed a formal method that expressed the information flow security semantics in CPS [50], Sabaliauskaite proposed an extended IFD by converting connections into information flows and adding flow direction, content, and transmission frequency.

The IFDs could be used for designing security countermeasures to achieve required level of security as well, especially since the SSM step 6 often requires use of additional information flows. While this strategy for comprehensive analysis of CPS safety and security can provide guidance for identifying possible failures and cyber attacks as well as selecting safety and security countermeasures, the applicability of the approach is only expanded upon using one example.

6.5.2 Formal Methods Transfer to CPS

Rapid integration is currently going on between the formal methods community and the control engineering community. The ERATO MMSD project, initiated in 2016 and proposed through 2022, funded by Japan Science and Technology Agency (JST) aims at enhanced quality assurance measures for CPS, and to contribute by taking a unique mathematical strategy to bridging gaps through finding a meta-level theory by transferring formal methods for quality assurance originally developed for software to quality assurance of the CPS in modern-day industry [51]. The researchers provide a list of concrete challenges that one is likely to encounter when solely following the V-model for software requirements. Some being the cost of hardware testing/simulation, correctness of designs and requirements, management of those designs and requirements, and optimization of complex systems. While these challenges are already present in computer system, and formal methods have already been developed to aid on the verification, synthesis, and specification, there are still challenges in the broader discipline of CPS. The methodology aims to provide a meta-mathematical transfer through nonstandard transfer from discrete to continuous/hybrid and from automata to coalgebras to move formal methods to heterogeneity. This includes coalgebraic model checking that unifies qualitative and quantitative aspects, quantitative semantics for enhanced expressivity, simulation and bi-simulation notions, compositionality, and collaboration and integration with control theory and robotics.

Several methodologies and frameworks have been proposed for system assurance test and evaluation in the design of resilient CPS. The amount of research into rigorous design of CPS has extensively increased over the past decade; however, a consistent and efficient model of the integration of the cyber and physical with the right level of fidelity for system design in its totality is still being researched [52]. The research states that we are far from reaching the desired degree of domain integration for the state-of-the-art CPS design. The authors believe that there is still the problem of writing "faithful and consistent models from networks of physical components." The paper attempted to perform a fair assessment for the CPS design and concluded that there exist some basic theoretical difficulties to be overcome by proposing methodologies adequately combining tool automation and designer ingenuity [52].

6.6   Summary

New approaches to address cybersecurity in development and operational use of CPS are an obvious concern and priority across many domains. Model-based approaches to requirements, design, and evaluation, especially those that can lead to use in the assurance process, must account for the complex interactions between threats, well-intended countermeasure patterns, and the resultant system behaviors that may or may not be what was intended. The overarching goal is to create methods that link to existing processes and answer how well the original functional capabilities of the cyber are preserved in the face of the threat(s) given the augmentation with the security design pattern(s). A dynamic graphical model will permit exploration and understanding of the structure–function relationships inherent in this ecosystem that produce those outcomes and link security choices to a trade-based decision process relating cost and level of security success. Additionally, such an approach provides the structure to visibly or quantifiably reveal the inherent diseconomies of scale that can result from overprotection.

The same functional model building activity dedicated to the analysis of counter-measure pattern effectiveness can also be used to build a test framework, essentially a separate functional harness that can be attached to portions of the functional model. Here, specific attack patterns are replaced by actual test functions (fuzzers) that evaluate vulnerabilities of the system. Threat patterns can be evaluated by test inputs that actually propagate through the system, and the test framework can simulate multiple attack inputs. In contrast to the system view, current methods typically test only one software component at a time. While this will not sufficiently address the inherent complexity of the system, it may provide a foundation from which the community can gradually build libraries of more complex system tests.

References

1. NSF. (2013). Cyber-physical systems. National Science Foundation (NSF) program solicitation 16-549, NSF document number nsf16549, March 4, 2016. [online] Retrieved June 1, 2017, from https://www.nsf.gov/publications/pub_summ.jsp?ods_key=nsf16549
2. NIST. (2016). National Institute for Standards and Technology (NIST) Framework for Cyber-Physical Systems Release 1.0: Cyber Physical Systems Public Working Group (Rep.). May 2016. Retrieved June 1, 2017, from https://pages.nist.gov/cpspwg/
3. Griffor, E. (Ed.). (2016). *Handbook of system safety and security: Cyber risk and risk management, cyber security, adversary modeling, threat analysis, business of safety, functional safety, software systems, and cyber physical systems.* Cambridge, MA: Syngress.
4. Avižienis, A., Laprie, J., Randell, B., & Landwehr, C. (2004). Basic concepts and taxonomy of dependable and secure computing. *IEEE Transactions on Dependable and Secure Computing, 1*(1), 11–22.
5. DoDI. (2014). Department of Defense Instruction (DoDI) 8500.01, Cybersecurity. March 14, 2014.
6. Reed, M. (2016). *DoD Strategy for Cyber Resilient Weapon Systems.* In Paper presented at the National Defense Industries Association, Annual Systems Engineering Conference, Alexandria VA, October 2016.

7. Boehm, B., & Kukreja, N. (2015). An initial ontology for system qualities. *INCOSE International Symposium, 25*(1), 341–356.
8. Newman, M., Barabasi, A., & Watts, D. (2011). *The structure and dynamics of networks.* Princeton, NJ: Princeton University Press.
9. Geard, N. (2010). In T. Gross & H. Sayama (Eds.), *Adaptive networks: Theory, models and applications.* Berlin: Springer-Verlag.
10. NATO. (2010). *North Atlantic Treaty Organization (NATO), engineering for system assurance in NATO programs.* Washington, DC: NATO Standardization Agency. DoD 5220.22M-NISPOM-NATO-AEP-67, February 2010.
11. Hilburn, T., Ardis, M., Johnson, G., Kornecki, A., & Mead, N. (2013). *Software assurance competency model.* Pittsburgh, PA: Software Engineering Institute, Carnegie Mellon University. Technical Note CMU/SEI-2013-TN-004, 2013. Retrieved October 1, 2018, from http://resources.sei.cmu.edu/library/asset-view.cfm?AssetID=47953
12. McDermott, T., & Horowitz, B. (2017). Human Capital Development – Resilient Cyber Physical Systems. Systems Engineering Research Center (SERC) Technical Report SERC-2017-TR-075, September 29, 2017. Retrieved October 1, 2018, from https://sercuarc.org/publication/?id=163&pub-type=Technical-Report&publication=SERC-2017-TR-113-Human+Capital+Development+%E2%80%93+Resilient+Cyber+Physical+Systems
13. Wan, J., Canedo, A., & Al Faruque, M. (2015). Security-aware functional modeling of cyber-physical systems. In *2015 IEEE 20th International Conference on Emerging Technology & Factory Automation (ETFA) 2015* (pp. 1–4).
14. Rashid, N., Wan, J., Quiros, G., Canedo, A., & Al Faruque, M. (2017). Modeling and simulation of cyberattacks for resilient cyber-physical systems. In *13th IEEE Conference on Automation Science and Engineering (CASE) 2017* (pp. 988–993).
15. Benner, L. (1975). Accident investigations: Multilinear events sequencing methods. *Journal of Safety Research, 7*(2), 67–73. 3.
16. Leveson, N. (2012). *Engineering a safer world: Systems thinking applied to safety* (p. 13). Cambridge, MA: MIT Press.
17. Goldman, H. (2010, November). *Building secure, resilient architectures for cyber mission assurance.* McLean, VA: The MITRE Corporation.
18. Young, W., & Leveson, N. (2013). Systems thinking for safety and security. In *Proceedings of the 29th Annual Computer Security Applications Conference (ACSAC '13)* (pp. 1–8). New York: ACM.
19. Lu, Y., Ferrese, F., & Labouliere, M. (2007) *Anti-threat mobile agent-based ship freshwater cooling system.* In Automation & Controls Symposium.
20. Lu, Y., Kuruganty, R., Al Faruque, M. A., Ren, Q., Zhang, W., & Scheidt, P. R. D. (2012). Risk based multi-agent chilled water control system for a more survivable naval ship. *International Journal of Intelligent Control and Systems, 17*(4), 102–112. 14.
21. Hirtz, J., Stone, R. B., Szykman, S., McAdams, D. A., & Wood, K. L. (2001). *Evolving a functional basis for engineering design.* In Proceedings of the ASME Design Engineering Technical Conference: DETC2001, Pittsburgh, PA.
22. Hirtz, J., Stone, R., McAdams, D., Szykman, S., & Wood, K. L. (2002). A functional basis for engineering design: Reconciling and evolving previous efforts. *Research in Engineering Design, 13*, 65. https://doi.org/10.1007/s00163-001-0008-3.
23. Wan, J., Canedo, A., & Al Faruque, M. (2014, December). Functional model-based design methodology for automotive cyber-physical systems. *IEEE Systems Journal, 11*(4), 2028–2039.
24. Wan, J., Canedo, A., & Al Faruque, M. (2015). Cyber-physical co-design at the functional-level for multi-domain automotive systems. *IEEE Systems Journal, 11*(4), 2949–2959.
25. Friedenthal, S., Moore, A., & Steiner, R. (2014). *A practical guide to SysML: The systems modeling language.* Amsterdam: Morgan Kaufmann.
26. Kruse, B., Gilz, T., Shea, K., & Eigner, M. (2014). Systematic comparison of functional models in SysML for design library evaluation. *Procedia CIRP, 21*, 34–39.

27. Weilkiens, T. (2011). *Systems engineering with SysML/UML: Modeling, analysis, design.* Burlington, MA: Morgan Kaufmann.
28. Li, L. (2007). *Topologies of complex networks: Functions and structures.* Pasadena, CA: California Institute of Technology.
29. Baresi, L., & Heckel, R. (2002). Tutorial introduction to graph transformation: A software engineering perspective. In *International Conference on Graph Transformation.* Berlin: Springer.
30. Ehrig, H., Rozenberg, G., & Kreowski, H. (1999). *Handbook of graph grammars and computing by graph transformation* (Vol. 3). London: World Scientific.
31. Karsai, G., Agrawal, A., Shi, F., & Sprinkle, J. (2003). On the use of graph transformation in the formal specification of model interpreters. *J. UCS, 9*(11), 1296–1321.
32. Plasmeijer, R., Van Eekelen, M., & Plasmeijer, M. (1993). *Functional programming and parallel graph rewriting* (Vol. 857). Reading, MA: Addison-Wesley.
33. Manadhata, P., Tan, K. M., Maxion, R. A., & Wing, J. M. (2007). *An approach to measuring a system's attack surface. No. CMU-CS-07-146.* Pittsburg, PA: Carnegie-Mellon University, School of Computer Science.
34. Sheyner, O., Haines, J., Jha, S., Lippmann, R., & Wing, J. (2002). Automated generation and analysis of attack graphs. In *Proceedings of the 2002 IEEE Symposium on Security and Privacy (SP '02).* Washington, DC: IEEE Computer Society.
35. Apvrille, L., & Roudier, Y. (2015). SysML-sec attack graphs: Compact representations for complex attacks. In *International Workshop on Graphical Models for Security.* Cham: Springer.
36. Luckett, B. (2013). Integration of graphical modeling techniques as a structural framework for system-aware cyber security architecture selection. Thesis from http://libra.virginia.edu/catalog/libra-oa:3720
37. Aguilar, J. (2009, June 4). Design assurance guide. aerospace.wpengine.netdna-cdn.com/wp-content/uploads/2015/04/TOR-20098591-11-Design-Assurance-Guide.pdf. Accessed online via DTIC, 12 Nov 2018.
38. Caslake, S. (1974). Quality assurance. *IEEE Transactions on Nuclear Science, 21*(1), 1974. https://doi.org/10.1109/TNS.1974.4327589.
39. Rachowitz, B., Maue, R. K., Angrisano, N. P., & Abramson, B. (1991). A guide to engineering workstations: Using workstations efficiently. *IEEE Spectrum, 28*(4), 38–40. https://doi.org/10.1109/6.76301.
40. Alberts, C, Ellison, R, & Woody, C (2009). Cyber assurance. 2009 CERT Research Report. Software Engineering Institute, Carnegie Mellon University. Available at http://resources.sei.cmu.edu/library/asset-view.cfm?assetid=77638
41. Brooks, T. (2018). *Cyber-assurance for the internet of things.* New York: Wiley. Accessed 2018.
42. Wolf, M., & Dimitrios, S. (2018). Safety and security in cyber-physical systems and internet-of-things systems. *Proceedings of the IEEE, 106*(1), 9–20. https://doi.org/10.1109/JPROC.2017.2781198.
43. Pothon, F. (2012). DO-178C/ED-12C versus DO-178B/ED-12B Changes and Improvements. www.adacore.com/uploads/books/pdf/DO178C-ED12C-Changes_and_Improvements-Sep2012.pdf. Report generated from ACG Solution on the new update to the standards.
44. Nakajima, S., Talpin, J. P., Toyoshima, M., & Yu, H. (Eds.). (2018). *Cyber-physical system design from an architecture analysis viewpoint: Communications of NII Shonan meetings* (Vol. 2017). Singapore: Springer.
45. Mitsch, S., & Platzer, A. (2016). Modelplex: Verified runtime validation of verified cyber-physical system models. *Formal Methods in System Design, 49*(1–2), 33–74. https://doi.org/10.1007/s10703-016-0241-z.
46. Sedjelmaci, H., Senouci, S. M., & Ansari, N. (2018). A hierarchical detection and response system to enhance security against lethal cyber attacks in UAV networks. *IEEE Transactions on Systems, Man & Cybernetics. Systems, 48*(9), 1594–1606.
47. Brissaud, F., Barros, A., Be'renguer, C., & Charpentier, D. (2009). Reliability study of an intelligent transmitter. In *15th IS- SAT International Conference on Reliability and Quality in Design.* (pp. 224–233). International Society of Science and Applied Technologies.

48. Modarres, M., & Cheon, S. (1999). Function-centered modeling of engineering systems using the goal tree–success tree technique and functional primitives. *Reliability Engineering & System Safety, 64*(2), 181–200.
49. Sabaliauskaite, G., & Adepu, S. (2017). Integrating six-step model with information flow diagrams for comprehensive analysis of cyber-physical system safety and security. In *Proceedings of IEEE International Symposium on High Assurance Systems Engineering* (pp. 41–48). https://doi.org/10.1109/HASE.2017.25.
50. Akella, R., Tang, H., & McMillin, B. (2010). Analysis of information flow security in cyber-physical systems. *International Journal of Critical Infrastructure Protection, 3*(3–4), 157–173.
51. Hasuo, I. (2017). Metamathematics for systems design: Comprehensive transfer of formal methods techniques to cyber-physical systems. *New Generation Computing, 1-35*, 1–35. https://doi.org/10.1007/s00354-017-0023-1.
52. Bliudze, S., Furic, S., Sifakis, J., & Viel, A. (2017). Rigorous design of cyber-physical systems. *Software & Systems Modeling, 2*(2), 1–24. https://doi.org/10.1007/s10270-017-0642-5.

Part III
Application-Specific Design Automation Methodologies and Tools

Chapter 7
Optimal Design of Distributed Controllers for Large-Scale Cyber-Physical Systems

Amer Mešanović, Xiaofan Wu, Simone Schuler, Ulrich Münz, Florian Dörfler, and Rolf Findeisen

7.1 Introduction

Cyber-Physical Systems (CPSs), such as power systems, can have thousands of components and span large geographical areas, e.g., the whole of Europe, or the western interconnection in the USA. Modern life as we know it today strongly depends on such systems, and they are critical for the operation of many other systems like transportation systems, factories, hospitals, etc. Centralized control of large power systems is impractical and challenging due to the need for fast and reliable communication between all components, possibly located far from each other. Additionally, a centralized entity would need to be trusted by all participants, thereby raising privacy concerns. Due to reliance on the centralized controller,

A. Mešanović (✉)
Siemens AG, Munich, Germany
e-mail: amer.mesanovic@siemens.com

X. Wu · U. Münz
Siemens Corp., Princeton, USA
e-mail: xiaofan.wu@siemens.com; ulrich.muenz@siemens.com

S. Schuler
GE Renewable energy, Munich, Germany
e-mail: simone@schulers-post.de

F. Dörfler
ETH Zurich, Zurich, Germany
e-mail: doerfler@control.ee.ethz.ch

R. Findeisen
Otto-von-Guericke University Magdeburg, Magdeburg, Germany
e-mail: rolf.findeisen@ovgu.de

© Springer Nature Switzerland AG 2019
M. A. Al Faruque, A. Canedo (eds.), *Design Automation of Cyber-Physical Systems*, https://doi.org/10.1007/978-3-030-13050-3_7

achieving satisfying robustness is also a challenge. Consequently, various distributed control methods are developed, which can cope with such large distributed systems.

We specifically focus on the control of frequency oscillations in power systems in the frequency range of 0.5–10 Hz between groups of power plants, which can occur across large distances. Sufficient damping of these oscillations is necessary for power system stability, resilience, as well as optimal operation [1]. Power oscillations in power systems result from a complex interplay between heterogeneous machine dynamics, sparsely connected grid topologies, and large inter-area power transfers [1, 2]. They are currently handled by special controllers, so-called Power System Stabilizers (PSSs). PSSs are local controllers, present in some power plants to improve Power Oscillation Damping (POD). Currently in practice, PSSs are tuned manually, without directly considering other power plants in the system. The process of tuning is iterative and can last up to several months for a single PSS. Typically, it is done during installation and not adapted until a problem in the system occurs. Such manual tuning is successful as long as the network and power plants do not change. Minor variations are present due to periodic load fluctuations, which are predictable.

This is challenged by the increasing amount of renewable, "low-inertia" generation [3], leading to large fluctuations in power systems operation. Depending on the weather conditions, renewable generation changes constantly and will shift geographically across different areas. Furthermore, if the weather conditions are not suitable for renewable generation, the percentage of conventional generation will need to be increased. The constantly shifting mixture of renewable and conventional generation leads to time-varying oscillatory modes in the system [4, 5]. If not handled accordingly, the PSSs, which are tuned today for fixed oscillatory modes, may become less and less effective, and the risk of a blackout increases [6]. Thus, new control and operation methods are necessary in order to improve the robustness of electric power systems.

Existing analysis and control methods of inter-area oscillations are based on the different modal approaches. Typical methods are $\mathcal{H}_\infty$ optimization, $\mathcal{H}_2$ optimization, and pole placement for controller synthesis, see, e.g., [7–12]. Other approaches are sensitivity analysis [13–15], sliding mode controller design [16], the use of reference models [17], coordinated switching controllers [18], genetic algorithms based tuning [19], and model predictive control [20]. An overview of different methods for power oscillation damping can, for example, be found in [21].

We focus on three approaches using $\mathcal{H}_\infty$ and $\mathcal{H}_2$ distributed controller synthesis for POD in power systems and compare their performance and resulting controller interconnections. Note that there are many other approaches for this purpose; however, they cannot be included due to length limitations. The first approach [22] is based on $\mathcal{H}_\infty$ optimization to tune the parameters of existing local structured controllers to the current system state, i.e., the local controllers do not exchange information. The second [23–26] and third approach [27] allow for communication between the controllers by introducing wide-area state-feedback controllers to the system which are designed using $\mathcal{H}_2$ and $\mathcal{H}_\infty$ optimization. They enable the consideration of the trade-off between system performance and sparsity of the required communication links for the control scheme.

The remainder of the chapter is organized as follows: first we introduce the considered distributed control structures for large CPS in Sect. 7.2, followed by a review of $\mathcal{H}_\infty$ and $\mathcal{H}_2$ norms in Sect. 7.3. Section 7.4 outlines the three approaches for distributed controller synthesis. Finally, the considered power system models are outlined in Sect. 7.5. The three approaches are then applied and compared on an exemplary power system in Sect. 7.6, before summarizing the chapter in Sect. 7.7.

7.2   Control Architectures in Large-Scale Cyber-Physical Systems

We first introduce an abstract representation of the considered control problem. This will be refined for power systems in Sect. 7.5. We consider linear time-invariant systems $H(s)$ consisting of N_s physical interconnected subsystems S_i. The model of the ith subsystem is given by

$$S_i : \quad \dot{\mathbf{x}}_i = A_{ii}(\mathbf{K}_{ti})\mathbf{x}_i + \sum_{\substack{j=1, \\ j\neq i}}^{N_s} A_{ij}(\mathbf{K}_{ti}, \mathbf{K}_{tj})\mathbf{x}_j + B_{wi}(\mathbf{K}_{ti})\mathbf{w} + B_{ui}(\mathbf{K}_{ti})\mathbf{u}_i$$

$$(7.1a)$$

$$\mathbf{y}_i = C_i\mathbf{x}_i + D_{wi}\mathbf{w}_i + D_{ui}\mathbf{u}_i,\tag{7.1b}$$

where $\mathbf{x}_i$ denotes the state vector of the ith subsystem, $\mathbf{w}$ is the disturbance input vector acting on all subsystems through the physical interconnection, $\mathbf{u}_i$ are the control inputs of the ith subsystem, and $\mathbf{y}_i$ is the performance output of the ith subsystem. We consider that the subsystems already possess local structured controllers, i.e., local controllers with a specified structure and which only exploit local measurements. While the structure is fixed, the parameters of these controllers can be tuned, and the vector of tunable controller parameters is denoted with $\mathbf{K}_{ti}$, leading to parameter-dependent state matrices $A_{ii}(\mathbf{K}_{ti})$. The interconnection matrices $A_{ij}(\mathbf{K}_{ti}, \mathbf{K}_{tj})$ between S_i and S_j are, in general, functions of the local controller parameter vectors $\mathbf{K}_{ti}$ and $\mathbf{K}_{tj}$. Note that the $A_{ij}(\mathbf{K}_{ti}, \mathbf{K}_{tj})$ are not necessarily zero or sparse matrices, leading to possibly many interconnections between the subsystems. Figure 7.1a shows an exemplary structure of the described system consisting of three subsystems S_1, S_2, and S_3. Combining the equations from each subsystem, one obtains the model of the whole system

$$\dot{\mathbf{x}} = A(\mathbf{K}_t)\mathbf{x} + B_w(\mathbf{K}_t)\mathbf{w} + B_u(\mathbf{K}_t)\mathbf{u}\tag{7.2a}$$

$$\mathbf{y} = C\mathbf{x} + D_w\mathbf{w} + D_u\mathbf{u}.\tag{7.2b}$$

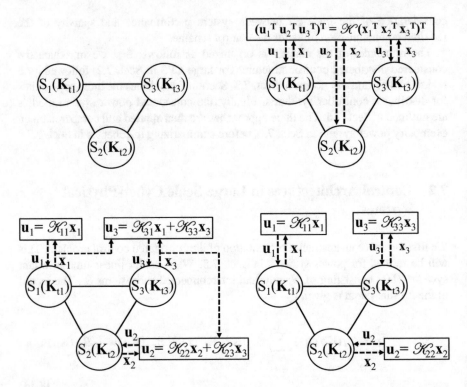

Fig. 7.1 Considered control architectures with different levels of decentralization. Solid lines between the systems refer to physical links. Dashed lines are communication links. (**a**) Decentralized control with sole local structured controllers. (**b**) Fully centralized static control with local control. (**c**) Partially decentralized (distributed) static control with local control. (**d**) Fully decentralized static control with local control

Distributed control of system (7.2) can be separated into several levels of decentralization. We explore these levels on the example of static state-feedback controller synthesis for the control input $\mathbf{u}$, i.e., only static controllers are introduced to the system. These controllers do not have any internal states, and take as input the state vector of the system $\mathbf{x}$, whereas the output of the controller is the control input vector to the system $\mathbf{u}$ given by

$$\mathbf{u} = \left(\mathbf{u}_1^T \dots \mathbf{u}_{N_s}^T\right)^T = \mathscr{K}\left(\mathbf{x}_1^T \dots \mathbf{x}_{N_s}^T\right)^T = \mathscr{K}\mathbf{x}, \qquad (7.3)$$

where $\mathscr{K} \in \mathbb{R}^{n_u \times n_x}$ is the distributed controller gain matrix, n_u is the number of controller inputs, and n_x is the total number of system states. Thereby, we use the calligraphic $\mathscr{K}$ to denote the gain matrix for distributed control, and the boldface $\mathbf{K}$ to denote the parameter vector of local controllers. The term "static" originates from the fact that $\mathscr{K}$ is a real-valued matrix, and not, e.g., a transfer matrix with internal states. Centralized controller synthesis using the local control inputs $\mathbf{u}_i$ is shown in Fig. 7.1b. One obtains a centralized controller which uses the states from all subsystems to calculate the control input for all subsystems, i.e., $\mathbf{u}_i = \sum_{i=1}^{N_S} \mathcal{K}_{ij}\mathbf{x}_j$, where $\mathcal{K}_{ij}$ is the appropriate sub-matrix of $\mathcal{K}$. If one wants to reduce the number of necessary communication links between the subsystems, it is desirable to SET some of the matrices $\mathcal{K}_{ij}$ to zero. In this way, the technical requirements for the realization of such a control law, in terms of required communication links, are lowered, c.f. Fig. 7.1c. Such sparse and communication-based controllers are often referred to as distributed controllers. In the limit, all off-diagonal elements $\mathcal{K}_{ij}$, with $i \neq j$, become zero, and one obtains a fully decentralized control law, as illustrated in Fig. 7.1d combining local controllers consisting of $\mathbf{K}_{ti}$ and $\mathcal{K}_{ii}$. To summarize, a distributed system can have decentralization levels ranging from fully distributed control, as shown in Fig. 7.1a, d, various levels of partially distributed control, as shown in Fig. 7.1c, to a fully centralized control solution, as shown in Fig. 7.1b.

The question arises, however, how the elements $[\mathcal{K}]_{ij}$ of the distributed controller $\mathcal{K}$ should be chosen and which should be set to zero to achieve a desired performance with reduced communication. In general, this is a challenging problem. Two methods presented in Sects. 7.4.2 and 7.4.3 give a (sub)optimal answer to this question. We also improve the system performance by tuning only the parameters of existing local controllers of the subsystems $\mathbf{K}_{ti}$. For this purpose, the method in Sect. 7.4.1 is introduced and all three methods are evaluated in Sect. 7.6.

7.3 Norms of Linear Systems

In order to objectively compare different methods for controller synthesis, we first need numerical performance metrics which can be used as a minimization objective. The methods presented subsequently in Sect. 7.4 minimize either the $\mathcal{H}_\infty$ or the $\mathcal{H}_2$ norm of a linear system. For this reason, we briefly revise the definition of these two norms on the linear time-invariant (LTI) system H defined as

$$H : \begin{cases} \dot{x} = A_H x + B_H w \\ y = C_H x + D_H w \end{cases}, \tag{7.4}$$

and $H(s) = C(sI - A)^{-1}B + D$.

The $\mathcal{H}_\infty$ norm of a linear system is defined as the maximal amplification of amplitude of any harmonic input signal in any output direction. Thus, minimizing the $\mathcal{H}_\infty$ norm minimizes the amplitude amplification of all oscillation frequencies after, e.g., a load step. Other interpretations of the $\mathcal{H}_\infty$ norm for LTI systems are that it represents the $\mathcal{L}_2$-gain or the power-gain of the system. It is defined as [28]

$$\|H(s)\|_\infty = \sup_{\omega \in \mathbb{R}} \overline{\sigma}\left(H(j\omega)\right), \tag{7.5}$$

where $\bar{\sigma}\,(H(j\omega))$ is the largest singular value of the transfer matrix $H(j\omega)$ for a given frequency ω. The so-called bounded real lemma provides a useful tool for the determination of the $\mathscr{H}_\infty$ norm of linear systems.

Lemma 7.1 ([29]) *Consider the continuous-time transfer function $H(s)$. The following statements are equivalent:*

- *The system $H(s)$ is asymptotically stable and $\|H(s)\|_\infty < \gamma$.*
- *There exists a symmetric positive definite solution $P \succ 0$ to the linear matrix inequality (LMI)*

$$\begin{pmatrix} A_H^T P + P A_H & P B_H & C_H^T \\ B_H^T P & -\gamma I & D_H^T \\ C_H & D_H & -\gamma I \end{pmatrix} \prec 0, \tag{7.6}$$

where $\prec$ and $\succ$ are used to denote the negative- and positive-definiteness of a matrix, respectively.

The $\mathscr{H}_2$ norm is the second considered cost function for oscillation damping. One interpretation of the $\mathscr{H}_2$ norm is that it represents the total output energy of the system response after an impulse in the input. Another interpretation is that the $\mathscr{H}_2$ represents the amplification of white stochastic disturbances from the input to the output. The $\mathscr{H}_2$ norm of a linear system is defined as

$$\|H(j\omega)\|_2 = \sqrt{\int_{-\infty}^{\infty} \text{trace}\,(H(j\omega)^* H(j\omega))\,\mathrm{d}\omega} = \sqrt{\text{trace}(B_H^T P B_H)}, \tag{7.7}$$

where P is the observability Gramian of H, see [28]. A more detailed discussion about the use of the $\mathscr{H}_2$ norm, and its use in power systems, can be found in [23, 28].

In case of single-input single-output (SISO) systems, the $\mathscr{H}_\infty$ norm is the maximum of the system Bode magnitude plot, i.e., the maximal amplitude amplification, and the $\mathscr{H}_2$ norm graphically represents the area underneath the Bode magnitude plot.

7.4 Distributed Controller Design Methods for Large-Scale CPS

This section describes three optimal design methods for distributed controllers, which are based on $\mathscr{H}_\infty$ and $\mathscr{H}_2$ optimal control synthesis.

The first method, presented in [22], minimizes the $\mathscr{H}_\infty$ norm of a linear system by tuning the parameters of structured local controllers. This corresponds to tuning the parameters $\mathbf{K}_{ti}$ in Fig. 7.1a. With this method, no new controllers are added to the system, only the existing local controllers are used more efficiently. We call this method "structured $\mathcal{H}_\infty$ controller synthesis of local controllers" subsequently (SHinf), and it is described in more detail in Sect. 7.4.1.

The second [23] and third [27] method introduce an additional higher-level control layer. They use $\mathcal{H}_2$ and $\mathcal{H}_\infty$ controller synthesis methods to design optimal static and distributed state-feedback controllers. This additional control layer requires fast communication of all system states to the controller, see Fig. 7.1b. This can be challenging in large CPSs, such as power systems, because fast, reliable, and time-synchronized communication across large distances is needed. In order to minimize the dependency on fast communication, these methods also minimize the number of communication links necessary for the state feedback. Depending on the desired performance, they achieve a control architecture similar to the one in Fig. 7.1c, or a fully decentralized architecture as in Fig. 7.1d. We call these methods subsequently "sparsity-promoting $\mathcal{H}_2$ controller synthesis" (SPH2), presented in Sect. 7.4.2, and "sparsity-promoting $\mathcal{H}_\infty$ controller synthesis" (SPHinf), presented in Sect. 7.4.3.

7.4.1 Structured $\mathcal{H}_\infty$ Controller Synthesis of Local Controllers (SHinf)

This section summarizes the controller tuning method presented in [22]. This method minimizes the $\mathcal{H}_\infty$ norm of system (7.2) by tuning the parameters of local structured controllers $\mathbf{K}_{ti}$, $i = 1...N_S$. For this method, the additional control input $\mathbf{u}$ is not used, and we can reduce (7.2) to

$$\mathbf{x} = A(\mathbf{K}_t)\mathbf{x} + B_w(\mathbf{K}_t)\mathbf{w} \tag{7.8a}$$

$$\mathbf{y} = C\mathbf{x} + D_w\mathbf{w}. \tag{7.8b}$$

We denote this system with $H_1(s, \mathbf{K}_t) = D_w + C(sI - A(\mathbf{K}_t))^{-1} B_w(\mathbf{K}_t)$. We drop the dependency on the complex variable s subsequently for notational ease and only write $H_1(\mathbf{K}_t)$. The controller tuning approach considered here is based on the bounded real lemma (Lemma 7.1). It enables us to formulate the theorem for the solution of the controller tuning problem of existing local controllers [22].

Theorem 7.1 *Given the system* $H_1(\mathbf{K}_t)$, *(7.8), and box constraints* $\mathbf{K}_{t,\min}$ *and* $\mathbf{K}_{t,\max}$ *on the controller parameters* $\mathbf{K}_t$. *Then* $\|H_1(\mathbf{K}_t)\|_\infty$ *is minimized, while keeping* $H_1(\mathbf{K}_t)$ *asymptotically stable, by solving the following optimization problem:*

$$\min_{P \in \mathbb{R}^{N_{st} \times N_S}, \mathbf{K}_t \in \mathbb{R}^{N_t}, \gamma \in \mathbb{R}} \gamma \tag{7.9a}$$

$$\text{s.t.} \quad \Phi(\gamma, \mathbf{K}_t, P_\mu) = \begin{pmatrix} A^T(\mathbf{K}_t)P + PA(\mathbf{K}_t) & PB_w(\mathbf{K}_t) & C^T \\ B_w^T(\mathbf{K}_t)P & -\gamma I & D_w^T \\ C & D_w & -\gamma I \end{pmatrix} \prec 0 \tag{7.9b}$$

$$\mathbf{K}_{t,\min} \le \mathbf{K}_t \le \mathbf{K}_{t,\max} \tag{7.9c}$$

$$P = P^T \succ 0. \tag{7.9d}$$

Note that (7.9) is non-convex due to the nonlinear parameter dependency in the system matrices $A(\mathbf{K}_t)$ and $B_w(\mathbf{K}_t)$, and the non-convex term (7.9b). Standard structured controller synthesis typically only considers linear parameter dependency with no box constraints or structural constraints on the controller parameters. The typical synthesis problem is solved by introducing a variable substitution [29]; however, this cannot be done in this case.

We now reformulate (7.9) in order to solve it iteratively in the linear matrix inequality framework. This is done in the μth iteration by first setting the parameter vector $\mathbf{K}_t$ in (7.9) to the value from the previous iteration $\mathbf{K}_{t,\mu-1}$. In this way, we obtain the Lyapunov matrix P_μ in the μth iteration

$$\min_{P_\mu, \gamma} \quad \gamma \tag{7.10a}$$

$$\text{s.t.} \quad \Phi(\gamma, \mathbf{K}_{t,\mu-1}, P_\mu) \prec 0 \tag{7.10b}$$

$$P_\mu = P_\mu^T \succ 0. \tag{7.10c}$$

In order to obtain $\mathbf{K}_\mu$, we need to fix P in (7.9) to P_μ. This is conceptually similar to the classic D-K-iteration. However, this problem is still non-convex due to the nonlinear parameter dependency in $A(\mathbf{K}_t)$ and $B(\mathbf{K}_t)$. For this reason, we linearize $A(\mathbf{K}_t)$ and $B(\mathbf{K}_t)$ around the parameter vector from the previous iteration $\mathbf{K}_{t,\mu-1}$

$$A_L(\mathbf{K}_t) = A(\mathbf{K}_{t,\mu-1}) + \left.\frac{\partial A(\mathbf{K}_t)}{\partial \mathbf{K}_t}\right|_{\mathbf{K}_{t,\mu-1}} (\mathbf{K}_l - \mathbf{K}_{t,\mu-1}), \tag{7.11}$$

and similarly for $B(\mathbf{K}_t)$ to obtain $B_L(\mathbf{K}_t)$. The following convex optimization problem results from these operations:

$$\min_{\mathbf{K}_{t,\mu}, \gamma} \quad \gamma \tag{7.12a}$$

$$\text{s.t.} \quad \Phi_L(\gamma, \mathbf{K}_{t,\mu}, P_\mu) \prec 0 \tag{7.12b}$$

$$\mathbf{K}_{t,\min} \le \mathbf{K}_{t,\mu} \le \mathbf{K}_{t,\max} \tag{7.12c}$$

$$\left| \mathbf{K}_{t,\mu} - \mathbf{K}_{t,\mu-1} \right| \le \Delta K, \tag{7.12d}$$

where $|.|$ is defined component-wise for vectors, and M_L is defined as

$$\Phi_L(\gamma, \mathbf{K}_{t,\mu}, P_\mu) = \begin{pmatrix} A_L^T(\mathbf{K}_t)P_\mu + P_\mu A_L(\mathbf{K}_t) & P_\mu B_{w,L}(\mathbf{K}_t) & C^T \\ B_{w,L}^T(\mathbf{K}_t)P_\mu & -\gamma I & D_w^T \\ C & D_w & -\gamma I \end{pmatrix}.$$

```
 1: procedure SHINF($H_1(\mathbf{K}_t), \mathbf{K}_{t,0}$)
 2:     $\mu = 1$
 3:     while stopping criteria not satisfied do
 4:         $(P_\mu, \gamma) \leftarrow$ Solution of (7.10)
 5:         $H_{1L}(\mathbf{K}_t) \leftarrow$ Linearization of $H_1(\mathbf{K}_t)$ around $\mathbf{K}_{t,\mu-1}$
 6:         $(\mathbf{K}_{t,\mu}, \gamma) \leftarrow$ Solution of (7.12)
 7:         $\mu \leftarrow \mu + 1$
 8:     end while
 9:     return $(\mathbf{K}_{t,\mu}, \gamma)$
10: end procedure
```

Fig. 7.2 Optimization algorithm for solving (7.9)

Constraint (7.12d) is added to (7.12) to define a trust region in which the linearization accuracy is preserved by limiting how much one can move away from the linearization point in one iteration.

Figure 7.2 presents the optimization algorithm for the iterative coordinate descent method. The inputs are the system $H_1(\mathbf{K}_t)$, and the initial parameter vector $\mathbf{K}_{t,0}$. The iteration is repeated until a stopping criterion is satisfied, e.g., the number of iterations reaches the specified limit, or the improvement of the $\mathscr{H}_\infty$ norm is smaller than the specified limit, etc. Results of the numerical evaluation for this system are presented in Sect. 7.6.

7.4.2 Sparsity-Promoting $\mathscr{H}_2$ Controller Synthesis (SPH2)

This section summarizes the sparsity-promoting optimal control design method presented in [23, 24]. This method aims to minimize the $\mathscr{H}_2$ norm of system (7.2) by introducing and designing the static state-feedback controller $\mathscr{K}_2$ using the sparsity-optimal control approach introduced in [24, 30]. The trade-off between sparsity and $\mathscr{H}_2$ performance is achieved by tuning of a sparsity emphasis parameter. When the sparsity emphasis parameter is set to zero, the algorithm recovers the optimal centralized controller, as shown in Fig. 7.1b. By appropriate selection of the sparsity emphasis parameter, control structures as shown in Fig. 7.1c, d can be obtained. For this method, we expand system (7.2) with the static state-feedback controller $\mathbf{u} = \mathscr{K}_2\mathbf{x}$, and we obtain

$$\dot{\mathbf{x}} = A\mathbf{x} + B_w\mathbf{w} + B_u\mathbf{u} \tag{7.13a}$$

$$\mathbf{y} = C\mathbf{x} + D_u\mathbf{u} \tag{7.13b}$$

$$\mathbf{u} = \mathscr{K}_2\mathbf{x}, \tag{7.13c}$$

where it is required that the matrix D_w is zero, because otherwise the $\mathscr{H}_2$ norm of the system would not be finite. We denote this system as

$$H_2(s) = (C + D_u \mathcal{K}_2) (sI - A - B_u \mathcal{K}_2)^{-1} B_w.$$

Note that A and B are also functions of the local controller parameters $\mathbf{K}_l$. These parameters, however, are not modified with this method and we drop this dependency in this chapter for notational convenience.

The $\mathcal{H}_2$ norm of H_2 can be calculated with [30]

$$J(\mathcal{K}_2) := \|H_2(s)\|_2 = \text{trace} \left(X (C^T C + \mathcal{K}_2^T D_u^T D_u \mathcal{K}_2) \right), \qquad (7.14)$$

where X is the solution to the equation

$$(A + B_c \mathcal{K}_2)X + X(A + B_2 \mathcal{K}_2)^T = -B_w B_w^T. \qquad (7.15)$$

A desired trade-off between the system's performance and the sparsity of $\mathcal{K}_2$ is achieved by solving the optimal control problem [30]

$$\min_{\mathcal{K}_2} \quad J(\mathcal{K}_2) + \alpha \|\text{vec}(\mathcal{K}_2)\|_0, \qquad (7.16)$$

where the notation $\text{vec}(\mathcal{K}_2)$ creates a vector from the elements of $\mathcal{K}_2$ by stacking all the columns of $\mathcal{K}_2$, and $\| \cdot \|_0$ denotes the 0-norm of a vector, i.e., the number of nonzero elements in the vector.[1] The first term of the objective function is smooth and we can use gradient-based methods to solve it. The second term of the objective function is non-smooth and non-convex, but has an explicit solution. The alternating direction method of multipliers (ADMM) takes advantage of different characteristics of the two terms. To obtain a smooth cost function, the problem is reformulated into

$$\min_{\mathcal{K}_2} \quad J(\mathcal{K}_2) + \alpha g(\mathcal{K}_2). \qquad (7.17)$$

The regularization term in (7.17) is given by the weighted ℓ_1-norm of $\mathcal{K}_2$,

$$g(\mathcal{K}_2) := \sum_{i,j} W_{ij} |\mathcal{K}_{2ij}|,$$

which is an effective proxy for inducing element-wise sparsity [31]. This problem is solved iteratively with ADMM. In each iteration, the weights W_{ij}'s are updated with the values of $\mathcal{K}_2$ from the previous iteration, see Fig. 7.3 and [31] for details. Problem (7.17) allows the synthesis of sparse state-feedback controllers. The sparsity level of the matrix $\mathcal{K}_2$ depends on the value of the sparsity emphasis parameter α. By varying α in (7.17), we can control the sparsity level which we want to achieve in the system. A larger value for α results in more elements of $\mathcal{K}_2$ being

---

[1]The formulation of (7.16) has been extended to promote block-sparsity in [24].

```
 1: procedure SPH2(H₂)
 2:     μ = 1
 3:     Wᵢⱼ = 1, ∀i, j
 4:     𝒦₂,₀ = dense feedback matrix
 5:     while stopping criteria not satisfied do
 6:         𝒦₂,μ ← solution of (1.17) via ADMM with 𝒦₂,μ₋₁ as initial value.
 7:         Wᵢⱼ ← 1/|𝒦₂,μ,ᵢⱼ|, ∀i, j
 8:         μ ← μ + 1
 9:     end while
10:     return (𝒦₂,μ)
11: end procedure
```

Fig. 7.3 Optimization algorithm of the SPH2 method for solving (7.17)

set to zero, however, at the cost of a worse $\mathcal{H}_2$ performance. With this method, we obtain a family of different distributed controllers with varying sparsity levels and performance. The control designer can then select one with an appropriate sparsity-performance trade-off.

The details of the sparsity-promoting optimal control algorithm can be found in [23]. We will perform the numerical evaluation of the SPH2 method in Sect. 7.6.

7.4.3 Sparsity-Promoting $\mathcal{H}_\infty$ Controller Synthesis (SPHinf)

In this section, we review the sparsity-promoting $\mathcal{H}_\infty$ controller synthesis presented in [27]. The objective of this method is to design a linear static feedback matrix $\mathcal{K}_\infty$ with as many zero entries as possible. This is similar to the SPH2 method; however, it achieves this with $\mathcal{H}_\infty$ optimization.

We consider again system (7.2), which is extended by a static state-feedback controller $\mathcal{K}_\infty$

$$\dot{\mathbf{x}} = A\mathbf{x} + B_w\mathbf{w} + B_u\mathbf{u} \tag{7.18a}$$

$$\mathbf{y} = C\mathbf{x} + D_u\mathbf{u} \tag{7.18b}$$

$$\mathbf{u} = \mathcal{K}_\infty\mathbf{x}. \tag{7.18c}$$

Note that the matrix D_w needs to be set to zero in order for the method presented here to be applicable to (7.2). We denote this system as

$$H_3(s) = (C + D_u\mathcal{K}_\infty)(sI - A - B_u\mathcal{K}_\infty)^{-1}B_w.$$

The centralized controller, i.e., the $\mathcal{K}_\infty$ matrix with all entries nonzero, shown in Fig. 7.1b, uses all possible degrees of freedom and can be designed via convex optimization [29]. In order to increase the sparsity of $\mathcal{K}_\infty$, the following theorem is presented in [27], which is also derived from Lemma 7.1.

Theorem 7.2 ([27]) *The following statements are equivalent:*

- *There exists a controller $\mathscr{K}_\infty$ which asymptotically stabilizes the system (7.18), such that the $\mathscr{H}_\infty$ norm of (7.18) is smaller than β.*
- *There exist matrices $P_1 \succ 0$, diagonal $P_2 \succ 0$, and full block matrices L_∞ and U_∞, such that*

$$
\Pi(P_1, P_2, L_\infty, U_\infty) := \begin{pmatrix} \Delta & P_1 B_u + D_u^T L^T & P_1 B_w & C^T \\ B_u^T P_1 + D_u L & -P_2 & 0 & D_u^T \\ B_w^T P_1 & 0 & -\beta I & 0 \\ C & D_u & 0 & -\beta I \end{pmatrix} \prec 0,
$$

$$(7.19)$$

with

$$
\Delta = A^T P_1 + P_1 A - U_\infty^T L_\infty - L_\infty^T U_\infty + U_\infty^T P_2 U_\infty.
$$

In this case, the controller is given by

$$
\mathscr{K}_\infty = P_2^{-1} L_\infty. \tag{7.20}
$$

We refer to [27] for proof of the given theorem. It gives the conditions when a static state-feedback controller $\mathscr{K}_\infty$ will achieve a defined $\mathscr{H}_\infty$ performance β. From (7.20), it follows that the sparsity pattern of $\mathscr{K}_\infty$ is the same as the sparsity pattern of L_∞. Consequently, the theorem leads to the following optimization problem for the sparsity improvement of $\mathscr{K}_\infty$ under a constant performance bound β:

$$
\min_{P_1, P_2, L_\infty, U_\infty} \quad \|\mathrm{vec}(L_\infty)\|_0 \tag{7.21}
$$

$$
\text{subject to} \quad \Pi(P_1, P_2, L_\infty, U_\infty) \prec 0, \tag{7.22}
$$

where the notation $\mathrm{vec}(L_\infty)$ creates a vector from the elements of L_∞ by stacking the columns of L_∞, and $\|\cdot\|_0$ denotes the 0-norm of a vector, i.e., the number of nonzero elements in the vector. The previous problem allows the synthesis of sparse state-feedback controllers with a guaranteed performance bound; however, due to the presence of the 0-norm in the cost function, it can be very challenging to solve. In order to improve the computation time, analogously as in SPH2, the 0-norm is relaxed into a weighted 1-norm, and we obtain the optimization problem

$$
\min_{P_1, P_2, L_\infty, U_\infty} \quad \|\mathrm{vec}(M_\infty \circ L_\infty)\|_1 \tag{7.23}
$$

$$
\text{subject to} \quad \Pi(P_1, P_2, L_\infty, U_\infty) \prec 0, \tag{7.24}
$$

where $\circ$ denotes the Hadamard (element-wise) product of two matrices, and $M_\infty = [m^{ij}]$ is a weighting matrix. If M_∞ is chosen to be the element-wise inverse of $L_\infty = [l^{ij}]$, i.e., $m^{ij} = \infty$ if $l^{ij} = 0$, and $m^{ij} = 1/|l^{ij}|$ otherwise, then the weighted 1-norm and the 0-norm coincide. The same reformulation is proposed in the SPH2 method as well. Since the weights depend on the solution of the optimization problem, this cost function cannot be used to obtain a convex optimization problem. Thus, an iterative solution, similar to the one presented for the SPH2 method, is proposed. However, even if the weights are set to a constant value in one iteration, the term Π in (7.24) is still non-convex, because U_∞ is multiplied with other optimization variables in $\Pi(P_1, P_2, L_\infty, U_\infty)$.

To solve this problem, the algorithm in Fig. 7.4 is proposed which gives the SPHinf method its final form. The small positive number v is introduced in the algorithm to avoid bad conditioning of the problem when $l_k^{ij} = 0$. Step 10 is introduced because the 1-norm penalizes large values in the cost term $M_k L_{\infty,k}$, resulting in a controller with small gain values. In Step 10, there is no term which minimizes these values, and the controller obtained previously can be further polished. With this algorithm, the problem is reduced to a series of convex linear matrix inequality problems. The convergence of weighted optimization with the 1-norm to a local minimum is proven [32]. However, in Fig. 7.4, the constraints are also changed in each step of the iteration. For this reason, convergence of the SPHinf

1: **procedure** SPHINF(G)
2:     $k = 1$
3:     $U_{\infty,1} = \mathcal{K}_{\infty,cent}$, where $\mathcal{K}_{\infty,cent}$ is the fully centralized controller which can be obtained with convex optimization.
4:     Choose m_0^{ij} and a sufficiently small number $v > 0$. Choose $\beta > 0$
5:     **while** $k \leq k_{max}$ or not converged **do**
6:         For fixed $U_{\infty,k}$ solve the following convex optimization problem

$$\{P_{1k}^*, P_{2k}^*, L_{\infty,k}^*\} = \underset{P_{1k} \succ 0, \text{ diagonal } P_{2k} \succ 0, L_{\infty,k}}{\arg\min} \left\| M_k L_{\infty,k} \right\|_1 \tag{7.25a}$$

$$\text{s.t. } \Pi(U_{\infty,k}, P_{1k}, P_{2k}, L_{\infty,k}) \prec 0. \tag{7.25b}$$

7:         Update $U_{\infty,k+1} = \left(P_{2k}^*\right)^{-1} L_{\infty,k}^*$ and $m_{k+1}^{ij} = (|l_k^{ij}| + v)^{-1}$.
8:         $k \leftarrow k + 1$
9:     **end while**
10:     Solve the feasibility problem for the fixed controller structure obtained in Step 7

$$\{P_{1k}^*, P_{2k}^*, L_{\infty,k}^*\} = \underset{P_{1k} \succ 0, \text{ diagonal } P_{2k} \succ 0, L_{\infty,k}}{\arg\min} \quad 0 \tag{7.26a}$$

$$\text{s.t. } \Pi(U_{\infty,k}, P_{1k}, P_{2k}, L_{\infty,k}) \prec 0 \tag{7.26b}$$

$$L_{\infty,k} \text{ has fixed controller structure from Step 7.} \tag{7.26c}$$

11:     $\mathcal{K}_\infty \leftarrow (P_{2k}^*)^{-1} L_{\infty,k}^*$.
12:     **return** $\mathcal{K}_\infty$
13: **end procedure**

Fig. 7.4 Optimization algorithm the SPHinf method [27]

method to a local minimum cannot be guaranteed. However, fast convergence for many numerical examples is reported in [27]. We perform the numerical evaluation of the presented method in Sect. 7.6.

7.5 Power System Model

We compare these controller design algorithms for large-scale CPS for a power system example. For this purpose, we first briefly summarize how power systems are modeled. In general, power systems consist of power plants (PP) which are interconnected through power flow equations. Figure 7.5 (left) illustrates the structure of a power system model, which consists of N power plants $PP_1, \ldots PP_N$. Each power plant can have an arbitrary structure. This figure has the same structure as Fig. 7.1b, where PP_i corresponds to subsystem S_i, and the power flow represents the interconnection between the subsystems.

7.5.1 Power Grid Model

Each power line and cable in a power grid has an inductance and capacitance which give itself internal dynamic states. The time constant of these states is in the order of milliseconds [1]. Since we consider interplant and inter-area oscillations, which are in the order of several Hz [1], we can neglect the power line dynamics [33], and

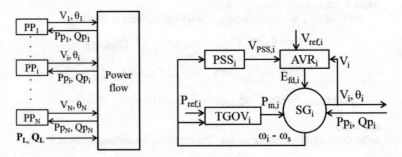

Fig. 7.5 *Left*: Power plants PP_i connected by the power flow equations. The vectors $\mathbf{P}_L$ and $\mathbf{Q}_L$ represent the active and reactive power of the loads, respectively. *Right*: Structure of one power plant model. Each power plant PP_i includes the synchronous generator SG_i, the turbine and governor ($TGOV_i$), as well as the automatic voltage regulator (AVR_i) and power system stabilizer (PSS_i)

we model the power grid, which consists of N_B buses, with algebraic power flow equations [1]

$$P_i = \sum_{j=1}^{N_B} |V_i||V_j|(G_{ij} \cos \Delta\theta_{ij} + B_{ij} \sin \Delta\theta_{ij}) \tag{7.27a}$$

$$Q_i = \sum_{j=1}^{N_B} |V_i||V_j|(G_{ij} \sin \Delta\theta_{ij} - B_{ij} \cos \Delta\theta_{ij}), \tag{7.27b}$$

where $P_i = P_{pi} - P_{Li}$ and $Q_i = Q_{pi} - Q_{Li}$ are the injected active and reactive net-power into the ith bus (node) in the grid by a power plant (P_{Pi}) or a load (P_{Li}), V_i and θ_i are the magnitude and angle of the voltage phasor at the ith bus, $\Delta\theta_{ij} = \theta_i - \theta_j$, and G_{ij} and B_{ij} are the elements of the conductance and susceptance matrix of the grid [1]. We assume that all buses with a zero net-power infeed are eliminated from the system with Kron reduction [34].

We gather all active and reactive powers of the loads: P_{Li}, and Q_{Li}, in the vectors $\mathbf{P}_L$ and $\mathbf{Q}_{Li}$. The load power can change at unknown times, and is thus considered as a disturbance input $\mathbf{w}$ for the system.

7.5.2 Power Plant Model

This section presents the model of a conventional power plant (PP), with the structure shown in Fig. 7.5 (right). Each power plant has a synchronous generator (SG_i), which converts the mechanical power of the turbine into electrical power. The rotational speed of the turbine and SG_i is controlled by the governor. The coupled governor and turbine model is denoted with $TGOV_i$. Additionally, the automatic voltage regulator (AVR_i) controls the power plant voltage via the exciter. In order to increase the system stability and improve power oscillation damping (POD), a power system stabilizer (PSS_i) is often connected to the AVR_i.

We present in the following subsections the models used for the different power plant components.

7.5.2.1 Synchronous Generator Model

The dynamics of the ith synchronous generator are given by 6th-order equations. The full synchronous generator model is not presented because it is not important for the subsequent numerical evaluation. The interested reader is referred to [1] for a detailed explanation of the model.

7.5.2.2 Controller and Actuator Models

In this subsection, we present exemplary models for the different PP controllers in Fig. 7.5 (right).

We use in the numerical example the simple AVR_i model shown in Fig. 7.7. This model emulates the controllers and hardware which control the generator terminal voltage V_i to a constant value via the field winding voltage $E_{fd,i}$. As inputs, it needs the reference terminal voltage $V_{ref,i}$, the actual/measured terminal voltage V_i, and the input from the power system stabilizer $V_{PSS,i}$. The gain parameter $K_{A,i}$ of this model can be adjusted, marked red in Fig. 7.7. We define an additional input u_i for this model, marked blue in Fig. 7.7, which is used as a control input for the SPH2 and SPHinf methods.

The AVR_i tries to achieve a (nearly) constant voltage $V_{t,i}$ at the PP terminal. However, this can degrade the POD in the system [1, 35]. This is the reason why power system stabilizers (PSS_i) are built into some PPs [36]. They are controllers which take as input, e.g., the frequency deviation from the nominal frequency ω_s at the PP terminals, or the PP power infeed, etc. We use the PSS model described in [1, 7], shown in Fig. 7.8. It consists of two lead-lag filters, a washout filter, and a gain. All parameters in Fig. 7.8 are red, because every parameter can be freely specified.

The governor controls the mechanical power $P_{m,i}$ which the turbine produces and transfers to the synchronous generator. It takes as input the frequency deviation of the PP from the nominal frequency ω_s. We model the standard IEEE turbine and governor model $TGOV_i$, c.f. Fig. 7.6. No parameter of the presented model can be changed.

Summarizing, the controllers, $TGOV_i$, AVR_i, and the PSS_i are described with a total of six states, the same as the synchronous generator model. Therefore, each power plant model comprises of a total of 12 states and 7 tunable controller parameters. Other parameters cannot be changed in the considered controller models.

Fig. 7.6 IEEE gas turbine and governor model TGOV1

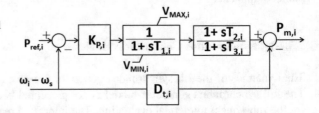

7.5.3 Coupled Linear Power System Model

When we combine the models of the N power plants, described in Sect. 7.5.2, with the power flow equations (7.27), we obtain the differential-algebraic nonlinear model

$$\dot{x} = f(\mathbf{x}, \mathbf{w}, \mathbf{K}_t, \mathbf{u}) \quad 0 = h(\mathbf{x}, \mathbf{w}, \mathbf{K}_t), \tag{7.28}$$

where $\mathbf{x} \in \mathbb{R}^{\cdot N_{st}}$ combines all power plant states in the model, $\mathbf{w}$ is the vector of disturbance inputs (here the active and reactive powers of the load), $\mathbf{K}_t \in \mathbb{R}^{\cdot N_t}$ is the vector of tunable parameters of all power plants, i.e., of parameters marked red in Figs. 7.7 and 7.8, $\mathbf{u} \in \mathbb{R}^N$ is the vector of wide-area control inputs for each PP, f describes the power plant dynamics, and h represents the algebraic equations in (7.27). We are interested in rejecting small changes in $\mathbf{w}$ in the performance output $\mathbf{y}$ (defined subsequently). Thus, we linearize the system (7.28) around the steady-state value $\mathbf{x}_0$ obtained with the known input $\mathbf{w}_0$. After eliminating the linearized algebraic equations in (7.28), we obtain a linear system of the form

$$\dot{x} = A(\mathbf{K}_t)\mathbf{x} + B_w(\mathbf{K}_t)\mathbf{w} + B_u(\mathbf{K}_t)\mathbf{u}. \tag{7.29}$$

Note that the choice of the tunable controller parameters $\mathbf{K}_t$ is made such that they do not change the steady-state of the system. This is done because the steady-state of power systems results from trade in the electricity market and any deviations cause additional costs for the power system operator. An additional consequence of this choice is that the linearization accuracy is unchanged when $\mathbf{K}_t$ is tuned, as the steady-state $\mathbf{x}_0$ is not a function of $\mathbf{K}_t$.

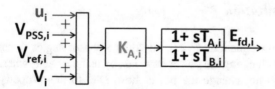

Fig. 7.7 Dynamic model of a simple AVR$_i$ used, e.g., in [1]. It consists of a gain $K_{A,i}$ and a transient gain reduction component with the time constants $T_{A,i}$ and $T_{B,i}$. The gain $K_{A,i}$, marked red, is tunable. The input u_i is additionally introduced, marked blue, which is used as a control input for the methods presented in Sect. 7.4

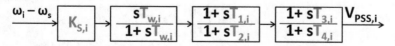

Fig. 7.8 Dynamic model of the simple power system stabilizer (taken from [1, 7]), where $K_{S,i}$ is the PSS gain, $T_{W,i}$ is the washout time constant, and $T_{1,i}$–$T_{4,i}$ are the lead-lag filters time constants. All of the PSS parameters are tunable

We minimize active power oscillations, which manifest as oscillations in the generator frequencies ω. Frequency oscillations, in turn, cause oscillations of generator rotor angles δ relative to each other. Thus, power plant frequencies ω_i, and rotor angles δ_i are a suitable outputs for observing power oscillations. Additionally, we want to avoid large control inputs into our system in order to avoid saturations and hardware limitations. We obtain the following performance output

$$\mathbf{y} = w_{\text{cond}} \left(\begin{pmatrix} L_{\text{perf}} \delta \\ \omega \end{pmatrix} + \mathbf{u} \right) = C\mathbf{x} + D_u \mathbf{u}, \tag{7.30}$$

where L_{perf} is a $N \times N$ Laplace matrix defined such that we obtain the difference between the generator rotor angles [23, 27] as one of the system outputs: $L_{\text{perf}} = I - \frac{1}{N} \mathbf{1} \mathbf{1}^T$. An additional multiplication factor of w_{cond} (e.g., $w_{\text{cond}} = 10$) is introduced to improve the conditioning of the optimization problems of the three methods. This definition of C and D quantifies the potential and kinetic energy stored in the electromechanical dynamics of the system [23, 27], leading to the overall system $G(\mathbf{K}_t, s)$

$$\dot{\mathbf{x}} = A(\mathbf{K}_t)\mathbf{x} + B_w(\mathbf{K}_t)\mathbf{w} + B_u(\mathbf{K}_t)\mathbf{u} \tag{7.31a}$$

$$\mathbf{y} = C\mathbf{x} + D_u \mathbf{u}. \tag{7.31b}$$

Note that the obtained system has the same form as (7.2). By setting $\mathbf{u} = 0$, or $\mathbf{u} = \mathcal{K}_2 \mathbf{x}$, or $\mathbf{u} = \mathcal{K}_\infty \mathbf{x}$, we obtain the setup for the SPinf, SPH2, and SPHinf method, respectively.

7.5.4 Adaptation of the Power System Model for $\mathcal{H}_2$ Optimization

In the power flow equations (7.27), all voltage phasor angles θ_i can be shifted by $\delta\theta \in \mathbb{R}$ without changing the power flow, i.e., by using $\theta_i + \delta\theta$, where $\delta\theta$ is identical for all i, we do not change the power flow. Due to this property, the A-matrix and L_{perf} have an eigenvalue at zero, which does not allow us to directly apply the SPH2 optimization method. We introduce in this subsection a transformation which eliminates the zero-eigenvalue and allows us to apply the SPH2 method. By expressing the state vector as

$$\mathbf{x}(t) := \begin{bmatrix} \delta(t) \\ \mathbf{r}(t) \end{bmatrix} \in \mathbb{R}^n,$$

where $\boldsymbol{\delta}(t) \in \mathbb{R}^N$ is the vector of generator rotor angles, and $\mathbf{r}(t) \in \mathbb{R}^{n-N}$ represents all other states in the system, we can define the structural properties of the matrices in (7.31)

$$A \begin{bmatrix} \mathbf{1} \\ 0 \end{bmatrix} = 0, \quad L_{\text{perf}} \, \mathbf{1} = 0, \quad \mathcal{K}_2 \begin{bmatrix} \mathbf{1} \\ 0 \end{bmatrix} = 0.$$

To eliminate the zero pole from (7.31) we introduce the following coordinate transformation [23]:

$$\mathbf{x} = \begin{bmatrix} \boldsymbol{\delta} \\ \mathbf{r} \end{bmatrix} = \underbrace{\begin{bmatrix} U & 0 \\ 0 & I \end{bmatrix}}_{T} \boldsymbol{\xi} + \begin{bmatrix} \mathbf{1} \\ 0 \end{bmatrix} \bar{\delta}, \tag{7.32}$$

where the matrix $U \in \mathbb{R}^{N \times (N-1)}$ is chosen such that its columns form an orthonormal basis that is orthogonal to span $(\mathbf{1})$ [23]. The $N - 1$ eigenvectors that correspond to the nonzero eigenvalues of L_{perf} in (7.30) can be used for the columns of U. In the new set of coordinates, $\boldsymbol{\xi}(t) = T^T \mathbf{x}(t) \in \mathbb{R}^{n-1}$, the closed-loop system takes the form

$$\dot{\boldsymbol{\xi}} = (\bar{A} + \bar{B}_u F)\boldsymbol{\xi} + \bar{B}_w \mathbf{w} \tag{7.33a}$$

$$\mathbf{y} = (\bar{C} + D_u F)\boldsymbol{\xi}, \tag{7.33b}$$

where

$$\bar{A} := T^T A T, \quad \bar{B}_w := T^T B_w, \quad \bar{B}_u := T^T B_u, \quad \bar{C} := C T,$$

and F is the controller matrix in the new ($\boldsymbol{\xi}$) coordinates. The controller matrices $\mathcal{K}_2$ and F are coupled by

$$F = \mathcal{K}_2 T \quad \Leftrightarrow \quad \mathcal{K}_2 = F T^T.$$

We can use this relation to minimize the number of elements in the $\mathcal{K}_2$ matrix, while at the same time using (7.5.4) to minimize the system $\mathcal{H}_2$ norm. This is achieved by applying the optimization Problem (7.17) to the system (7.5.4), where we introduce the additional constraint

$$\begin{aligned} \underset{F, \, \mathcal{K}_2}{\text{minimize}} \quad & J(F) + \alpha \, g(\mathcal{K}_2) \\ \text{subject to} \quad & F T^T - \mathcal{K}_2 = 0, \end{aligned} \tag{7.34}$$

where the equality constraint ensures the validity of the transformation, and the term $g(\mathcal{K}_2)$ increases the sparsity of the $\mathcal{K}_2$ matrix. This problem can be solved using ADMM, as described in [23].

7.6 Numerical Comparison

We now compare the three methods presented in Sect. 7.4 on a numerical example. Since SPH2 uses the $\mathscr{H}_2$ norm as the performance metric, whereas SHinf and SPHinf use the $\mathscr{H}_\infty$ norm, a direct comparison between all three methods is not straightforward. For this reason, we compare the methods in several aspects on a numerical example:

- The achieved $\mathscr{H}_\infty$ and $\mathscr{H}_2$ norms and how they correlate in the power system example.
- The sparsity patterns in the $\mathscr{H}_2$ and $\mathscr{H}_\infty$ matrices achieved by SPH2 and SPHinf, respectively.
- Impact on the maximal singular value of the transfer function for all methods.
- Time-domain simulations achieved by a disturbance (load) step.

For the comparison, we consider the grid from [1, Example 12.6, p. 813] presented in Fig. 7.9. The parameters of the transmission grid, as well as the parameters of power plants, can be found in [1]. All power plants have controllers described in Sect. 7.5.2.2. The parameters of the TGOV1 controller are $T_1 = 0.04$ s, $T_2 = 1$ s, $T_3 = 2$ s, $k_P = 150$, and $D_t = 0$, while the voltage regulator time constants are set to $T_A = 1$s and $T_B = 10$s, as in [1]. The initial value for the tunable voltage regulator gain is $K_A = 200$, and for all PSSs: $K_S = 50.5$, $T_1 = 0.0037$ s, $T_2 = 0.0079$ s, $T_3 = 40.9$ s, $T_4 = 2.1386$ s, and $T_W = 3.9604$ s.

As disturbance input **w** for the optimization, we consider the loads in buses 7 and 9, marked red in Fig. 7.9. All generators are equipped with PSSs in this example because we consider them as equivalent models for several generators.

The system consists of 48 states and 28 tunable parameters described in Sect. 7.5.2.2.

7.6.1 Analysis of the System with Initial Parameters

Figure 7.10 shows the simulation of generator frequencies with the initial parameters after a load step in bus 7. In the first 20 s after a load step, poorly dampened oscillations with a frequency of approx. 3 Hz are present. Afterwards, slow oscilla-

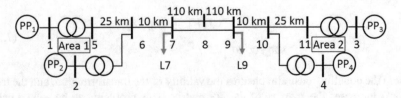

Fig. 7.9 A two-area system from [1, p. 813, Example 12.6]

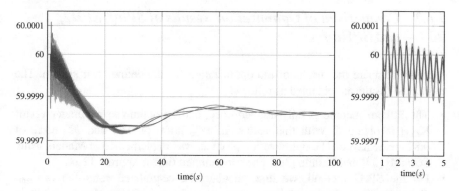

Fig. 7.10 (left) Generator frequency response after a 100 MW load step in bus 9. Strong oscillations are present in the system. (right) Zoomed fast oscillations after the load step

Table 7.1 Poorly dampened modes of the considered system

Mode	Frequency (rad/s)	Damping ratio (%)
1	9.74	1.7
2	9.42	2.5
3	7.85	0.9
4	0.13	2.8
5	0.13	3.1

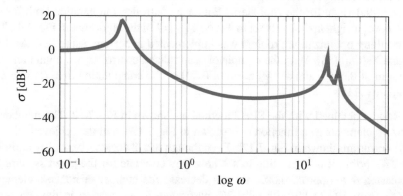

Fig. 7.11 Largest singular value of $G(\mathbf{K}_I, \omega)$

tions are noticeable, which are not completely dampened even after 100 s. Table 7.1 presents the weakly dampened oscillatory modes of the system, with damping below 5%, which confirm the analysis of the step response plot.

The singular value plot of the system is shown in Fig. 7.11. The frequencies at which the peaks occur in Fig. 7.11 correspond to the oscillatory modes in Table 7.1. The initial system $\mathcal{H}_\infty$ norm is 7.3, and the $\mathcal{H}_2$ norm is 1. Thus, the considered test system is complex with slow and fast badly dampened oscillatory modes on which we can test the methods presented in Sect. 7.4.

7.6.2 Comparison of Optimization Results of SPinf, SPH2, and SPHinf

We now compare the results of the optimization for the defined test system. The numerical results are obtained as follows:

- The SPHinf method is only applied once, and an optimized parameter vector $\mathbf{K}_{t,\text{opt}}$ is obtained. With this vector, an $\mathcal{H}_\infty$ norm of 0.81 and $\mathcal{H}_2$ norm of 0.82 are achieved. The optimization problem was implemented in Matlab, using SeDuMi [37] and Yalmip [38]. The computation time is approx. 120 s.
- For the SPH2 method, we first calculate the centralized controller $\mathcal{K}_{2,\text{cent}}$. Then, we vary the weighting parameter α from 0.001 to approx. 1000 in 40 logarithmically spaced steps and obtain the matrices $\mathcal{K}_{2,a}$, $a = 1...40$ with increasing sparsity-degrees. With the completely centralized controller, we obtain an $\mathcal{H}_\infty$ norm of 0.6824, and an $\mathcal{H}_2$ norm of 0.4685. The computation time for one α is approx. 4 s.
- For the SPHinf method, we also first calculate a centralized controller $\mathcal{K}_{\infty,\text{cent}}$ which achieves an $\mathcal{H}_\infty$ norm β_{cent}. We then vary the performance bound β from 1% degradation to approx. 32% degradation in 30 steps, and from 32% to 1000% compared to β_{cent} in 10 steps. We obtain the matrices $\mathcal{K}_{\infty,b}$, $b = 1...40$ matrices with increasing sparsity-degrees. For the calculation of $\mathcal{K}_{\infty,b}$, we use $\mathcal{K}_{\infty,b-1}$ as an initial value for the "hot-start." For each controller, the iteration limit for the 1-norm weight update was set to 4 (i.e., $k_{\max} = 4$ in the algorithm in Fig. 7.4). The resulting optimization problem was implemented in Matlab, using SeDuMi [37] and Yalmip [38]. With the completely centralized controller, we obtain an $\mathcal{H}_\infty$ norm of 0.6653, and an $\mathcal{H}_2$ norm of 0.5018. The computation time for one β is approx. 115 s.

Results obtained with the SPH2 method are shown in Fig. 7.12. The number of nonzero elements as a function α is shown in Fig. 7.12a, and the achieved $\mathcal{H}_2$ and $\mathcal{H}_\infty$ norms are shown in Fig. 7.12b. Even though SPH2 does not explicitly optimize the $\mathcal{H}_\infty$ norm, it is interesting to see how they correlate for the given system. By increasing α to approx. 0.003, we can decrease the number of nonzero elements in $\mathcal{K}_2$ from 192 to 89 with only 3% performance degradation in the $\mathcal{H}_2$ norm compared to the initial controller. We denote this controller with $\tilde{\mathcal{K}}_2$. Afterwards, the performance degradation is relatively steep, and by setting the α to 1.3, we obtain an 85% performance degradation with 32 nonzero elements.

Results obtained with the SPHinf method are shown in Fig. 7.13. The number of nonzero elements as a function β compared to β_{cent} is shown in Fig. 7.13a, and the achieved $\mathcal{H}_2$ and $\mathcal{H}_\infty$ norms are shown in Fig. 7.13b. SPHinf reduces the number of nonzero elements to 26 with a 7% performance degradation in the $\mathcal{H}_\infty$ norm; however, the $\mathcal{H}_2$ norm is increased by 470%. We denote this controller as $\tilde{\mathcal{K}}_\infty$. Thus, even though with the SPH2 optimization, the two norms seem to correlate, this is not the case for the SPHinf method. The method, however, does not use the full performance degradation available through the constraints, e.g., if a 1000% degradation of the $\mathcal{H}_\infty$ norm is allowed, the optimization achieves a degradation of

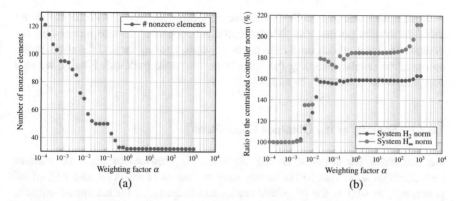

Fig. 7.12 Nonzero elements in $\mathcal{K}_2$ and the system norms as functions of the weighting γ with the SPH2 method. (**a**) Number of nonzero elements in $\mathcal{K}_2$. The full controller has 192 elements. (**b**) The $\mathcal{H}_2$ and $\mathcal{H}_\infty$ system norms relative to their respective values with the completely centralized controller

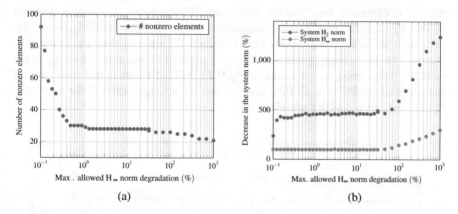

Fig. 7.13 Nonzero elements in $\mathcal{K}_\infty$ and the system norms as functions of performance limit with the SPHinf method. (**a**) Number of nonzero elements in $\mathcal{K}_\infty$. The full controller has 192 elements. (**b**) The $\mathcal{H}_2$ and $\mathcal{H}_\infty$ system norms relative to their respective values with the completely centralized controller

Table 7.2 Comparison of norms of the methods SPHinf, SPH2, and SPHinf

Norm	Initial	$\mathbf{K}_{t,opt}$	$\mathcal{K}_{2,cent}$	$\tilde{\mathcal{K}}_2$	$\mathcal{K}_{\infty,cent}$	$\tilde{\mathcal{K}}_\infty$
$\mathcal{H}_\infty$	7.3	0.81	0.6824	0.6905	0.6653	0.7137
$\mathcal{H}_2$	1	0.82	0.4685	0.4825	0.5018	2.3513

300%. This suggests that it does not converge to a global optimum, as already noted in [27]. The optimized norms for the different controller parameters are shown in Table 7.2. Not surprisingly, the best $\mathcal{H}_\infty$ norm is achieved with $\mathcal{K}_{\infty,cent}$, and the best $\mathcal{H}_2$ norm with $\mathcal{K}_{2,cent}$. Interestingly, the degradation of the $\mathcal{H}_2$ norm from $\mathcal{K}_{2,cent}$ to $\tilde{\mathcal{K}}_2$ and of the $\mathcal{H}_\infty$ norm from $\mathcal{K}_{\infty,cent}$ to $\tilde{\mathcal{K}}_\infty$ is very small, and $\tilde{\mathcal{K}}_2$ achieves a better $\mathcal{H}_\infty$ norm than $\tilde{\mathcal{K}}_\infty$.

In summary, the two methods achieve very different results with regard to the achieved system norms. The SPHinf optimization causes a large increase in the system $\mathcal{H}_2$ norm, which the SPH2 method avoids, causing large differences in the optimization results regarding the sparsity-performance trade-off. Finally, we also compare the structures of $\mathcal{K}_2$ and $\mathcal{K}_\infty$.

7.6.2.1 Sparsity of the Obtained Controllers

We first show the structure of the $\tilde{\mathcal{K}}_2$ and $\tilde{\mathcal{K}}_\infty$ matrices in Fig. 7.14. Further investigations show that SPH2 mostly uses the states of the AVR and PSS of all generators, as well as the generator angles and frequencies for the state-feedback. On the other hand, SPHinf mostly uses the states of the same PP for the controller: e.g., the controller of PP1 uses mostly the states of PP1, with two additional states from PP2 and PP4.

We also analyze the structures of $\mathcal{K}_2$ and $\mathcal{K}_\infty$ when they both have a similar number of elements. We chose the case where SPH2 achieves 32 nonzero elements in the $\mathcal{K}_2$ matrix, a performance degradation of the $\mathcal{H}_2$ norm of 60%, and of the $\mathcal{H}_\infty$ norm of 85%. The SPHinf method achieves 33 nonzero elements in the $\mathcal{K}_\infty$ matrix, a performance degradation of the $\mathcal{H}_2$ norm of 446%, and of the $\mathcal{H}_\infty$ norm of 0.2%. The results are presented in Fig. 7.15. SPH2 focuses in this case only on the states of the PSSs of all PPs, whereas for WHinf, the choice of the necessary states varies with the PP controller.

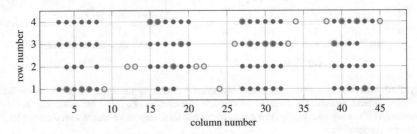

Fig. 7.14 Comparison of nonzero entries of the $\tilde{\mathcal{K}}_2$ (blue) and $\tilde{\mathcal{K}}_\infty$ (red) matrices

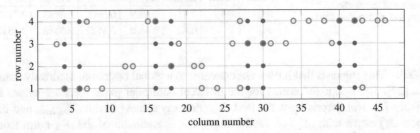

Fig. 7.15 Comparison of nonzero entries of the $\mathcal{K}_{2,23}$ (blue) and $\mathcal{K}_{\infty,8}$ (red) matrices which have 32 and 33 nonzero elements, respectively

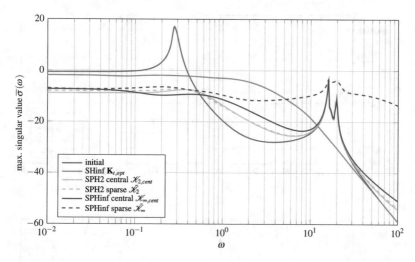

Fig. 7.16 Comparison of the singular value plots obtained with the initial system, $\mathbf{K}_{t,\text{opt}}$, $\mathcal{K}_{2,\text{cent}}$, $\mathcal{K}_{\infty,\text{cent}}$, $\tilde{\mathcal{K}}_{\infty}$, and $\tilde{\mathcal{K}}_{2}$

7.6.2.2 Singular Value Plot

We now analyze the impact of the methods on the singular value plot of the system, which is shown for the initial system in Fig. 7.11. We show the results obtained with the initial system, $\mathbf{K}_{t,\text{opt}}$, $\mathcal{K}_{2,\text{cent}}$, $\mathcal{K}_{\infty,\text{cent}}$, $\tilde{\mathcal{K}}_{\infty}$, $\tilde{\mathcal{K}}_{2}$ in Fig. 7.16. We see that all methods were able to eliminate the largest peak in the low frequency range. However, only SPHinf is able to eliminate the peak in the 3 Hz area as well. This may be due to the fact that SPHinf only tunes the parameters of controllers which were already designed that specific purpose. All other methods focused on reducing the max. singular value in the very low frequency area (i.e., in the area of below 0.1 Hz), which cannot be achieved without significant control effort. Interestingly, all controllers achieve different DC gains (in the low frequency area). Another interesting observation is that the curve obtained with $\tilde{\mathcal{K}}_{\infty}$, which has the worst $\mathcal{H}_{2}$ degradation of 446%, has also the largest area below the curve, which also corresponds to the SISO interpretation of the $\mathcal{H}_{2}$ norm.

7.6.2.3 Time-Domain Comparison

Finally, we also compare the time-domain step responses after a load step in bus 7 for the parameterizations $\mathbf{K}_{t,\text{opt}}$, $\mathcal{K}_{2,\text{cent}}$, and $\mathcal{K}_{\infty,\text{cent}}$, shown in Fig. 7.17. Other time-domain results with SPH2 and SPHinf are visually similar and are thus not included here due to length limitations. Surprisingly, the SPHinf method achieved the best POD result for the tested system. A possible cause for this is that external controllers $\mathcal{K}$ cannot provide enough compensation if the internal controllers are

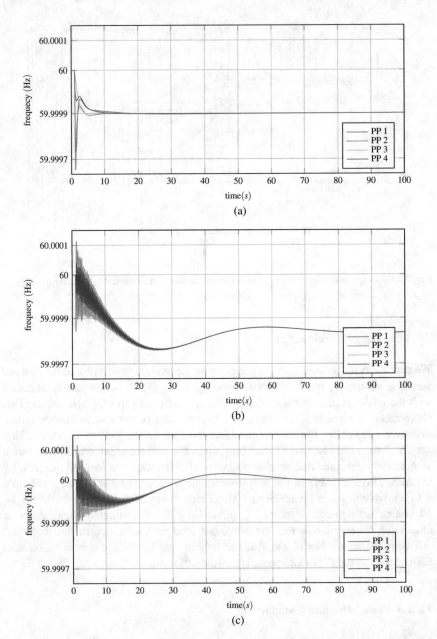

Fig. 7.17 Time-domain simulation after a load step in bus 7 with the parameterizations $\mathbf{K}_{t,\mathrm{opt}}$, $\mathcal{K}_{2,\mathrm{cent}}$, and $\mathcal{K}_{\infty,\mathrm{cent}}$. (a) $\mathbf{K}_{t,\mathrm{opt}}$ parameterization. (b) $\mathcal{K}_{2,\mathrm{cent}}$ parameterization. (c) $\mathcal{K}_{\infty,\mathrm{cent}}$ parameterization

Table 7.3 Overshoot and settling time to within ±0.004 mHz, which corresponds to a $\pm2\%$ bound of the initial steady-state deviation

	Initial	SHinf	SPH2	SPHinf
Overshoot (mHz)	0.2762	0.336	0.2535	0.19
Settling time (s)	138	13	93	85

badly tuned. However, SPHinf also has the largest overshoot after a load step. Better results with the SPH2 and SPHinf methods could possibly be obtained by introducing shaping functions; however, this is out of the scope of this chapter. The overshoot (in %) and the settling time are shown in Table 7.3. Even though SHinf has the smallest settling time to within a 2% bound of the steady-state deviation, it achieves the largest overshoot. On the other hand, SPHinf achieves the smallest overshoot.

7.6.2.4 Computational Complexity

The three methods also have different computational complexities. Overall, the SPH2 method achieves the fastest solution times, taking approx. 4 s for one α, as it can be solved in a distributed fashion via ADMM. The SHinf and SPHinf methods, on the other hand, need to find a positive definite matrices in each iteration using a centralized optimization, thereby increasing the computation. Consequently, the SHinf method needs approx. 120 s to find the optimal parameters, and the SPHinf method needs approx. 115 s for one β.

7.6.2.5 Discussion

Overall, all three methods are able to improve the system behavior. As visible from the singular value plot in Fig. 7.16, all methods eliminated the resonant peak at approx. 1 Hz, whereas only SHinf also eliminated the second peak at approx. 3 Hz. These results indicate that, in the considered system, external controllers may not be able to compensate all internal resonances in a system, and that tuning of existing controllers in the system may be necessary.

When comparing SPH2 and SPHinf, they achieved very different results regarding the sparsity-performance trade-off. The cause of the difference is that the SPHinf method, while minimizing the system $\mathcal{H}_\infty$ norm, significantly increases the $\mathcal{H}_2$ norm of the considered system. On the other hand, the SPH2 method, while minimizing the $\mathcal{H}_2$ norm, also achieved a good $\mathcal{H}_\infty$ norm. Consequently, the sparsity patterns of the obtained controller matrices are also very different.

In the time domain, SHinf achieved the best settling time, however, at the cost of an increased overshoot. SPH2 and SPHinf decreased both the settling time and overshoot; however, the 3 Hz oscillation is still present. The presented methods can also be combined. Best results could possibly be obtained by combining SHinf with SPH2 or SPHinf, this is, however, part of future work.

7.7 Conclusions

In this chapter, we reviewed distributed control concepts of cyber-physical systems and presented three methods which create (sub-)optimal controllers for distributed systems with varying degrees of decentralization. The first method [22] tunes the parameters of local structured controllers with $\mathcal{H}_\infty$ optimization, whereas the second [23] and third [27] method create static state-feedback controllers using $\mathcal{H}_2$ and $\mathcal{H}_\infty$ optimization methods, respectively, with varying degrees of decentralization. In the second part, we applied these methods for power oscillation damping in power systems on a test system with four power plants. The chapter is concluded with the comparison of the numerical results in the time and frequency domain with the three presented methods, as well as the obtained sparsity results. Overall, the SPH2 method outperformed the SPHinf method regarding the computation time, while the time-domain results look similar. On the other hand, the SHinf method achieved the best results in the time domain, while having the longest computation time.

References

1. Kundur, P. (1993). *Power system stability and control*. New York: McGraw-Hill.
2. Sauer, P., & Pai, M. (2007). *Power system dynamics and stability*. Champaign, IL: Stipes Publishing LLC.
3. REN21. (2018) Renewables 2018 global status report. [Online]. Available: http://www.ren21.net/gsr-2018/
4. Al Ali, S., Haase, T., Nassar, I., & Weber, H. (2014). Impact of increasing wind power generation on the north-south inter-area oscillation mode in the European ENTSO-E system. *IFAC Proceedings Volumes, 47*(3), 7653–7658.
5. Preece, R., & Milanovic, J. (2014). Tuning of a damping controller for multiterminal VSC-HVDC grids using the probabilistic collocation method. *IEEE Transactions on Power Delivery, 29*(1), 318–326.
6. Milano, F., Dörfler, F., Hug, G., Hill, D. J., & Verbič, G. (2018). Foundations and challenges of low-inertia systems. In *Power Systems Computation Conference (PSCC)* (pp. 1–25). Piscataway: IEEE.
7. Raoufat, M., Tomsovic, K., & Djouadi, S. (2016). Virtual actuators for wide-area damping control of power systems. *IEEE Transactions on Power Systems, 31*(6), 4703–4711.
8. Pipelzadeh, Y., Chaudhuri, N., Chaudhuri, B., & Green, T. (2017). Coordinated control of offshore wind farm and onshore HVDC converter for effective power oscillation damping. *IEEE Transactions on Power Systems, 32*(3), 1860–1872.
9. Zhu, C., Khammash, M., Vittal, V., & Qiu, W. (2003). Robust power system stabilizer design using $\mathcal{H}_\infty$ loop shaping approach. *IEEE Transactions on Power Systems, 18*(2), 810–818.
10. Befekadu G., & Erlich, I. (2005). Robust decentralized structure-constrained controller design for power systems: An LMI approach. In *Conference on Power Systems Computation*.

11. Mahmoudi, M., Dong, J., & Tomsovic, K. (2015). Application of distributed control to mitigate disturbance propagations in large power networks. In *North American Power Symposium (NAPS), 2015* (pp. 1–6). Piscataway: IEEE.
12. Preece, R., Milanovic, J., Almutairi, A. M., & Marjanovic, O. (2013). Damping of inter-area oscillations in mixed AC/DC networks using WAMS based supplementary controller. *IEEE Transactions on Power Systems, 28*(2), 1160–1169.
13. Marinescu, B., Mallem, B., Bourles, H., & Rouco, L. (2009). Robust coordinated tuning of parameters of standard power system stabilizers for local and global grid objectives. In *PowerTech, Bucharest*. Piscataway: IEEE.
14. Rouco, L. (2001). Coordinated design of multiple controllers for damping power system oscillations. *International Journal of Electrical Power & Energy Systems, 23*(7), 517–530.
15. Borsche, T., Liu, T., & Hill, D. J. (2015). Effects of rotational inertia on power system damping and frequency transients. In *54th Annual Conference on Decision and Control (CDC)* (pp. 5940–5946). Piscataway: IEEE.
16. Liao, K., He, Z., Xu, Y., Chen, G., Dong, Z., & Wong, K. (2017). A sliding mode based damping control of DFIG for interarea power oscillations. *IEEE Transactions on Sustainable Energy, 8*(1), 258–267.
17. Yaghooti, A., Buygi, M., & Shanechi, M. (2016). Designing coordinated power system stabilizers: A reference model based controller design. *IEEE Transactions on Power Systems, 31*(4), 2914–2924.
18. Liu, Y., Wu, Q. H., & Zhou, X. X. (2016). Coordinated switching controllers for transient stability of multi-machine power systems. *IEEE Transactions on Power Systems, 31*(5), 3937–3949.
19. Taranto, J., do Bomfim, A., Falcao, D., & Martins, N. (1999). Automated design of multiple damping controllers using genetic algorithms. In *Proceedings of the IEEE Power Engineering Society. Winter Meeting* (pp. 539–544). Piscataway: IEEE.
20. Fuchs, A., Imhof, M., Demiray, T., & Morari, M. (2014). Stabilization of large power systems using VSC-HVDC and model predictive control. *IEEE Transactions on Power Delivery, 29*(1), 480–488.
21. Obaid, Z. A., Cipcigan, L., & Muhssin, M. T. (2017). Power system oscillations and control: Classifications and PSSs' design methods: A review. *Renewable and Sustainable Energy Reviews, 79*, 839–849.
22. Mešanović, A., Unseld, D., Münz, U., Ebenbauer, C., & Findeisen, R. (2018). Parameter tuning and optimal design of decentralized structured controllers for power oscillation damping in electrical networks. In *American Control Conference (ACC)* (pp. 3828–3833). Piscataway: IEEE.
23. Wu, X., Dörfler, F., & Jovanović, M. R. (2016). Input-output analysis and decentralized optimal control of inter-area oscillations in power systems. *IEEE Transactions on Power Systems, 31*(3), 2434–2444.
24. Wu, X., & Jovanović, M. R. (2017). Sparsity-promoting optimal control of systems with symmetries, consensus and synchronization networks. *Systems & Control Letters, 103*, 1–8.
25. Dörfler, F., Jovanović, M., Chertkov, M., & Bullo, F. (2014). Sparsity-promoting optimal wide-area control of power networks. *IEEE Transactions on Power Systems, 29*(5), 2281–2291.
26. Dörfler, F., Jovanović, M., Chertkov, M., & Bullo, F. (2013). Sparse and optimal wide-area damping control in power networks. In *American Control Conference (ACC)* (pp. 4289–4294). Piscataway: IEEE.
27. Schuler, S., Münz, U., & Allgöwer, F. (2014). Decentralized state feedback control for interconnected systems with application to power systems. *Journal of Process Control, 24*(2), 379–388.
28. Scherer, C., & Weiland, S. (2015). Linear matrix inequalities in control, lecture notes. [Online]. Available: http://www.mathematik.uni-stuttgart.de/opencms/opencms/fak8/imng/lehrstuhl/lehrstuhl_fuer_mathematische_systemtheorie/lehre_/lecurenotes/LectureNotes.pdf
29. Gahinet, P. & Apkarian, P. (1994). A linear matrix inequality approach to $\mathcal{H}_\infty$ control. *International Journal of Robust and Nonlinear Control, 4*(4), 421–448.

30. Lin, F., Fardad, M., & Jovanović, M. R. (2013). Design of optimal sparse feedback gains via the alternating direction method of multipliers. *IEEE Transactions on Automatic Control, 58*(9), 2426–2431.
31. Candès, E. J., Wakin, M. B., & Boyd, S. P. (2008). Enhancing sparsity by reweighted ℓ_1 minimization. *Journal of Fourier Analysis and Applications, 14*, 877–905.
32. Fazel, M., Hindi, H., & Boyd, S. P. (2003). Log-det heuristic for matrix rank minimization with applications to Hankel and Euclidean distance matrices. In *American Control Conference (ACC)* (Vol. 3, pp. 2156–2162). Piscataway: IEEE.
33. Schiffer, J., Zonetti, D., Ortega, R., Stanković, A. M., Sezi, T., & Raisch, J. (2016). A survey on modeling of microgrids-from fundamental physics to phasors and voltage sources. *Automatica, 74*, 135–150.
34. Dörfler, F. & Bullo, F. (2013). Kron reduction of graphs with applications to electrical networks. *IEEE Transactions on Circuits and Systems I: Regular Papers, 60*(1), 150–163.
35. Hanson, O., Goodwin, C., & Dandeno, P. (1968). Influence of excitation and speed control parameters in stabilizing intersystem oscillations. *IEEE Transactions on Power Apparatus and Systems, 87*(5), 1306–1313.
36. Kundur, P., Berube, G., Hajagos, L., & Beaulieu, R. (2003). Practical utility experience with and effective use of power system stabilizers. In *Power Engineering Society General Meeting, IEEE* (Vol. 3, pp. 1777–1785). Piscataway: IEEE.
37. Sturm, J. F. (2017). Using SeDuMi 1.05, A Matlab Toolbox for Optimization Over Symmetric Cones. [Online]. Available: https://sedumi.ie.lehigh.edu/?page_id=58
38. Lofberg, J. (2004). YALMIP: A toolbox for modeling and optimization in MATLAB. In *IEEE International Conference on Robotics and Automation* (pp. 284–289). Piscataway: IEEE.

Chapter 8
Model-Driven Software Design Automation for Complex Rehabilitation

Pranav Srinivas Kumar and William Emfinger

8.1 Introduction

Powered wheelchairs (PW) [14] are widely used around the world [68] by people with motor, sensory, or cognitive impairments for their everyday mobility needs. The speed and direction of the PW drive mechanism are controlled using integrated joystick controllers. Traditional joystick-based PW control has numerous drawbacks, e.g., along corridors, joystick jerks induced by uncontrolled motions are a source of wall collisions. To address such issues, researchers have leveraged technologies developed in the domain of mobile robotics to create smart and autonomous wheelchairs using alternative control systems such as vision [59], head joysticks [56], smart wearable devices [70], and natural language [72]. There is an evident trend towards increased and higher-level autonomy in PW designs. The trend is evident in mobile robotics in general, e.g., self-driving cars, unmanned aerial vehicles, and warehouse robots. These robots are tasked with understanding the world around them, planning actions, and in the case of PW, interacting with the humans that also occupy that world.

Mobile robots cover many layers of the software stack: from low-level physical dynamics, sensing, actuation, and control to high-level requirements such as goal specification, coordination, and fault tolerance. While existing tools provide support at different layers, e.g., Simulink [69], ROS [63], Gazebo [44], etc., designers are often left to navigate a large set of independent tools, systems, and methods. The ultimate challenge with software development is managing complexity, so that

P. S. Kumar (✉)
Siemens Corporation, Princeton, NJ, USA
e-mail: pranav.kumar@siemens.com

W. Emfinger
Permobil, Lebanon, TN, USA
e-mail: william.emfinger@permobil.com

© Springer Nature Switzerland AG 2019
M. A. Al Faruque, A. Canedo (eds.), *Design Automation of Cyber-Physical Systems*, https://doi.org/10.1007/978-3-030-13050-3_8

artifacts that satisfy some objectives, e.g., functionality, performance, resilience, etc., can be produced with acceptable levels of effort. Increased autonomy in mobile robotic systems has created increased complexity in software design and engineering. These robots operate in homes, on roads, and in warehouses, performing safety-critical missions such as transportation and medical surgery. The algorithms that govern these mobile robots are also pushing the limits of theories and tools at our disposal for design, implementation, and verification. Software implementation is prompting the use of new and novel software architectures and components, e.g., neural networks. In order to reduce deployment surprises and enable rigorous testing of robotic software, integrated robotics development *platforms* are being utilized.

Many of these platforms (discussed in Sect. 8.2) apply model-driven engineering (MDE) [66]. *Models* represent requirements, structure, behavior, deployment, etc., of the engineered system. Communicating design decisions via models helps engineers to develop a common understanding of the system. Following a model-driven workflow for software has its challenges though, as there needs to be consistency between models and implementation code. A number of platforms offer code generation from models, but the generated code is just a skeleton for a full implementation, i.e., a human needs to provide further details. Nevertheless, MDE is useful as models can express properties of the software that are hard to discern from the implementation. Understanding how a complex system operates is essential to future changes, maintenance, and analysis. Thus, even if models are disconnected from the implementation code, they are a valuable resource in the design process.

In this chapter, we present our experiences using MDE for the software engineering of a domain-specific use-case: distributed, real-time embedded software for the Permobil SmartDrive MX2+ PushTracker, shown in Fig. 8.1. This is a power-assist drive system for manual wheelchairs and our system under test (SUT). We present two MDE tools: ROSMOD [47] and HFSM Design Studio [18], used for

Fig. 8.1 The SmartDrive MX2+ attached to a rigid-frame manual wheelchair

engineering this SUT. In the next section, we discuss various state-of-the-art tools and frameworks for MDE, focusing on robotic CPS. This is followed by sections on ROSMOD and HFSM Design Studio. The next section introduces two example complex rehabilitation systems and presents the MDE workflow. The final sections summarize the observations and outline our vision for the next stage in design automation.

8.2 Background and Related Research

The background and related research is divided into two subsections covering different aspects of robotics software development: (1) MDE tools in robotics and (2) middleware technologies in robotics. MDE tools cover the design-time modeling aspect of robotics research and middleware technologies cover the run-time distributed communications aspects of mobile robots. Both these areas are equally important for PW and active areas of research. In later sections, we discuss how both these aspects are considered in our own MDE tools.

8.2.1 Model-Driven Engineering in Robotics

In the last decade, the robotics community has seen a large number of libraries and tools, developed by research labs and universities, providing modeling [69], simulation (e.g., Gazebo [44]), and distributed run-time frameworks (e.g., ROS [63]). To a large extent, frameworks like these help to rapid prototype robotic systems. More recently, integrated system-level analysis methodologies are being integrated into such frameworks [22, 45] that can help engineers and system integrators at design-time to estimate properties such as trigger-to-response times, deadline violations, and synchronization issues. MDE has been widely applied in domains such as avionics and automotive engineering. Motivated by the gains in workflow efficiency, correct-by-construction modeling, and rapid prototyping that MDE provides, the software engineering community in robotics has gradually moved in this direction [65].

MDE, however, cannot be directly applied to CPS like robotic systems due to the highly changing, dynamic environments in which robots are deployed. The unpredictability and non-determinism span various layers of the software development life-cycle, from requirements, specification, design, implementation, and deployment. The system cannot be realized fully with a unidirectional workflow from requirements to deployment since neither the design space nor the run-time environments can be modeled in its entirety as in traditional embedded systems. So, the current trend in MDE for robotics is to generate containers where engineers can insert hand-written code called *business logic*. The business logic provides the core logic of the state machine of the robot, where engineers interface with external libraries to interact with sensors and actuators. The rest of the robot software, e.g., code for distributed communication, the build system, time-triggered operation, etc., are fully generated based on the requirements specification of the engineers. In this section, we describe a few model-driven development approaches, such as ROSMOD, that have been developed for robotics.

The BRICS component model [6] is based on model-driven development and separation of concerns. The syntax of the system model is represented by a connector-port-connector (CPC) meta-model, and the semantics is mapped to communication, computation, configuration, coordination, and composition. BRICS components represent computation and can be composed. A composite component contains a coordinator, a piece of software responsible for starting and stopping the computational components. Ports represent types of communication and BRICS allows ports for dataflow, events, service calls, etc. Connectors connect two compatible ports.

RobotML [12] is a domain-specific modeling language for designing, simulating, and deploying robotic applications. It is developed and integrated as part of the PROTEUS [54] project. The domain model consists of architecture, communication, behavior, and deployment meta-models. The architectural model defines (1) the system structure using a CPC model, and (2) data types, mission, and platform-specific properties. The communication model specifies the ports and port types—dataflow or services. The behavioral model uses state machines. Specific tasks are associated with states and transitions are mapped to specific algorithms. The deployment model specifies the assignment of each component to a target robotic middleware or simulator.

Similarly, the V^3CMM [13] component model consists of different views of a robotic system—structural, coordination, and algorithmic views. The structural view provides a static view of the components in the system, the coordination view describes event-driven behavior of the components, and the algorithmic view describes the algorithms executed by each component based on its current state. The coordination model is defined using UML state machines [48] and the algorithmic view is specified using UML activity diagrams [16].

8.2.2 Middleware in Robotics

Middleware is a software that connects software components or applications. Middleware consists of a set of services that allow multiple processes running on one or more devices to interact with each other. This technology has evolved to provide inter-operability in distributed applications. Middleware sits in the "middle" between the applications that may be running on multiple operating systems and could be imagined as a layer that operates between the application code and the run-time infrastructure. Middleware generally consists of a library of functions, and enables applications to use these functions from the common library rather than re-implement them for each application. Middleware, for this reason, provides reusability, portability, reliability, and simplifies the development process by managing the complexity.

Robotic systems middleware are a set of projects that all share the same basic concepts and goals, help improve the development process for robotic software—scenarios where many hardware and software components need to communicate and collaborate to reach a common goal. Robotic software applications, through the use of middleware technologies, have moved to a separation-of-concerns approach, e.g., "get sensor reading," "apply edge detection algorithm," "drive motor at speed x," etc. Components can exchange data using the common middleware communication framework and remain consistent across applications. This enables code reuse among different projects and applications.

Robot operating system (ROS) [63] is an open-source meta-operating system, the most popular robotics middleware. ROS provides services expected from an operating system, including hardware abstraction, low-level device control, implementation of commonly used functionality, message passing between processes and package management. ROS is aimed at primarily enabling code reuse in robotics for the development and release of software packages that perform a large set of common robotic tasks. The ROSMOD communication middleware is based on ROS.

The OROCOS [4] project aims at becoming a general-purpose, robot control software package that provides open-source software where developers can build sub-systems and modules without needing to deal with the code for the entire system. OROCOS is composed of C++ libraries provided through the OROCOS real-time toolkit (RTT), enabling the development of real-time component-based applications. This library allows designers to build configurable and interactive control systems and scripting interfaces. The communication interface is CORBA [67] and components can be configured using XML files. The OROCOS kinematics and dynamics library (KDL) [5] provides an application-independent framework for modeling and computation of kinematic chains, bio-mechanical human models, machine tools, etc. Orca [49] is an open-source framework for the development of component-based robotic software. Orca and OROCOS differ in the underlying communication framework. While OROCOS uses CORBA, Orca uses ZeroC [76], an Internet Communications Engine. Orca also provides some libraries for common applications and tools to simplify component development.

BRICS [6] is a middleware initiative to identify and document best practices in the development of complex robotic systems, refactoring existing components to achieve reusability and robustness with support for well-structured tool chains and configurable component code. This project aims to provide structure and formalization for developing robots and for increasing inter-operability of robot hardware and software components through well-defined interfaces on its component model and by providing a BRICS integrated development environment (BRIDE) based on the Eclipse [11] platform. BRICS, through BRIDE, supports model-based and component-based code generation for various robotics libraries such as OROCOS and ROS.

8.3 ROSMOD

Component-based software engineering (CBSE) and development [8, 36, 60] has
become an accepted practice for tackling software complexity in large-scale embedded robotic software. CBSE tackles escalated demands with respect to requirements
engineering, high-level design, error detection, tool integration, verification, and
maintenance. The widespread use of component technologies in the market has
made CBSE a focused field of research in the academic sectors. Applications are
built by assembling together small, tested component building blocks that implement a set of services. These building blocks are typically built from design models,
class diagrams, or imported from other projects/vendors and connected together via
exposed interfaces, providing a *black box* approach to software assembly.

ROSMOD [20, 21, 47] is a CBSE platform for distributed robotics using ROS.
ROSMOD consists of two parts: a modeling toolsuite and a run-time platform. The
modeling toolsuite, developed in WebGME [52], enables developers to graphically
model and implement their ROS applications. Coupling this with run-time tools
allows for cross-compilation, automated deployment on both desktop and embedded
platforms, and monitoring of these applications. In prior work, we have used
ROSMOD for design automation and analysis of various distributed CPS, e.g.,
fractionated spacecraft [15, 57], robotics [47], etc.

8.3.1 Component Model

The ROSMOD component model defines the basic software units that can be used
to assemble applications. An application can be distributed across several processes,
and each process has one or more *components*. Figure 8.2 shows the anatomy
of an application containing two processes created using ROSMOD components.

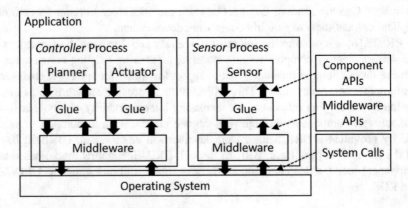

Fig. 8.2 Anatomy of a ROSMOD application

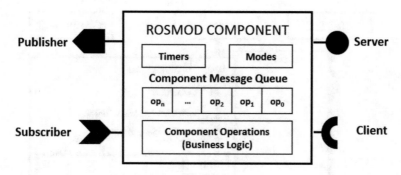

Fig. 8.3 ROSMOD component

The first process hosts two components: planner and actuator, the second process contains a single component: sensor. The components interact with each other via the generated "glue" code that connects them to the middleware layer.

Figure 8.3 provides an overview of a component. A component can have four different kinds of ports: publisher, subscriber, server (provided interface), and client (required interface). Publisher ports publish messages, without blocking, on a message topic. Subscriber ports subscribe to such topics and receive all messages published on the topic. This interaction implements an anonymous topic-driven publish–subscribe message passing scheme [26]. Server ports provide an interface to a service. Client ports can use this interface to call such services, exercising a blocking peer-to-peer synchronous remote method invocation [23]. Lastly, components can be triggered periodically or sporadically with timers.

A ROSMOD component is single-threaded; at most one thread can be active in a component at any time. All component operations are serially scheduled based on the associated triggers. Figure 8.4 shows the interaction between a periodic trigger, the ROS middleware, and the component. The middleware launches the operations associated with the trigger when the trigger becomes active. A time-based trigger will generate timeout events at a specified rate. If the operation is not completed within the time deadline specified, an anomaly is detected and reported.

Triggers may be used to realize the following interaction patters:

* *Periodic Publisher (Time Trigger with Publisher)*: This pattern relies on an operation in a component that publishes data at a periodic rate.
* *Periodic Client (Time Trigger with Client)*: Similarly, a time-triggered client can be specified. This enables a component to periodically refresh itself with new data from another component.
* *Periodic Subscriber (Time Trigger with Subscriber)*: This pattern can be used to implement a "pull" subscriber operation that is triggered periodically.
* *Aperiodic Subscriber (Event Trigger with Subscriber)*: This pattern will allow the implementation of a "push" subscriber operation, which gets triggered only when data is available.

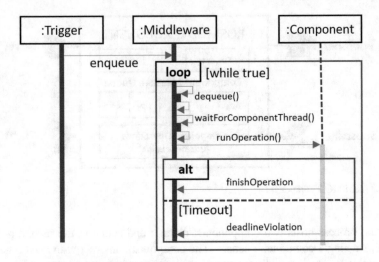

Fig. 8.4 Interaction of a time trigger and component

Inside a component, once a trigger has started an operation, it is possible to daisy-chain the calls by directly invoking other operations as functions. However, care should be taken because the deadline of the first operation must account for the longest chain of other operations that can be called as part of the first operation [45].

8.3.2 Model-Driven Development Toolsuite

The development process for ROSMOD applications can be realized in two ways:

1. *The conventional process*: The application developer constructs all the software using an implementation language (e.g., C++) and using middleware libraries to access the services provided by ROS. Technically, the developer can develop applications using the core interfaces provided by ROS libraries, but this involves tedious low-level coding and re-building functionality that is provided by the middleware libraries. The developer delivers the application (as source or object code) and the meta-data, as required by the system integrator.
2. *The model-driven process*: The application developer performs the system design and the high-level specification of the application using model-based tools (e.g., an architecture modeling language with graphical tool support), uses the tools to generate the infrastructure ("glue") code needed to integrate the application logic with ROS libraries, and adds the "business logic" of the application using an interactive development environment. In this scheme, the developer uses the well-established, conventional code development style for implementing the core application functionality, and all the low-level, glue code is auto-generated from

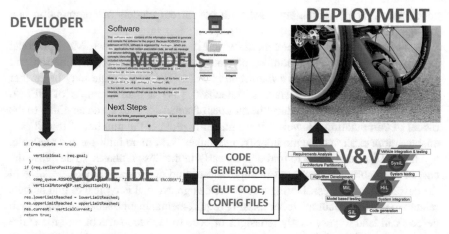

Fig. 8.5 Model-driven development for ROSMOD. The developer icon in this figure was made by *Vectors Market* [30] from www.flaticon.com [29]

the models. The developer delivers the application (as source or object code) and the models of the application (from which the glue code and the meta-data will be re-generated by the system integrator).

The model-driven process is outlined in Fig. 8.5.

The ROSMOD toolsuite includes tools for MDE, including an extensible domain-specific modeling language and its supporting visual modeling environment, the various software generators that produce code and other implementation artifacts, and model-driven tools for model analysis and verification. Other model-driven tools can be used as well. For instance, the business logic of applications can be developed using Simulink/Stateflow and the resulting models (and the code generated from them) integrated into the final application. In the next section, we briefly discuss one such tool: a design environment for modeling hierarchical finite state machines, called the *HFSM Design Studio*.

8.3.3   HFSM Design Studio

Finite state machines (FSM) [32] have long been used to model and analyze event-driven reactive systems, e.g., embedded devices. Devices react to some kind of external or internal stimuli which leads to an action and, eventually, to a change of state. Due to their finite nature, FSM are more amenable to analysis and synthesis than alternative control models, e.g., with FSM, a designer can enumerate the set of reachable states of the system to determine that a particular unsafe state cannot be reached. Most modern computer systems have both complex control flow and concurrency. Thus, combining FSM with concurrent models of computation is a popular design choice [1, 34, 50, 51, 74].

Plain FSM are flat and sequential, which is a major weakness. Most practical systems have a very large number of states and transitions. Representation and analysis become difficult. In a *hierarchical* FSM (HFSM), a state may be further refined into another FSM. Harel's statecharts [34] are an example of HFSM. Composing FSM does not reduce the number of states nor does it add anything to the model of computation. However, it significantly reduces the number of transitions, enables more explicit code modularity, and makes FSM more intuitive and readable.

HFSM allow engineers to find defects early in the design phase. To decrease the cost of poor software quality, it is important to find defects as early as possible in the development process. During a software design cycle, it is typical to find defects related to unclear, incomplete, or missing requirements. In embedded software such defects can lead to very costly redesigns or even to the reconstruction of the entire system. There is a definite need for early, integrated testing and simulation to identify architectural and behavioral defects in embedded software.

It is easily possible to execute a state machine in a simulator and allow the user to send events to the machine and observe how the HFSM reacts to the sent stimuli. This way, the user can interactively test the model and improve it where necessary. Depending on the embedded system, this type of testing may not be possible with the actual target platform. Also, for UML state machines [48], the OMG has specified a set of well-formedness rules within the UML specification [62], e.g., *"Final states must only have incoming transitions."* These rules as well as a number of additional rules [33] can be automatically checked by a model checker during the design phase.

Once a HFSM has been checked, the implementation can start. HFSM can be tricky to code by hand, especially in highly composite cases. When new transitions or states have been added to the HFSM, one wishes to have a generator at hand taking over all the error-prone placement of entry, exit, and action code associated with states or transitions. Automatic code generation has many benefits, especially if a model checker is integrated in the generator and can perform a large number of checks automatically. It must be said that automatic code generation does not make source code analysis needless. Static code analysis tools have been evolving for more than two decades [17]. Both model checking and code analysis complement each other quite well and tools should make sure that the generated code is not in conflict with code analysis tools.

ROSMOD has an integrated *HFSM Design Studio* [18, 19], a WebGME application for constructing UML state machines. The design studio includes a visualizer, called HFSMViz, and a simulation environment. Figure 8.6 shows a toy example. WebGME *plugins* are available for generating executable code directly from these HFSM models. We discuss more about how this environment is used for our PW software in the next section.

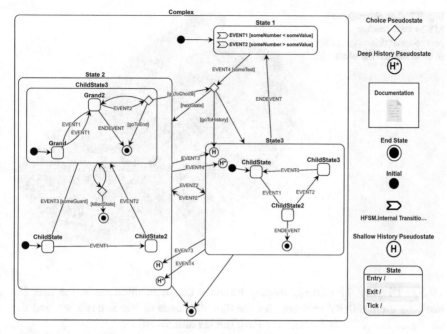

Fig. 8.6 Example HFSM modeled in HFSM Design Studio

8.4   Case Studies

8.4.1   PushTracker

The PushTracker is a Bluetooth low-energy (BLE) wearable activity tracker for manual wheelchair users. It is designed to be usable by people with varying degrees of ability and allows the users to control a light-weight power-assist drive system for their wheelchair—the SmartDrive. The SmartDrive easily attaches within the footprint of both folding and rigid-frame manual wheelchairs. These two products, collectively the SmartDrive MX2+ PushTracker (Fig. 8.7), operate together through a BLE connection over which they transmit control and status information, and over-the-air (OTA) software updates.

The PushTracker is comprised of the following functional components:

Bluetooth Radio/Processor   The BLE radio is the only micro-processor on-board the PushTracker and is the architecture for which the software is developed. It provides the main function of the PushTracker—the connectivity and communication with the SmartDrive. Additionally, it also provides the connectivity to mobile applications for data and OTA updates.

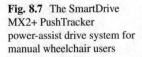

Fig. 8.7 The SmartDrive
MX2+ PushTracker
power-assist drive system for
manual wheelchair users

OLED Display The OLED display provides the main interface to the user for showing their activity measures for the day, the status of the SmartDrive, and the configuration controls for both the PushTracker and SmartDrive.

Status LEDs Status LEDs are used to provide secondary visual feedback to the user that can be viewed from different angles of the PushTracker. These lights are designed to be visible while the PushTracker is worn and the user's hands are on the hand-rim of their PW. This allows users to maintain a position of control over their PW while still being able to see important information, shown by different patterns of color and blinks.

Push Buttons The two push buttons provide the primary means of input to the PushTracker, controlling the PushTracker's power, the connection to and control of the SmartDrive, and the auxiliary interfaces to the configuration menu and connectivity to the mobile applications. In addition to the tactile feedback from pressing the buttons, status LEDs on the face of the PushTracker illuminate while the user is pressing the buttons to inform them that the PushTracker is registering their input.

Accelerometer The accelerometer is responsible for detecting the various actions of the user, including when they (1) push their PW, (2) perform different types of tap gestures, and (3) reorient their wrist. All of these actions make up the different gestures that the user can use as various control inputs to the PushTracker, e.g., if the user holds the PushTracker level, the status LEDs will turn blue indicating that the user can then perform a double tap gesture to toggle the power-assist system. While the PushTracker is running, it is constantly using the accelerometer to look for these gestures and to count the number of pushes the user is performing. These gestures provide a convenient control interface for the PushTracker for users with limited hand and finger dexterity.

Vibration Motor The vibration motor allows the PushTracker to provide non-visual feedback to the user during critical periods of activity where the user must pay attention, e.g., when they are engaging the SmartDrive while maneuvering their PW. This feedback provides yet another level of support to the user and enables safe and effective operation of the SmartDrive.

Real-Time Clock Finally, the real-time clock (RTC) provides a low-power reliable time source for use with data logging and display to the user.

From the onset, the PushTracker software has been developed traditionally (without MDE) and using an off-the-shelf proprietary scripting language called BGScript (BGS) [3]. This language lends itself well to rapid development of event-based software but has many limitations, e.g., limited stack size, limited reusability, lack of tool-chain support, and advanced debugging. To keep up with the evolving design of the PushTracker, the software design team transitioned into an MDE approach, integrating these proprietary tools [3, 37] with open-source modeling frameworks [18, 47, 52]. We discuss this design process in more detail in the following subsections.

8.4.1.1   Structured Code Generation

The PushTracker software is designed as a FSM. BGS, however, lacks high-level language features such as custom data types or object-oriented programming. The language also has a very limited call stack size; factoring the code into a large set of functions is not an option. As such, code reviews are tedious and the code does not lend itself to readability.

To address these issues, we have integrated the HFSM Design Studio directly into the design process. Figure 8.8 shows the HFSM model of the PushTracker, as modeled in ROSMOD. Collaboratively editable, version-controlled, high-level modeling like this facilitates robust and rapid software design and implementation. The PushTracker code base, i.e., BGS, and all required execution files are now fully generated through model transformation from these HFSM models. The "glue code" generated from HFSM models is around 16,000 lines of code. The engineer implements the core business logic for the HFSM states (9000 lines of code). This hand-written code includes input/output and interrupt handling, and data manipulation, e.g., *enable_acc()* to toggle power to the accelerometer component.

8.4.1.2   Switching Languages

Decoupling the system design from the implementation, even if not completely, allows engineers to study the bottlenecks in the current implementation and consider changes at the implementation level without any breaking changes at the design level. HFSM-based modeling and code generation means that developers can transition their implementation to a more suitable embedded programming

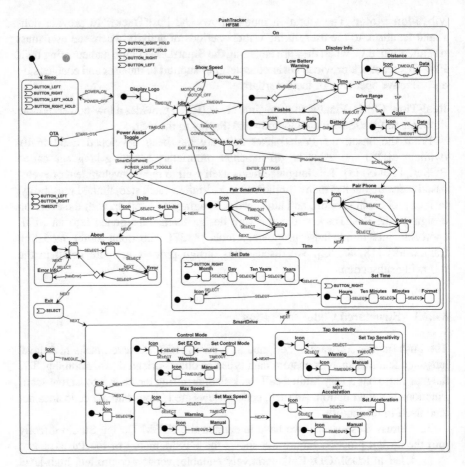

Fig. 8.8 HFSM for PushTracker, modeled in ROSMOD

language, such as C, and only require changes to the code generators. The initial engineering effort required for this transition may be high but the reward over the long-term justifies the effort. In practice, our team had to develop 900 lines of code transformation software to generate C code from HFSM models. Comparing the model transformation from HFSM to BGS, this is a negligible amount of effort. The reduction in code re-implementation allows the developers to focus their efforts on properly translating the state machine business logic without having to worry about aberrant HFSM behavior that might have come about from manually translating the HFSM glue code.

8.4.1.3 Process Control and Verification

Since the PushTracker controls the SmartDrive MX2+, it is classified in the USA by the Food and Drug Administration (FDA) as a Class II medical device, just like a PW. This classification imposes strict requirements on the pre-market approval required to bring the device to market, the design control procedures for its development, and the standards and tests to which it must conform. Currently, the FDA approval process involves data tables linking risks, hazards, device requirements for software and hardware, and tests performed to check the deployed software and hardware for these risk scenarios. The controller software is not model-checked, although code analysis is performed for quality control.

One of the planned improvements to the HFSM-based modeling is to integrate with risk and system-level assurance analysis. For example, SEAM [38], also developed with WebGME, supports the goal structuring notation (GSN) [43] standard to build assurance case models. SEAM uses hierarchical models, as well as cross-referencing to manage complexity in GSN models. SEAM also allows linking assurance cases to system models to provide context to assurance case arguments.

8.4.2 Drive Assistance for Powered Wheelchairs

PW users want to live their lives without being defined by their injury. Unfortunately, many activities of daily living (ADL) are difficult for complex PW users [28, 73]. These include navigating complex environments, maneuvering through tight spaces, and docking the chair in transport vehicles. Despite these challenges, many users do not want a fully autonomous solution [58].

Mobility solutions on the market consist of traditional direct-drive joysticks and a variety of alternative drive controls for users who lack the hand functions required for operating a traditional joystick. Such products give the user full, direct control over the chair. However, such direct control puts all of the navigation and environmental awareness burden onto the user, causing fatigue [24, 27, 28] which can impact the user's mobility and increase risk of fall or injury [28].

Autonomous powered wheelchairs (AWC) comprise multiple functional requirements from chair status and health monitoring, user interaction and intent determination, and environmental awareness and navigation. Such requirements naturally lend themselves to the design and development of a suite of software and hardware components that implement a subset of these features and which can be composed together through standard interfaces to produce the desired autonomous system.

One important requirement for the AWC software is the inter-operation of ROSMOD with Unity [9]. In the AWC project, see Fig. 8.9, much of the high-level path planning and environmental sensing, localization, and mapping are performed on an off-the-shelf Microsoft HoloLens [7], using custom software developed in Unity. The HoloLens is the SLAM device [39] for the AWC. It communicates with the user interface wirelessly and controls a Unity application running on a

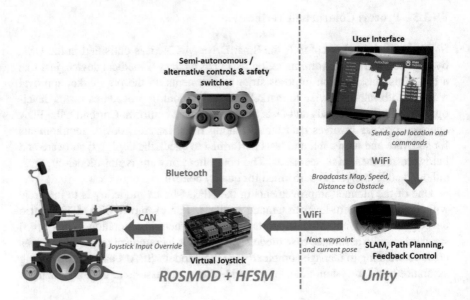

Fig. 8.9 Overall design of autonomous powered wheelchair prototype. The virtual joystick provides a wireless joystick interface to the powered wheelchair, translating the wireless control commands from both the hand-held controller (manual) and the Microsoft HoloLens (autonomous)

tablet mounted to the AWC chair. To control the AWC base, the HoloLens software communicates over WiFi with a Raspberry Pi [61]. The Raspberry Pi executes code, generated from ROSMOD and HFSM, and controls the AWC via a CAN bus [25]. The control is either manual or autonomous. Manual control is via a Playstation DualShock 4 Controller [10] over Bluetooth. Autonomous control is via the HoloLens. Additionally, a semi-autonomous mode is available whereby the user controls the chair trajectory with the controller's joystick and the Raspberry Pi modulates the speed according to the proximity of obstacles that it receives from the HoloLens.

8.4.2.1 Overview of System Design

The AWC is comprised of these interconnected sub-systems:

Environment Detection and Segmentation The most critical aspect of the AWC platform is its capability to detect and segment the environment. In our prototype, these functions are carried out by code developed in Unity, running on the HoloLens. The HoloLens provides the ability to detect and remember surrounding geometry, without any semantic meaning or segmentation—all data is stored as a single mesh of points connected together by triangles. From this data, it is the responsibility of the segmentation sub-system to determine what the salient features of the environment are for the AWC: the floor, drivable surface, the walls, bounds of the drivable surface, and the objects or humans present on the drivable surface.

This segmentation task is performed by customizing and composing several Unity components in a model-driven fashion. For performance reasons, only the volume within a 20-m rectangle that is 5-m tall centered around the HoloLens is considered for environment segmentation and path planning. Therefore, a simple segmentation method is chosen whereby an orthographic camera is placed as a child of the user (HoloLens camera) looking down the world Y-axis (in Unity coordinate frame) at the HoloLens and placed vertically 2.5-m above the HoloLens camera. This orthographic camera renders its view to a render texture which runs a shader pipeline as a post-process technique to perform the environment segmentation in parallel on the graphics processing unit (GPU). The main criteria for segmentation are both the per-pixel depth from the orthographic camera and the per-vertex normal of the geometry being rendered. Using these criteria, we can project the three-dimensional (3D) geometry into two dimensions and color them according to whether they belong to the floor, boundary, or obstacle categories. The output of this pipeline is a two-dimensional (2D) map of the surrounding area that can be viewed by the user and used for path planning. Because this map is generated using this Unity's render-texture pipeline, it is updated automatically every time Unity's entity execution framework runs its *Update* event, which in this case is approximately 10–20 Hz.

The use of Unity for this development allows all of the rendering, containment, shader setup, and GPU communications code to be generated, leaving only the actual segmentation code (the business logic of this component) to be written by hand according to the Unity software interfaces.

Path Planning Similarly to the segmentation sub-system, the path planning sub-system consists of a collection of inter-operating components which receive inputs from both the segmentation sub-system and the user interface sub-system. Creating these as Unity components allows us to leverage Unity to generate the interaction and communication code between these components as well as the event handling and entity execution code for integrating with Unity's run-time scheduling framework. The path planning is composed of two different components, one which runs every time the execution framework updates and re-renders and the other which runs only when the user specifically inputs a desired goal location. The periodic component, such as the segmentation component, is executed as a post-process pixel shader to automatically perform a distance transform on the output of the segmentation (the 2D segmented map). It does this by implementing the jump flooding algorithm (JFA) [64] to first compute the Voronoi diagram of the input and then apply the distance transform. This distance transform can then be used immediately by a traditional path finding algorithm such as A* [35] when the user provides a goal location as input (the start location is always the center of the map since the map follows the AWC).

The glue code required to receive these inputs and execute on user input were all generated according to the model defined in Unity, allowing much of the code outside of the actual distance transform and path finding business logic to be generated entirely from Unity's modeling environment.

Hand-held User Interface The hand-held user interface provides a 2D touch-enabled wireless control and display to the user to inspect the AWC's segmentation of its environment, manage control over the autonomy, and inspect the generated path to goal. Since much of the data being shown on the hand-held interface is the same as what the HoloLens uses, much of the interface code is shared between these components.

Augmented-Reality User Interface To prototype the system in both a safe and decoupled manner, an augmented reality (AR) interface is developed and deployed on the HoloLens. This interface allows the engineer to walk around their environment and evaluate how well the environmental segmentation and path planning works from both a functional and a performance perspective. Because Unity is a game development engine, integrating the map and distance transformation becomes painless. Additionally, the Unity-based collision detection used during autonomous navigation is able to be visualized in 3D in the real world which expedites the iterative development process. Many of these components, especially the user interface, are pre-built models from Unity's and the HoloLens' library which are easily instantiated and configured to produce a rich user interface and debugging experience with minimal effort.

Communications In order to tie the user interface, the segmentation/path planning, and the chair controls together, we needed a communications sub-system which can interface between the Unity components running in the user interface/segmentation/path planning components and the ROSMOD components that handle manual control overrides and embedded interfaces to the PW. Since the HoloLens does not support wired data connections and the user interface is designed to be mobile, all communications are designed to use low-overhead wireless protocols. Much of the code for the communications components across the different sub-systems involve a mix of custom and library code for encoding/decoding and compressing/decompressing the map image data to maximize the number of image frames we could send/receive per second. To solve this challenge, all Unity communications code resided in a single Unity component which could be instantiated in all the Unity models. Similarly, a ROSMOD component is responsible for acting as the adapter from the wireless communications to the ROS communications between the low-level control components of the chair interface module.

Manual and Alternative Control Input Since this is a semi-autonomous PW, it supports manual control overrides and safety switches, in addition to chair-level configuration controls. Additionally, by leveraging an off-the-shelf consumer gaming controller, we are able to prototype other alternative manual drive controls for the PW, e.g., a touchpad-based joystick and gyroscopic controls for the PW. The interface to this controller is implemented as a ROSMOD component, acting as an adapter from the driver-specific Bluetooth interface to a ROS message topic containing the input data.

Motion Planning and Execution   Another ROSMOD component is used to combine the manual and autonomous control inputs and plan the actual trajectory for the PW. This component ensured the safety of the user by ensuring that manual control inputs always had higher priority than the autonomous control inputs. Finally, this component is responsible for transforming these inputs into a single virtual joystick command.

PW Interface   The final component for interfacing with the chair is a ROSMOD component acting as an adaptation layer between the ROS communications and the CAN bus on the chair.

The use of both ROSMOD and Unity for modeling the different components of the AWC system allowed a team of only four software and hardware developers to create a functional prototype of an autonomous powered wheelchair in 3 months. Such design automation allowed the team to focus on the core algorithms they wanted to evaluate for use in AWC, without having to focus on platform or glue code—which as prototype code with off-the-shelf hardware would not be relevant to the final system.

8.5   Future Work

The work we have shown thus far regarding design automation for both the PushTracker and the AWC presents a work-in-progress for both a toolsuite and a fully integrated MDE design process. Much of the work that has been done thus far has been for the modeling of glue and interaction code in order to leverage code generators and model checkers for correctly transforming abstract models into executable code. However, much of the design process is still performed traditionally, e.g., the creation of design, device, and hardware requirements, the analysis of the system, and the test procedures required for verification and validation. ROSMOD has been developed as a design studio which can function within and alongside other design studios that will enable such systems engineering and more formalized design procedures. The rest of this section will discuss some of the possible extensions to this work along such a path.

8.5.1   Controller Synthesis

It is evident from our discussion in Sect. 8.4.1.1 that ROSMOD and HFSM do not synthesize complete controllers. The HFSM Design Studio generates glue code for the HFSM structure and it is still the responsibility of the engineer to hand-write the business logic for these applications. This still leaves thousands of lines of software to be written by a human; software that is neither generated from "correct by construction" models nor easily verifiable. Frameworks such as GASPARD [31] have, in prior work, transformed OMG standard profile MARTE models [53] of embedded applications into synchronous programs from which controllers can be synthesized using formal techniques. Such synchronous models are very close to source-code level in abstraction while constituting a verifiable representation. One of our planned projects is to study state-of-the-art controller synthesis methods and integrate such tools into ROSMOD if applicable.

8.5.2 Learning-Enabled Systems

One aspect where our current tools fall short, and where design specification as well becomes more challenging is for a newer class of devices: learning-enabled systems [2, 40, 55]. These systems are forming the core of new technologies which allow embedded systems to adapt to unknown environments and to their user for a more personalized experience. Such technologies have the potential to provide great benefit to users in complex rehabilitation but the specification, development, verification, and validation of such systems pose a significant challenge to ensuring that these systems remain safe and effective for their intended use throughout the life-cycle of the device.

To address these concerns, current modeling, synthesis, development, and testing frameworks need to be extended to accommodate learning-enabled systems. These are systems with learning-enabled components such as deep neural networks [71]. This is an active field of research [41, 42, 75] and so far no single paradigm has shown such promise as to be the accepted standard. However, tools which allow the designer to more formally describe the system and its environment provide a path towards componentization of learning-enabled systems into compositions of traditional software components and learning-enabled components, allowing for the analysis and verification of such a composed system.

8.5.3 Simulation

As complex rehabilitation systems become autonomous, traditional software model checking may prove an intractable problem for showing safety and effectiveness. Such complex systems require holistic analysis including the addition of comprehensive simulated-based testing with high-fidelity hardware-in-the-loop simulation environments [46]. These testbeds allow for automated, large-scale testing of a battery of different environments, use-cases, and failure modes which would be infeasible or potentially dangerous to perform in situ.

The integration of simulation tools for performing automated testing is not new, but only recently have high-fidelity simulations of many complex physical systems, actors, and locations at once been readily available. The ROSMOD team has already started working on integrating with various physical simulation tools, many of which come from the video gaming community. But recently, vendors such as Unity and NVIDIA have started releasing their own specific simulation tools to address these problems. These new tools provide unparalleled simulation capability with open, extensible interfaces which can dramatically help the automation of iterative software development and testing.

8.6   Conclusions

In this work, we have covered many aspects of design automation for software development in robotics, specifically as they relate to devices in complex rehabilitation such as power-assist devices for manual wheelchairs and drive-assistance devices for powered wheelchairs. These devices were used to showcase the current state of design automation for software engineering of complex rehabilitation systems, as well as to motivate the need for continued research in this area. The benefit of model-based software engineering was described with motivating examples from different toolchains provided for free from different vendors. Finally, some promising future work was presented which helps address current limitations of existing tools for developing state-of-the-art complex rehabilitation systems.

Acknowledgements ROSMOD was partly supported by DARPA under contract NNA11AB14C and USAF/AFRL under Cooperative Agreement FA8750-13-2-0050, and by the National Science Foundation (CNS-1035655). The authors would like to thank Ben Hemkens, Liyun Guo, Kennth Shafer, Dexter Watkins, and Devon Doebele for their work on the projects mentioned in this chapter.

References

1. Benveniste, A., & Berry, G. (1991). The synchronous approach to reactive and real-time systems. *Proceedings of the IEEE, 79*(9), 1270–1282.
2. Beom, H. R., & Cho, H. S. (1995). A sensor-based navigation for a mobile robot using fuzzy logic and reinforcement learning. *IEEE Transactions on Systems, Man, and Cybernetics, 25*(3), 464–477.
3. BlueGiga. (2017). BGScript from Silicon Labs/BlueGiga. https://bluegiga.com. Accessed July 2018.
4. Bruyninckx, H. (2001). Open robot control software: The OROCOS project. In *Proceedings 2001 ICRA. IEEE International Conference on Robotics and Automation, 2001* (Vol. 3, pp. 2523–2528). Piscataway: IEEE.
5. Bruyninckx, H., Soetens, P., & Koninckx, B. (2003). The real-time motion control core of the OROCOS project. In *Proceedings. ICRA'03. IEEE International Conference on Robotics and Automation, 2003* (Vol. 2, pp. 2766–2771). Piscataway: IEEE.
6. Bruyninckx, H., Klotzbücher, M., Hochgeschwender, N., Kraetzschmar, G., Gherardi, L., & Brugali, D. (2013). The BRICS component model: A model-based development paradigm for

complex robotics software systems. In *Proceedings of the 28th Annual ACM Symposium on Applied Computing* (pp. 1758–1764). New York: ACM.

7. Chen, H., Lee, A. S., Swift, M., & Tang, J. C. (2015). 3D collaboration method over HoloLens and Skype end points. In *Proceedings of the 3rd International Workshop on Immersive Media Experiences* (pp. 27–30). New York: ACM.

8. Clemens, S., Dominik, G., & Stephan, M. (1998). *Component software: Beyond object-oriented programming*. Boston, MA: Addison-Wesley.

9. Creighton, R. H. (2010). Unity 3D game development by example: A seat-of-your-pants manual for building fun, groovy little games quickly. Birmingham: Packt Publishing Ltd.

10. Davies, M., Read, H., Xynos, K., & Sutherland, I. (2015). Forensic analysis of a Sony playstation 4: A first look. *Digital Investigation, 12*, S81–S89.

11. des Rivières, J., & Wiegand, J. (2004). Eclipse: A platform for integrating development tools. *IBM Systems Journal, 43*(2), 371–383.

12. Dhouib, S., Kchir, S., Stinckwich, S., Ziadi, T., & Ziane, M. (2012). RobotML, a domain-specific language to design, simulate and deploy robotic applications. In *International Conference on Simulation, Modeling, and Programming for Autonomous Robots* (pp. 149–160). Berlin: Springer.

13. Diego, A., Cristina, V. C., Francisco, O., Juan, P., & Bárbara, Á. (2010). V3cmm: A 3-view component meta-model for model-driven robotic software development. *Journal of Software Engineering in Robotics, 1*(1), 3–17.

14. Ding, D., & Cooper, R. A. (2005). Electric powered wheelchairs. *IEEE Control Systems, 25*(2), 22–34.

15. Dubey, A., Emfinger, W., Gokhale, A., Karsai, G., Otte, W. R., Parsons, J., et al. (2012). A software platform for fractionated spacecraft. In *2012 IEEE Aerospace Conference* (pp. 1–20). Piscataway: IEEE.

16. Dumas, M., & Ter Hofstede, A. H. (2001). UML activity diagrams as a workflow specification language. In *International Conference on the Unified Modeling Language* (pp. 76–90). Berlin: Springer.

17. Emanuelsson, P., & Nilsson, U. (2008). A comparative study of industrial static analysis tools. *Electronic Notes in Theoretical Computer Science, 217*, 5–21.

18. Emfinger, W. (2018). HFSM design studio. https://cps-vo.org/group/hfsm.

19. Emfinger, W. (2018). HFSM design studio on github. https://github.com/finger563/webgme-hfsm.

20. Emfinger, W. (2018). ROSMOD design studio. https://cps-vo.org/group/rosmod.

21. Emfinger, W. (2018). ROSMOD github. https://github.com/rosmod.

22. Emfinger, W., Karsai, G., Dubey, A., & Gokhale, A. (2014). Analysis, verification, and management toolsuite for cyber-physical applications on time-varying networks. In *Proceedings of the 4th ACM SIGBED International Workshop on Design, Modeling, and Evaluation of Cyber-Physical Systems* (pp. 44–47). New York: ACM.

23. Emmerich, W., & Kaveh, N. (2001). Component technologies: Java Beans, COM, CORBA, RMI, EJB and the CORBA Component Model. In *ACM SIGSOFT Software Engineering Notes* (Vol. 26, pp. 311–312). New York: ACM.

24. Erdogan, A., & Argall, B. D. (2017). The effect of robotic wheelchair control paradigm and interface on user performance, effort and preference: An experimental assessment. *Robotics and Autonomous Systems, 94*, 282–297.

25. Etschberger, K. (2001). *Controller area network: Basics, protocols, chips and applications*. Weingarten: IXXAT Automation GmbH.

26. Eugster, P. T., Felber, P. A., Guerraoui, R., & Kermarrec, A. M. (2003). The many faces of publish/subscribe. *ACM Computing Surveys (CSUR), 35*(2), 114–131.

27. Ezeh, C., Trautman, P., Devigne, L., Bureau, V., Babel, M., & Carlson, T. (2017). Probabilistic vs linear blending approaches to shared control for wheelchair driving. In *2017 International Conference on Rehabilitation Robotics (ICORR)* (pp. 835–840). Piscataway: IEEE.

28. Fehr, L., Langbein, W. E., Skaar, S. B. (2000). Adequacy of power wheelchair control interfaces for persons with severe disabilities: A clinical survey. *Journal of Rehabilitation Research and Development, 37*(3), 353–360.
29. Flaticon. (2018). Flaticon. https://www.flaticon.com.
30. Flaticon. (2018). Vectors market. https://www.flaticon.com/authors/vectors-market.
31. Gamatié, A., Yu, H., Delaval, G., & Rutten, É. (2009). A case study on controller synthesis for data-intensive embedded systems. In *2009 International Conference on Embedded Software and Systems* (pp. 75–82). Piscataway: IEEE.
32. Gill, A. (1962). *Introduction to the theory of finite-state machines*. New York: McGraw-Hill.
33. Gomez, M. (2004). Embedded state machine implementation: Turning a state machine into a program can be straightforward if you follow the advice of a skilled practitioner. In *The firmware handbook* (pp. 101–109). Burlington: Elsevier.
34. Harel, D. (1987). Statecharts: A visual formalism for complex systems. *Science of Computer Programming, 8*(3), 231–274.
35. Hart, P. E., Nilsson, N. J., & Raphael, B. (1968). A formal basis for the heuristic determination of minimum cost paths. *IEEE Transactions on Systems Science and Cybernetics, 4*(2), 100–107. https://doi.org/10.1109/TSSC.1968.300136.
36. Heineman, G. T., & Councill, W. T. (2001). *Component-based software engineering. Putting the pieces together* (p. 5). London: Addison-Wesley.
37. IAR. (2017). Embedded Workbench from IAR. https://iar.com/products/iar-embedded-workbench. Accessed July 2018.
38. ISDE. (2018). Systems engineering and assurance modeling. https://modelbasedassurance.org/.
39. Ishikawa, R., Oishi, T., & Ikeuchi, K. (2018, preprint). Offline and online calibration of mobile robot and slam device for navigation. arXiv:180404817.
40. Kaelbling, L. P., Littman, M. L., & Moore, A. W. (1996). Reinforcement learning: A survey. *Journal of Artificial Intelligence Research, 4*, 237–285.
41. Katz, G., Barrett, C., Dill, D. L., Julian, K., & Kochenderfer, M. J. (2017). Reluplex: An efficient SMT solver for verifying deep neural networks. In *International Conference on Computer Aided Verification* (pp. 97–117). Berlin: Springer.
42. Katz, G., Barrett, C., Dill, D. L., Julian, K., & Kochenderfer, M. J. (2017, preprint). Towards proving the adversarial robustness of deep neural networks. arXiv:170902802.
43. Kelly, T., & Weaver, R. (2004). The goal structuring notation–A safety argument notation. In *Proceedings of the Dependable Systems and Networks 2004 Workshop on Assurance Cases*, Citeseer (p. 6).
44. Koenig, N. P., & Howard, A. (2004). Design and use paradigms for Gazebo, an open-source multi-robot simulator. In *IROS*, Citeseer (Vol. 4, pp. 2149–2154).
45. Kumar, P. S. (2016). *Integrated Timing Analysis and Verification of Component-Based Distributed Real-Time Systems*. PhD thesis, Vanderbilt University.
46. Kumar, P. S., Emfinger, W., & Karsai, G. (2015). A testbed to simulate and analyze resilient cyber-physical systems. In *2015 International Symposium on Rapid System Prototyping (RSP)* (pp. 97–103). Piscataway: IEEE.
47. Kumar, P. S., Emfinger, W., Karsai, G., Watkins, D., Gasser, B., & Anilkumar, A. (2016). ROSMOD: A toolsuite for modeling, generating, deploying, and managing distributed real-time component-based software using ROS. *Electronics, 5*(3), 53.
48. Lilius, J., & Paltor, I. P. (1999). *The semantics of UML state machines*. Turku: Turku Centre for Computer Science.
49. Makarenko, A., Brooks, A., & Kaupp, T. (2006). Orca: Components for robotics. In *International Conference on Intelligent Robots and Systems (IROS)* (pp. 163–168).
50. Maraninchi, F. (1991). The Argos language: Graphical representation of automata and description of reactive systems. In *IEEE Workshop on Visual Languages*, Citeseer (Vol. 3).

51. Maraninchi, F. (1992). Operational and compositional semantics of synchronous automaton compositions. In *International Conference on Concurrency Theory* (pp. 550–564). Berlin: Springer.
52. Maróti, M., Kecskés, T., Kereskényi, R., Broll, B., Völgyesi, P., Jurácz, L., et al. (2014). Next generation (meta) modeling: Web-and cloud-based collaborative tool infrastructure. *MPM@ MoDELS, 1237*, 41–60.
53. MARTE, O. (2007). Profile for modeling and analysis of real-time and embedded systems (MARTE). Beta1 Google Scholar.
54. Martinet, P., & Patin, B. (2008). PROTEUS: A platform to organise transfer inside French robotic community. In *3rd National Conference on Control Architectures of Robots (CAR)*.
55. Matarić, M. J. (1997). Reinforcement learning in the multi-robot domain. In *Robot colonies* (pp. 73–83). New York: Springer.
56. Mazumder, O., Kundu, A. S., Chattaraj, R., & Bhaumik, S. (2014). Holonomic wheelchair control using EMG signal and joystick interface. In *2014 Recent Advances in Engineering and Computational Sciences (RAECS)* (pp. 1–6). https://doi.org/10.1109/RAECS.2014.6799574.
57. Otte, W. R., Dubey, A., Pradhan, S., Patil, P., Gokhale, A., Karsai, G., et al. (2013). F6COM: A component model for resource-constrained and dynamic space-based computing environments. In *2013 IEEE 16th International Symposium on Object/Component/Service-Oriented Real-Time Distributed Computing (ISORC)* (pp. 1–8). Piscataway: IEEE.
58. Padır, T. (2015). Towards personalized smart wheelchairs: Lessons learned from discovery interviews. In *2015 37th Annual International Conference of the IEEE Engineering in Medicine and Biology Society (EMBC)* (pp. 5016–5019). Piscataway: IEEE.
59. Pasteau, F., Krupa, A., & Babel, M. (2014). Vision-based assistance for wheelchair navigation along corridors. In *2014 IEEE International Conference on Robotics and Automation (ICRA)* (pp. 4430–4435). Piscataway: IEEE.
60. Pastor, O., España, S., Panach, J. I., & Aquino, N. (2008). Model-driven development. *Informatik-Spektrum, 31*(5), 394–407.
61. Pi, R. (2015). Raspberry Pi model B.
62. Prochnow, S., Schaefer, G., Bell, K., & von Hanxleden, R. (2006). Analyzing robustness of UML state machines. In *MARTES 2006 at MoDELS 2006, LCAV-CONF-2006-030* (pp. 61–80).
63. Quigley, M., Conley, K., Gerkey, B., Faust, J., Foote, T., Leibs, J., et al. (2009). ROS: An open-source robot operating system. In *ICRA Workshop on Open Source Software*, Kobe (Vol. 3, p. 5).
64. Rong, G., & Tan, T. S. (2006). Jump flooding in GPU with applications to Voronoi diagram and distance transform. In *Proceedings of the 2006 Symposium on Interactive 3D Graphics and Games* (pp. 109–116). New York: ACM.
65. Schlegel, C., Haßler, T., Lotz, A., & Steck, A. (2009). Robotic software systems: From code-driven to model-driven designs. In *International Conference on Advanced Robotics, 2009. ICAR 2009* (pp. 1–8). Piscataway: IEEE.
66. Schmidt, D. C. (2006). Model-driven engineering. *Computer IEEE Computer Society, 39*(2), 25.
67. Siegel, J., & Frantz, D. (2000). *CORBA 3 fundamentals and programming* (Vol. 2). New York, NY: Wiley.
68. Simpson, R. C., LoPresti, E. F., & Cooper, R. A. (2008). How many people would benefit from a smart wheelchair? *Journal of Rehabilitation Research & Development 45*(1), 53–71.
69. Simulink, M., & Natick, M. (1993). The mathworks.
70. Sun, W., Liu, J., & Zhang, H. (2017). When smart wearables meet intelligent vehicles: Challenges and future directions. *IEEE Wireless Communications, 24*(3), 58–65.
71. Szegedy, C., Toshev, A., & Erhan, D. (2013). Deep neural networks for object detection. In *Advances in neural information processing systems* (pp. 2553–2561).
72. Tellex, S., Kollar, T., Dickerson, S., Walter, M. R., Banerjee, A. G., Teller, S. J., et al. (2011). Understanding natural language commands for robotic navigation and mobile manipulation. In *AAAI* (Vol. 1, p. 2).

73. Torkia, C., Reid, D., Korner-Bitensky, N., Kairy, D., Rushton, P. W., Demers, L., et al. (2015). Power wheelchair driving challenges in the community: A user's perspective. *Disability and Rehabilitation: Assistive Technology, 10*(3), 211–215.
74. von der Beeck, M. (1994). A comparison of statecharts variants. In *International Symposium on Formal Techniques in Real-Time and Fault-Tolerant Systems* (pp. 128–148). Berlin: Springer.
75. Yang, F., & Paindavoine, M. (2003). Implementation of an RBF neural network on embedded systems: Real-time face tracking and identity verification. *IEEE Transactions on Neural Networks, 14*(5), 1162–1175.
76. ZeroC I. (2003). The Internet communications engine.

Chapter 9
Design Automation Using Structural Graph Convolutional Neural Networks

Sujit Rokka Chhetri, Jiang Wan, Arquimedes Canedo, and Mohammad Abdullah Al Faruque

9.1 Introduction

A system design process consists of various steps such as problem definition, background research, requirement specification, brainstorming solutions, selecting the best solution, developing a prototype, testing, and finally redesigning [19]. During each of these steps, a wide variety of high volume and continuous data is generated. These design steps are repeated throughout various systems such as mechanical, electronic, and software. Due to the repetitive nature of these tasks, engineers can save a large amount of time if they can compare existing similar designs that closely match to the desired functionality while designing a new system. Rather than having to redesign, they can find functionally similar designs and modify such designs to fit their needs. Moreover, this can further lead to the creation of artificial intelligent assistants that assist human experts to design new systems faster.

The engineering design data varies from domain to domain. In electronic design, it consists of high-level design descriptions, register transfer level descriptions in Verilog or VHDL, schematics, etc. In mechanical design, it consists of data regarding structural designs, modeling, and analysis of components. Moreover, there is a wealth of data generated throughout the supply chain of engineering including computer-aided design (CAD) and computer-aided manufacturing (CAM) tools. In order to perform meaningful learning from these data, we need to utilize non-Euclidean or graph learning algorithms that are able to extract, categorize, and label these sparse data.

S. Rokka Chhetri (✉) · J. Wan · M. A. Al Faruque
University of California, Irvine, Irvine, CA, USA
e-mail: schhetri@uci.edu; jiangwan@uci.edu; alfaruqu@uci.edu

A. Canedo
Siemens Corporate Technology, München, Germany
e-mail: arquimedes.canedo@siemens.com

© Springer Nature Switzerland AG 2019
M. A. Al Faruque, A. Canedo (eds.), *Design Automation of Cyber-Physical Systems*, https://doi.org/10.1007/978-3-030-13050-3_9

237

In this chapter, we propose to utilize a structural graph learning algorithm to abstract the detailed engineering data (such as configuration, code, hybrid equations, geometry, and sensor data). This chapter expands the general view presented in [43]. To achieve this, we first represent the engineering data using a knowledge graph and then perform semi-supervised learning to be able to classify the subgraphs based on their structural property and the corresponding attributes. This will allow engineers to compare and cluster functionally similar designs, configurations, codes, geometry, sensor data, etc. This in return will allow engineers to automate the process of quickly searching the required engineering designs available in their library that meets the functional requirements presented in the specification.

9.2   Related Work

Design automation of engineering designs such as electronics and mechanics has seen an influx in the usage of machine learning and artificial intelligence algorithms [8, 23, 28, 33, 35]. These learning algorithms have helped the design automation by making the design process easier, faster, and efficient. However, current approaches are mostly limited in the utilization of Euclidean domain data and algorithms.

Research on more general non-Euclidean domain-based learning algorithms has recently gained momentum [12]. Moreover, a significant amount of work has been done in implementing the convolutional neural network on non-Euclidean structure data and manifolds of 3D objects [5, 7, 10, 18, 21, 25, 37, 41]. These works can be divided into two general directions in which the learning algorithms have been implemented on non-Euclidean data (such as a graph). The first direction is in the spectral domain, and the second direction is in the spatial or vertex domain. In spectral domain-based analysis, just like how filter weight is learned in the traditional 2-dimensional convolutional neural network, filter kernels are learned. In order to do this, first, the graph is transformed to Fourier domain by projecting the high-dimensional vertex domain graph to low-dimensional space using Eigenbasis of the graph Laplacian operation [9]. There are works where the various form of graph Laplacian operator and approximation methods for reducing the size of the graph kernel and the Fourier transformation of the graphs have been proposed [1, 10, 25, 27, 44]. In the second direction, first, the neighborhood information of a vertex is gathered using various techniques. This aggregated neighborhood information is then treated as features and different transformations on these features are proposed [15–17, 20, 40]. The major contribution of this work comes from the fact that the sampling and aggregation can focus on a node's neighborhood, thus not requiring the whole graph to be seen during the sampling and aggregation steps. The sampling and aggregation can be used by either using breadth-first [20] or utilizing both breadth-first and depth-first search [17]. These algorithms can effectively extract the node features based on their neighborhood which can be used to perform clustering and classification. Moreover, it has been shown to be effective in generating the graph embedding using the auto-encoders [16]. Furthermore, some work [1] have even proposed to combine the vertex domain and spectral domain approaches to utilize the strengths of each domain.

Both the spectral domain and vertex domain-based approaches have shown tremendous potential in generalizing the graph learning algorithms. However, each of these algorithms has a major weakness. The spectral domain-based approach relies highly on the Laplacian matrix. It filters the graph by using the eigenvalues of the Laplacian operator. However, this Laplacian matrix is dependent on the graph and the filter weights trained in one graph cannot be applied to another one. On the other hand, the vertex domain approaches have shown high efficiency in node-level clustering and classification. Although, there are algorithms such as sub-graph2vec [31], struct2vec [36], a more general CNN-based approach for learning rich features from a sub-graph is lacking. In engineering design automation, we would require a more general graph learning algorithm that is able to learn sub-graph or whole-graph property for providing more intuitive functional classification, clustering, or even generation.

In addition, custom graph kernel modeling for mining the graph by measuring the structural similarity between the pairs of graphs has also been proposed [42]. The major limitation of this approach is that they do not consider the features of the individual nodes and only rely on the structural similarity of the graph. Hence, it may be useful in mining structures from design engineering data; however, it will not help if the similar structure have different node features (which normally is the case in the engineering data). To address the existing limitation in the graph learning algorithms, in this chapter we introduce the structural graph convolutional neural network that is graph invariant (unlike spectral domain approaches), can learn rich features from nodes, sub-graph, or the whole graph, and is able to use both the structure and node features to learn rich features to classify and cluster different engineering domain graphs to aid in design automation.

9.3 Graph Learning Using Convolutional Neural Network

Let us define some preliminary notations for explaining the graph structure. The graph is denoted as $\mathcal{G} = (\mathcal{V}, \mathcal{E})$, where the set of vertices of the graph is denoted as $\mathcal{V}$ and the set of edges of the graph is denoted as $\mathcal{E}$. The graph edges can be weighted $w_{ij}, i, j \in \mathcal{V}$ and be directed. In this chapter, we will consider unweighted graphs with $w_{ij} = 1$. However, we may easily expand the graph learning algorithm for weighted graphs as well. When the engineering data is represented using such graph structure, each of the vertices will have some features (such as design version and mechanical properties). We represent such features for each vertices using the symbol f_i, where $i \in \mathcal{V}$. f_i consists of a vector whose dimension depends on the amount of information present in the engineering data. The raw format of features may vary from being a text, image, continuous or discrete signals, etc. In such situation, these features need to be converted to its corresponding vector

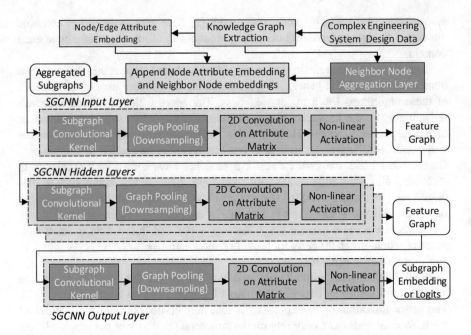

Fig. 9.1 Proposed graph learning algorithm for engineering design automation [43]

representation using auto-encoders. The adjacency matrix of the graph $\mathcal{G}$ is denoted by $\bar{A}$. The sub-graph of $\mathcal{G}$ is denoted as $\mathcal{G}_s = (\mathcal{V}_s, \mathcal{E}_s)$, where $\mathcal{V}_s \subseteq \mathcal{V}$ and $\mathcal{E}_s \subseteq \mathcal{E}$.

The proposed graph learning algorithm's architecture is shown in Fig. 9.1. Before the algorithm can be used, the raw engineering design data needs to be converted into a knowledge-based graph [39]. The structure of the knowledge-based graph should be tailored towards the engineering domain. The next step will involve converting the high-dimensional information present in each of the vertices and edges to a lower dimensional vector space embedding. These embedding will form the features f_i for all the vertices. The major contributions of the proposed graph learning algorithm are: (1) neighbor node aggregation layer, (2) sub-graph convolutional kernel, and (3) graph pooling algorithm. However, there are various components of the proposed algorithm which enable it to function. Each of these components is explained in the following subsections.

9.3.1 Knowledge Graph Extraction

The first task for utilizing the proposed graph learning algorithm is converting the complex engineering domain data into a meaningful data structure which can concisely and precisely represent the engineering domain data. For this purpose, we propose to utilize a knowledge graph. The knowledge graph stores information between the various uniquely identifiable entities and their corresponding relationship. These relationships are stored in a form of a triple (node–edge–node) relationship. This type of triples has previously been used to create knowledge graphs such as DBPedia [3]. The main advantage of using such knowledge graphs is that it can store rich engineering domain information and continuously evolve, grow, and be linked to other engineering domain data as well.

9.3.2   Attribute Embedding

After the knowledge graph has been created, we will have the structure of the graph ready to be utilized in the graph learning algorithm. However, in a knowledge graph the node and edge may be in different data format (such as text and images). This high-dimensional attribute of the nodes and edges needs to be converted into a low-dimensional feature embedding. To embed such attributes we utilize various state-of-the-art auto-encoders. For example, for encoding the text we will utilize word2vec [29]. For embedding images, we will utilize existing deep auto-encoder [11]. These vector embedding generated from the attribute of the nodes and edges form the feature which is utilized in the attribute matrix described in Sect. 9.3.4.1.

9.3.3   Neighbor Nodes Aggregation

One of the fundamental tasks in performing graph learning in engineering domain data is being able to capture the features of the vertices with respect to its location in the graph. Each of the vertices not only has special topology but also share a set of attributes across the knowledge graph. Hence, it is necessary to capture each of the unique structural and feature-based relation of vertices with respect to its neighbor. In order to do this, in the proposed graph learning algorithm, we utilize concepts similar to the vertex domain approach [20] (see Fig. 9.2). Before the neighbor node aggregation is performed, a user-defined query or a *schema* is used to induce a sub-graph. The process of defining the schema and inducing the sub-graph is domain specific. Depending on the type of engineering data, the schema can be tailored to extract only meaningful engineering design information. For example, the schema can be used to induce graphs that contain certain keywords (such as engines, piston, etc.) and has certain relations (such "has sub-components"). Based on this schema various instances are generated by the sample generator. These induced samples are then passed to two blocks in parallel. One block converts the node/edges attribute to its corresponding vector form, whereas the other block is the neighbor node aggregation layer. In the neighbor node aggregation layer, for the induced sub-graph $\mathcal{G}_t = (\mathcal{V}_t, \mathcal{E}_t)$, our algorithm performs both breadth-first and depth-first search to

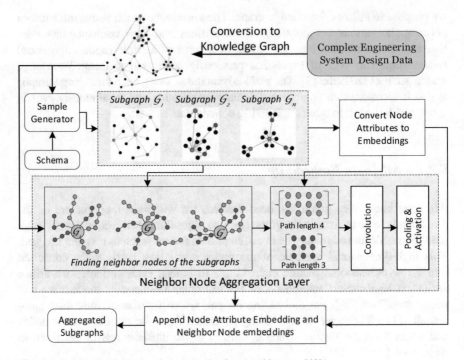

Fig. 9.2 Proposed neighbor node aggregation layer architecture [43]

collect the neighbor nodes of the given sub-graph. We utilize the parameter d as the depth to search in the graph, and n as the number of paths to be searched. The neighbor nodes borrow concept from node2vec [17], where a balance is maintained between the depth and the breadth for context generation. This balance helps us to maintain a balance between the local and non-local neighbors to the sub-graph.

In order to find the neighbors, for all $v_i \in \mathcal{V}_t$, we search the original knowledge graph $\mathcal{G}$ to find n paths of length d. Let us denote each of these paths by $\mathcal{P}_i^d$, where d denotes the path length, and i denoted the ith path of length d. For the path to be considered for aggregation, the nodes lying in the path must satisfy the condition $v_i \notin \mathcal{V}_t$. We acquire vector of paths $\mathcal{P}^d = \{\mathcal{P}_1^d \mathcal{P}_2^d, \ldots\}$ for each of the sub-graph $\mathcal{G}_t$. However, finding and using all of such paths is non-trivial, hence we then sample s paths of length d for each of the sub-graph. This sampling is done randomly in the proposed algorithm. Hence, s is another input parameter to the proposed graph learning algorithm. Using these randomly sampled paths, we form a neighbor feature matrix $\bar{N}$. The bar is s by d matrix with each element having a feature vector having extracted earlier. The number of paths found in $\mathcal{P}^d$ may be smaller than s, $\mathcal{P}^d$ can be padded to make $\bar{N}$ with at least s number of rows/paths. The algorithm for neighbor node aggregation is presented in Algorithm 1.

The input to Algorithm 1 is the knowledge graph $\{\mathcal{G}_1, \mathcal{G}_2, \ldots, \mathcal{G}_t\}$ using the user-defined schema, list of depth to search $D = \{d_1, d_2, \ldots, d_n\}$, list of sample numbers per depth $S = \{s_1, s_2, \ldots, s_n\}$, and feature vector x_n for the vertices. The output of

---

Algorithm 1: Neighbor node aggregation [43]

Input: Induced Sub-graphs: $\mathcal{G}_1, \mathcal{G}_2, \ldots, \mathcal{G}_t$
Input: A list of depth to search: $D = d_1, d_2, \ldots, d_n$
Input: A list of sample numbers per depth: $S = s_1, s_2, \ldots, s_n$
Output: A feature vector: x_n
1 **foreach** $d_i \in D$ **do**
2   **foreach** $v_j \in \mathcal{V}_t$ **do**
3   Find all length d_i paths $P_j^{d_i}$
4   Remove paths containing nodes in $\mathcal{V}_t$ from $P_j^{d_i}$
5   Add $P_j^{d_i}$ into P^{d_i}
6   Construct $\bar{N}$ by randomly selecting s_i paths from P^{d_i}
7   Extract feature x_{di} from $\bar{N}$
8 $x_n = f_{pool}^d(x_{d1}, x_{d1}, \ldots, x_{dn})$
9 **return** x_n

Algorithm 1 is the extracted feature to be appended to the vertices of the sub-graph G_t. One of the challenges in extracting the features from the neighborhood nodes as discussed in [20] is the fact that the non-Euclidean data has no natural ordering. Which means that the feature extraction should be applied over un-ordered set of paths to make sure that the arbitrary change in the order of rows of matrix $\bar{N}$ still results in the same feature being extracted. In order to achieve this in Algorithm 1, we first apply a general 1-D convolution operation with trainable 1 by d weight matrix $\bar{W}$ on $\bar{N}$ and then utilize a symmetric pooling function to extract the neighbor nodes' feature vector x_n as follows:

$$x_n = \sigma(f_{pool}(\bar{W} \circledast \bar{N}) + b) \tag{9.1}$$

This equation is used in Line 8 of Algorithm 1. The b in Eq. (9.1) is a bias variable and σ is an activation function (e.g., ReLU, LeakyRelu, Sigmoid, Tanh, etc.), and f_{pool} is a pooling function. As mentioned, the pooling function has to be invariant to permutation of rows in $\bar{N}$. To achieve this, pooling function such as a mean operator can be applied over all the rows in the matrix, or we may utilize a max pool operator that extracts max values out of all the rows. The use of specific pooling function is treated as a hyper-parameter in the graph learning algorithm and later configured in the training and hyper-parameter tuning process. As shown in Fig. 9.2, we extract various path length $\{d_1, d_2, d_3, \ldots, d_k\}$. This path length will integrate various topological and localized attributes of the engineering knowledge graph. Hence, Algorithm 1 will return feature vector $\{x_{d1}, x_{d2}, x_{d3}, \ldots, x_{dk}\}$ depending on total number of path lengths extracted from the graph. Each of these path lengths is aggregated to the sub-graph. If the extracted path length number is large, we may perform pooling to reduce the dimension of the extracted features as:

$$x_n = f_{pool}^d(\{x_{d1}, x_{d2}, x_{d3}, \ldots, x_{dk}\}) \tag{9.2}$$

The returned neighborhood feature vector is then concatenated to all the feature vectors of the vertices $v_i \in V_t$ as $x_{agg_i} = \{x_i, x_n\}$. Since the neighborhood aggregation layer is part of the graph learning algorithm, all the parameters of the neighborhood aggregation layer (such as weights of the 1-D convolution neural network) are learned during training. This allows the algorithm to automatically focus on relevant neighborhood node features to aggregate during the training.

9.3.4 Structural Graph Convolutional Neural Network Layers

The structural graph convolutional neural network (SGCNN) layers are shown in Fig. 9.3. The input of the SGCNN are the aggregated sub-graphs that were generated by the neighbor node aggregation layer described in Sect. 9.3.3 earlier. Each of these aggregated sub-graphs is passed in batches with individual vertices having a feature matrix. The SGCNN layers consists of input, hidden and output layers. These layers abstract the aggregated graph's feature in each layer just like the traditional 2D-convolutional neural networks. Each of the SGCNN consists of various components: (1) sub-graph convolution kernel, (2) graph pooling, (3) 2D convolution on adjacency matrix, (4) new adjacency matrix calculation, and (5) non-linear activation. We will discuss each of these components in detail in the following sections.

9.3.4.1 Sub-graph Convolution Kernel

The main block of the SGCNN is the sub-graph convolutional kernel. The task of the sub-graph convolutional kernel is to abstract meaningful feature vectors from the aggregated graph. The kernel receives the aggregated graph $\mathcal{G}_t$ with the

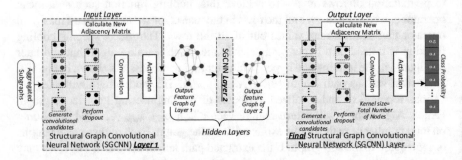

Fig. 9.3 Proposed structural graph convolutional neural network architecture [43]

corresponding feature matrix $\bar{X}$, which consists of the aggregated features $x_{agg_i} = \{x_i, x_n\}$ for $v_i \in \mathcal{V}_t$. It also receives the adjacency matrix $\bar{A}$ of the aggregated graph $\mathcal{G}_t$. With the adjacency matrix and the feature matrix, the first step the sub-graph convolutional kernel block will do is to create the adjacency matrix $\bar{Ar}$ by taking the Hadamard product between $\bar{X}$ and $\bar{A} + \bar{I}$ as follows:

$$\bar{Ar} = \bar{X} \circ (\bar{A} + \bar{I}) \tag{9.3}$$

We have added the identity matrix $\bar{I}$ to make sure that the vertices do not lose their feature information. This self-loop to each of the vertices in the sub-graph makes sure that each vertex retain their own information while calculating the Hadamard product. An example of an attribute matrix is shown in Fig. 9.4.

The attribute matrix $\bar{Ar}$ is used to define a graph convolutional kernel. The kernel consists of a k by k weight matrix $\bar{W}^k$. The size of the k can be varied just like the 2D-convolutional kernels. The convolutional is applied between the $\bar{W}^k$ and $\bar{Ar}$. In traditional 2D convolution, a kernel is slide from left to right and top-down in an order to perform the 2D convolution. However, since there is no notion of 2D-grid structure in the graph, we cannot perform convolution in a similar manner. Hence, in our graph convolution, we propose a new method to perform convolution operation for the graph data structure.

The algorithm for performing the graph convolution is presented in Algorithm 2. The first step in performing the graph convolution involves generating candidate graph kernels with size k by k, where we select the k vertices at a time. This candidate graph is then convoluted with the weight matrix $\bar{W}^k$. In order to generate the candidate graph kernels, we use the fact that removing the ith row and i^{th} column in $\bar{Ar}$ is equivalent to removing the vertex i from $\mathcal{G}_t$. Hence, assuming that the total vertices in the induced aggregated sub-graph $\mathcal{G}_t$, let it be denoted by n, is greater than the size of the kernel k, we will remove $n - k$ number of vertex from $\mathcal{G}_t$. The new sub-graph will be left with a new k by k attribute matrix $\bar{Ar}^k$. The drawback of generating the attribute matrix like this is that there are $O\binom{n}{k}$ possible ways to generate the attribute matrix $\bar{Ar}^k$. This might be okay for an induced graph with a

$$\begin{bmatrix} X_1 & X_2 \dots & X_{n-1} & X_n \\ X_1 & X_2 \dots & X_{n-1} & X_n \\ \vdots & \vdots & \vdots & \vdots \\ X_1 & X_2 \dots & X_{n-1} & X_n \\ X_1 & X_2 \dots & X_{n-1} & X_n \end{bmatrix} \circ \begin{bmatrix} 1 & 1 \dots & 1 & 0 \\ 1 & 1 \dots & 0 & 0 \\ \vdots & \vdots & \vdots & \vdots \\ 1 & 1 \dots & 1 & 0 \\ 0 & 0 \dots & 0 & 1 \end{bmatrix} = \begin{bmatrix} X_1 & X_2 \dots & X_{n-1} & 0 \\ X_1 & X_2 \dots & 0 & 0 \\ \vdots & \vdots & \vdots & \vdots \\ X_1 & X_2 \dots & X_{n-1} & 0 \\ 0 & 0 \dots & 0 & X_n \end{bmatrix}$$

Fig. 9.4 Example of an attribute matrix calculated using Hadamard product

---

Algorithm 2: Graph convolution kernel [43]

Input: An input graph: $\mathcal{G}_t$ with n vertices
Input: A convolution kernel: $\bar{W}^k$
Input: Sample size: s
Output: An output feature graph: G'
1 Generate adjacency matrix $\bar{A}$ from G
2 Using the same vertices order to generate list of features $\mathcal{X}$
3 Create feature matrix $\bar{X}$ with n rows, and each row being $\mathcal{X}$
4 $\bar{A}r = \bar{X} \circ (\bar{A} + \bar{I})$
5 $m = \binom{n}{k}$
6 $CombList$ = Enumerating choice of $n - k$ elements from $1, 2, \ldots, n$
7 **foreach** $comb \in CombList$ **do**
8 $\quad$ $\bar{A}r_{comb}$ = remove rows and columns list in $cand$ from $\bar{A}r$
9 $\quad$ Add $\bar{A}r_{comb}$ into $CandList$
10 **if** $m > s$ **then**
11 $\quad$ Down-sample $CandList$ to s elements
12 **else if** $m < s$ **then**
13 $\quad$ Pad $CandList$ to s elements
14 **foreach** $cand \in CandList$ **do**
15 $\quad$ $x^k = \bar{W}^k \circledast cand + b$
16 $\quad$ Add new vertex v_k into G'
17 $\quad$ Add feature vector x^k on v_k
18 $\quad$ Connect v_k based on $cand$'s connection in G
19 **return** G'

---

lower number of nodes, however for engineering data the induced graph size can have a large number of nodes. And normally the size of the $k \ll n$, making the complexity of generating the $\bar{A}r^k$ very high. Hence, to tackle this impractical $\bar{A}r^k$ generation step, we propose to relax it by only picking s number out of $O\binom{n}{k}$ as a convolution candidate. Hence, the total number of the possible $\bar{A}r^k$ is thus reduced to $O(s)$. By doing this, we make it feasible to generate the $\bar{A}r^k$ from the $\mathcal{G}_t$ as a potential candidate of graph kernels to be convoluted with the k by k weight filter matrix $\bar{W}^k$. The procedure of down-sampling of $O\binom{n}{k}$ possible $\bar{A}r^k$ values to just s is explained in Sect. 9.3.4.2.

Algorithm 2 presents the graph convolution steps in detail. The input to the graph is the induced aggregated graph $\mathcal{G}_t$. The output of the algorithm is the new graph where each node represents the merged vertices present in $\bar{A}r^k$. In Line 1, we first generate the adjacency matrix from the graph. In Line 4, the attribute matrix is generated by performing the Hadamard product between the feature matrix and the adjacency matrix. In Line 7, possible combination of $\bar{A}r^k$ is listed and down-sampled in Line 10. In Line 14, a 2D convolution is performed between the $\bar{A}r^k$ and the filter weight matrix W^k. Finally, the new graph is returned in Line 19. Each of the SGCNN layers will generate a new graph which progressively abstracts and fuses the topological and attributes of the previous graph.

9.3.4.2 Graph Pooling Algorithm

When we generate $\bar{A}r^k$, there are $O\binom{n}{k}$ possible in the beginning. After down-sampling it to s number of $\bar{A}r^k$, the next stage will have $O\binom{s}{k}$ possible combination of $\bar{A}r^k$. One of the desired properties of convolutional neural networks is to be able to perform the convolution over a deep number of layers. Hence, to be able to perform the graph convolution in deeper layers, we need to perform down-sampling or pooling at each layer to manage the size of possible $\bar{A}r^k$ matrices generated from the graph. Without down-sampling, it will be unfeasible to perform a large number of convolutional operation between $\bar{A}r^k$ and W^k. Hence, we propose to perform the pooling operation before the convolution operation in each of the SGCNN layers. The proposed down-sampling/pooling operation utilizes the topology of the graph to remove samples from combinations of $\bar{A}r^k$. For each of the possible $\bar{A}r^k$, we calculate the corresponding total degrees. The intuition behind the proposed down-sampling algorithm is that out of $O\binom{n}{k}$, due to the sparsity of the aggregated sub-graph, combination of various nodes will not have any edges or lower number of edges among them. Hence, combining them together to perform convolution will be less fruitful than selecting the combination of $\bar{A}r^k$ that have higher connectivity among the vertices. The proposed down-sampling algorithm is presented in Algorithm 3.

The input to Algorithm 3 is the graph $\mathcal{G}$, the possible list of candidate combination $Comb$, the pooling sample size s, and the dropout rate d. The output of the algorithm is the list of combination $Comb''$ which is down-sampled. To achieve this, first in Line 1, it randomly samples $Comb'$ by using the dropout rate d. This is

---

Algorithm 3: Graph pooling algorithm [43]

Input: An input graph: $\mathcal{G}$
Input: The list of the candidate nodes combinations: $Comb$
Input: Sample size: s
Input: Dropout Rate: d
Output: The list of down-sampled candidate nodes: $Comb'$

1 Randomly sample $Comb'$ from $Comb$ using d
2 Generate adjacency matrix $\bar{A}$ from G for only $Comb'$
3 **foreach** $c \in Comb'$ **do**
4      $d_c = 0$
5   **foreach** $n \in c$ **do**
6   Calculate Degree of n as d_n
7          $d_c = d_c + d_n$
8   **foreach** $c' \in Comb'$ **do**
9   **if** c' *is connected with c in* $\bar{A}$ **then**
10              $d_c = d_c + 1$
11 Keep the s number of nodes with the highest degrees and store in $Comb''$
12 **return** $Comb''$;

necessary because it is computationally unfeasible to calculate the adjacency matrix in the next step for all the possible candidates in $Comb$. In Line 2, we generate the adjacency matrix $\bar{A}$ for the new candidate combination of nodes. This adjacency matrix carries the graph structural data and passes it through the deeper layers. This step is important, as the different layers will be able to abstract the graph data in a hierarchical manner. Lines 3–10 compute the total degrees of each candidate nodes combination $Comb'$, which are combinations of the nodes in G that are generated by the convolution kernel and these combinations will serve as the new nodes of the output feature graph after the graph convolution kernel. Specifically, Lines 3–7 compute the total degrees inside the combination, and Lines 8–10 compute the degrees in between different combinations. Finally, we keep the s number of nodes combinations which have the highest degree and remove the rest. We significantly reduce the size of the graph convolution kernel by dropping the combinations in the calculated degrees using the max pooling. Nevertheless, we ensure that the convolution is performed on the graph structures with higher connectivity.

9.3.4.3 2D Convolutions on Attribute Matrix

After we have down-sample s possible candidate $\bar{A}r^k$ matrices, we apply simple 2D convolution operation to extract the feature vector for the give combination of the $\bar{A}r^k$. The convolution operation can be written as follows:

$$x^k = \phi(\bar{W}^k \circledast \bar{A}r^k + b) \tag{9.4}$$

where $\phi(.)$ is a non-linear activation function. We will have s extracted feature vectors as: $x_1^k, x_2^k, \ldots, x_s^k$. We consider the extracted feature vectors $x_1^k, x_2^k, \ldots, x_s^k$ as a new feature graph $\mathcal{G}'$ with s number of vertices, and x_i^k as the feature vector for node i. An example of this process is shown in Fig. 9.5.

9.3.4.4 New Adjacency Matrix Calculation

For the deep SGCNN architecture to work, we have to keep track of the adjacency of the new $\mathcal{G}'$ generated at each of the layers. This $\mathcal{G}'$ is used as input to another sub-graph convolution layer to form a deep SGCNN model. The constructed adjacency matrix will allow the next SGCNN layer to recalculate the new attribute matrix and the potential candidates of $\bar{A}r^k$ to be selected for convolution and down-sampling. To calculate the new adjacency for the new graph, we check the edges between inter and intra nodes of the graph convolution kernels. The edge between these intra and inter nodes allows the graph structure to be propagated through the SGCNN layers, making sure the topological information is being utilized to learn the filter weight matrix W^k at each layer.

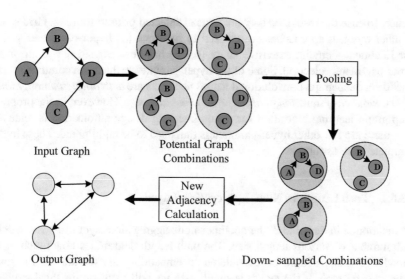

Fig. 9.5 Convolution kernel example: given the input graph with four vertices, a 3-by-3 convolution kernel is selected. As a result, four convolution candidates are generated, down-sampled, and adjacency recalculated resulting in a new graph with three vertices

9.3.5 Classification for Engineering Design Abstraction

Given a large graph and labeled sub-graphs representing engineering design, the SGCNN can be used to classify them. While classifying these sub-graphs, various features regarding the design are learned by the graph convolutional kernels. Once the design and the labels (which can be the function served by the design) are trained, the design is abstracted based on their functions. This classification is done by using a softmax function and cross-entropy of the logits. In addition, the feature vectors (sub-graph embedding) generated by the final SGCNN can also be used by clustering algorithms to identify nearest neighbors sub-graphs that have an equivalent function in the graph based on their node attributes and structure. In engineering, there are several use-cases for sub-graph embedding including the identification of functionally equivalent structures that engineers are unaware of, and to identify structures that mislabeled.

9.3.6 Graph Learning Algorithm Hyper-Parameters

In deep learning convolutional neural networks, hyper-parameters play a crucial role in improving the accuracy and convergence of the algorithm. These hyper-parameters are difficult to derive during the training as it requires large resources.

Hence, to ease the resources they are instead selected prior to training. Grid search or other efficient approaches are utilized to fine-tune the hyper-parameter values. For Euclidean domain, extensive study [6, 13] has been carried out to aid in the hyper-parameter selection. Some of the hyper-parameters that even common to non-Euclidean deep graph convolutional neural networks are *activation function, hidden layers, number of iteration, learning rate, and batch size*. However, as the proposed deep graph learning algorithm for the engineering design automation is relatively new, there are few other hyper-parameters that need to be highlighted. These hyper-parameters are as follows.

9.3.6.1 Path Length in Node Aggregation Layer

As mentioned in Sect. 9.3.3, the neighbor node aggregation layer utilizes a specific path number of various length size. The path length determines how much of the knowledge graph should be considered in embedding the feature for the given induced sub-graph. If the length is small, then we will focus on the local structure and attributes of the sub-graph, while selecting longer path length will allow the sub-graph to be embedded with global features. Hence, this value needs to be optimized and fine-tuned according to the specific engineering domain data.

9.3.6.2 Graph Convolution Kernel Size

The graph convolutional kernel size determines how many vertices to be considered for convolution with the filter weight matrix at each layer. The smaller kernel size means that each of the layers will abstract sub-graph feature by considering its immediate neighbors. However, if the kernel size is small, while the number of vertices in the induced sub-graph is large than deeper SGCNN layers may need to be implemented. However, if we select larger kernel, the SGCNN layer may be shallow. The size of the kernel may depend on the induced graph's size and structure and will require tuning before the training is performed.

9.3.6.3 Dropout of Candidate Kernels

As presented in Sect. 9.3.4.2, we have used dropout to make the deep graph learning feasible. We have combined the random and degree-based dropout. If permitted by the resource, taking a large number of candidate attribute matrix may be helpful to better abstract the induced sub-graph. Due to sparse nature of the engineering design data taking large combinations of attribute matrix for convolution may also be a waste of resources. Hence, a number of candidate kernels to drop out before the convolution is performed need to be tuned as a hyper-parameter.

9.4   GrabCAD Dataset

To demonstrate the applicability of the proposed graph learning algorithm in engineering design automation, we have selected the 3D engineering CAD models as training and testing dataset for engineering design functionality classification.

We have extracted the dataset from GrabCAD,[1] which is an online repository of 3D CAD models shared and maintained by an online community of designers, engineers, and manufacturers. It consists of over 4 million members with over 2 million engineering design models. From this vast online dataset, we have extracted meta-information from six functional categories of 3D CAD models. These functional categories are *Car*, *Engine*, *Robotic arm*, *Airplane*, *Gear*, and *Wheel* (see Fig. 9.6). Since these are engineering designs from the mechanical domain, we have tailored a custom schema consisting of the properties such as 3D model's name, author of the design, description of the model, parts names, tags, likes, time-stamps, and comments on the engineering design. With this schema we have induced sub-graphs for each of the categories with 2271 samples for *Car*, 1597 samples for *Engine*, 2013 samples for *Robotic arm*, 2114 samples for *Airplane*, 1732 samples for *Gear*, and 2404 samples for *Wheel*. The induced sub-graph consists of 17 nodes consisting of both social network data (such as user-to-user interaction through comments and likes) and engineering data (such as model-to-tags relationship and model-to-model relationships). By inducing a sub-graph

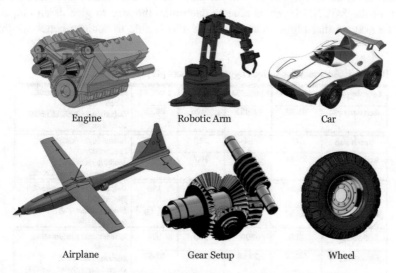

Engine   Robotic Arm   Car

Airplane   Gear Setup   Wheel

Fig. 9.6 Sample of the 3D CAD models (Engine [26], Robotic Arm [22], Car [30], Airplane [32], Gear Setup [4], and Wheel [38]) extracted from GrabCAD repository

[1]https://grabcad.com/.

in GrabCAD dataset, we aim to demonstrate that similar designs with functional description represented by a knowledge graph can be efficiently classified using the proposed supervised graph learning algorithm.

9.5 Results

In our experiment, the total number of induced graph from all the engineering design is 14,131. From this sample, we have used 11,304 as training samples and 2827 samples for testing if the similar functional designs get classified accurately. In order to make sure that proper hyper-parameters are selected and tuned, we have used a grid search approach over the possible hyper-parameter values. The result of hyper-parameter selection is shown in Table 9.1. In the table, we have tested the accuracy for various hyper-parameters. The first hyper-parameter is the learning rate of the optimization algorithm. We have used ADAM optimizer [24] to adjust the filter weights. From the table, it can be seen that the learning rate of 0.01 is able to achieve higher accuracy. The second hyper-parameter in the table is the batch size. The batch size determines how many of the induced sub-graph are passed together once for calculating the loss and updating the gradient.

It can be seen the batch size of 32 is able to obtain the highest accuracy of 79.22% classification accuracy. The next hyper-parameter is the hidden layer feature size. Each of the SGCNN layers is able to determine the size to give as an output. It can be noticed that higher feature size of 150 is able to achieve better accuracy.

Table 9.1 Accuracy for various hyper-parameters [43]

Learning Rate	1.00E-04	1.00E-03	1.00E-02	1.00E-01	Hidden Layer Feature=100, Epoch =100, Output Layer Kernel Size=25, Batch size =64
Accuracy (%)	45.70	66.051	79.26	80.15	
Batch size	32	64	128	256	Hidden Layer Feature=100, Epoch=100, Learning rate=1e-2, Output Layer Kernel Size=25
Accuracy (%)	79.22	78.65	76.7	72.01	
Hidden Layer Feature Size	20	50	100	150	Output Layer Kernel Size=25, Epoch=100, Learning rate=1e-2, Batch size =64
Accuracy (%)	82.28	83.10	85.54	87.89	
Epochs	100	200	300	500	Hidden Layer Feature=100, batch=64, Learning rate=1e-2, Output Layer Kernel Size=25
Accuracy (%)	78.87	79.68	80.43	81.12	
Output Layer Kernel	25	30	40	45	Hidden Layer Feature=100, Epoch=100, Learning rate=1e-2, Batch size=64
Accuracy (%)	85.61	85.22	86.43	87.92	

Three Layers (Aggregate, SGCNN Input, SGCNN Output), Random Dropout

With a larger feature size, we also increase the number of parameters in the filter weight matrix. The nest hyper-parameter is the epoch (the total number of times the training goes over the whole dataset). It can be seen that a higher epoch number of 500 is able to achieve higher classification accuracy. The last hyper-parameter shown in Table 9.1 is the output layer kernel size. The final output layer of the SGCNN acts like a dense layer, which means that it will try to generate probability values for each of the categories. The final layer of the kernel depends on the dropout carried out earlier. From the table, it can be seen that lower dropout or larger output kernel produces higher accuracy. Beside these hyper-parameters, we have also tested other hyper-parameters which are presented in the following sections.

9.5.1 Activation Functions

Activation functions are used before the output in each of the SGCNN layers. Most of the operation before the activation function are mostly linear. However, the activation function increases the capacity of the SGCNN layer by making it non-linear. We have explored various activation functions which are well studied in Euclidean domain. These activation functions are *sigmoid, softplus, tanh, rectifier linear unit*, and *leaky rectifier linear unit*.

The training loss and engineering design classification accuracy during testing is shown in Fig. 9.7. It can be seen that out of all activation functions, leaky rectifier linear unit ($f(x) = \alpha x$ for $x < 0$ and $f(x) = x$ for $x >= 0$) activation function (with $\alpha = 0.2$) is able to achieve lower loss and higher classification accuracy compared to other activation functions. Although the accuracy is higher, it introduces some noise in both the loss and the accuracy values. One of the parameters that can be used in *LeakyRelu* is the alpha value. It determines the slope to be used to cut off the values when $x < 0$. Further analysis is required to tune the value of the α.

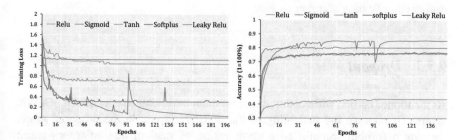

Fig. 9.7 (a) Training loss (Left) and (b) accuracy for different activation functions (layers = 2, aggregate and graph embedding layers) (Right) [43]

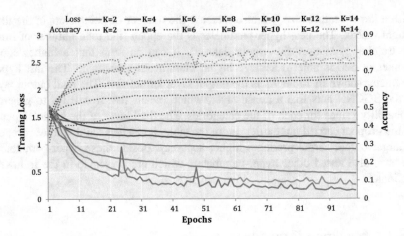

Fig. 9.8 Training loss and accuracy for different size of kernels (with random dropout, last layer kernel size = 5, layers = 3, hidden layer activation = relu, final layer activation = leaky relu) [43]

9.5.2 Kernel Size

One of the important hyper-parameters for the SGCNN layers is the size of the kernel used to perform the convolution with the attribute matrix $\bar{A}r^k$. The value of k determines the total number of vertices that are considered at a time to perform the convolution with the filter weights W^k. In our experiment, the total of nodes available in the induced sub-graph is 17, hence we have selected kernel size as 2, 4, 6, 8, 10, 12, and 14. It can be seen from Fig. 9.8 that the kernel size has a drastic effect on the classification accuracy of the engineering design data. For three layers of SGCNN, larger kernel size is able to achieve higher testing accuracy and lower training loss. However, having a larger kernel size in every hidden layer is not feasible, as it increases the complexity of the training algorithm. In Sect. 9.5.4, it can be seen that with deeper layers, a smaller kernel size is able to obtain higher testing classification accuracy as well.

9.5.3 Dropout

As mentioned earlier, without dropout, it becomes unfeasible to create deep SGCNN layers. To improve the scalability of the SGCNN layers, in the proposed graph learning algorithm we utilized a combination of random and adjacency-based dropout. In the adjacency-based dropout, we utilize the degree of the vertices to select the candidate combination of $\bar{A}r^k$ for the convolution. The dropout rate determines the size of the final kernel size. If the dropout rate is higher, the size

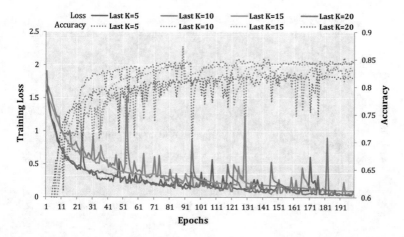

Fig. 9.9 Training loss and accuracy for various random dropouts to match the last layer's kernel size (hidden layer kernel size=14, layers=3, hidden layer activation=relu, final layer activation=leaky relu) [43]

of the kernel at the output SGCNN layer will be smaller and vice versa. The result of the various size of the final layer's kernel size as a result of changing the dropout is shown in Fig. 9.9. It maybe noticed that if the dropout is less (resulting in larger kernel size in the output layer), the testing accuracy is higher and training loss is lower. The result is shown for just three layers of the SGCNN. The total number of possible combination of $\bar{Ar}^k$ with the kernel size of 14 for the first layer is $\binom{17}{14} = 680$. We can notice that even when the total candidates have been drastically dropped to just 5, 10, 15, and 20, the graph learning algorithm is able to perform quite well in classifying the engineering designs. This may be due to the fact that the induced graphs are sparse in nature and that the initial kernel size of 14 is able to capture all of the node's features during convolution.

In addition to random dropout, we have implemented the adjacency-based dropout as well. The down-sampling algorithm first uses random dropout to initially reduce the possible candidates of $\bar{Ar}^k$, and out of the remaining selects the ones with higher connectivity. The result for the down-sampling combined with the random dropout is shown in Fig. 9.10. It maybe noticed that adjacency-based dropout is able to achieve the highest accuracy of around 90%.

9.5.4 Layers

The most advantageous property of the proposed graph learning algorithm is able to have deeper layers that are able to abstract the features of the induced graph in each iteration. To demonstrate this capability, we have selected a kernel size of 2 for the filter weights, and measure the accuracy of the graph learning algorithm for

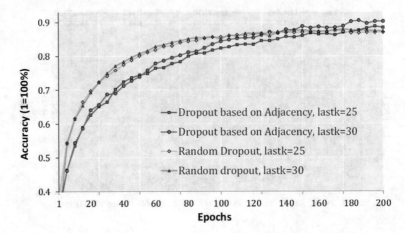

Fig. 9.10 Comparison between random dropout and adjacency-based dropout (hidden layer kernel size=14, layers=3, hidden layer activation=relu, final layer activation=leaky relu) [43]

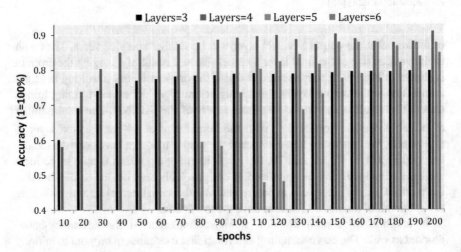

Fig. 9.11 Performance comparison between various layer sizes (hidden layer kernel size=2, activation =leaky relu, lastk=30, Dropout=Random) [43]

layer size of 3, 4, 5, and 6. It can be noticed from Fig. 9.11 that layer size of 4 and 5 is able to achieve higher testing classification accuracy compared to shallow three layers and deeper six layer size. The highest accuracy achieved was ≈91% with four layers. The deep SGCNN layers are able to abstract the structural and attribute properties of the induced sub-graph in a hierarchical manner by using a smaller kernel size of 2. This means that in each layer smaller node size is fused and the new combined features are learned. This property can be helpful in engineering design data, where there is some form of hierarchy in terms of design.

9.6   Discussion

The SGCNN's capability to learn sub-graph structure embedded with attributes was demonstrated in Sect. 9.5. It achieves very positive results on functional lifting using the GrabCAD dataset. We have presented the GrabCAD dataset as an engineering dataset to show the applicability of design automation. However, in our future work, we will demonstrate the use of the SGCNN to electronic design dataset as well. Although SGCNN was created to address the functional lifting problem in engineering, we believe this is broadly applicable to other domains. In our future work, we will compare the performance of SGCNN against the latest work on graph convolutional networks targeting sub-graph-level embedding [2, 31, 36]. Currently, we have shown that SGCNN is able to perform supervised learning in abstracting the design features and classifying them based on their functions. For robust design automation, designers would be interested not only in classifying the designs but being able to reproduce or generate designs that slightly vary in functionality. In order to do this, the same structure of the SGCNN may be used for generative learning algorithms such as variational auto-encoder (VAE) [34] or generative adversarial networks (GAN) [14].

9.7   Conclusions

This chapter presents a novel structural graph convolutional neural network which can be used to abstract the non-Euclidean graph or sub-graph dataset. These graphs can be used to represent the engineering design data and allow designers to effectively perform design automation by effectively clustering engineering designs with similar functionality. This allows designers to search for designs that are similar to the required functionality and aid in design automation.

References

1. Abu-El-Haija, S., Kapoor, A., Perozzi, B., & Lee, J. (2018). N-GCN: Multi-scale graph convolution for semi-supervised node classification. arXiv preprint arXiv:1802.08888.
2. Adhikari, B., Zhang, Y., Ramakrishnan, N., & Prakash, B. A. (2018). Sub2vec: Feature learning for subgraphs. In *Pacific-Asia Conference on Knowledge Discovery and Data Mining* (pp. 170–182). Cham: Springer.
3. Auer, S., Bizer, C., Kobilarov, G., Lehmann, J., Cyganiak, R., & Ives, Z. (2007). Dbpedia: A nucleus for a web of open data. In *The semantic web* (pp. 722–735). Berlin: Springer.
4. Bana. Differential gearbox. https://grabcad.com/library/differential-gearbox-speed-reducer-worm-gear-with-spider-gears-1. Retrieved November 14, 2018.
5. Belkin, M., & Niyogi, P. (2003). Laplacian eigenmaps for dimensionality reduction and data representation. *Neural Computation, 15*(6), 1373–1396.

6. Bergstra, J., & Bengio, Y. (2012). Random search for hyper-parameter optimization. *Journal of Machine Learning Research, 13*, 281–305.
7. Cao, S., Lu, W., & Xu, Q. (2015). GraRep: Learning graph representations with global structural information. In *Proceedings of the 24th ACM International on Conference on Information and Knowledge Management* (pp. 891–900). New York: ACM.
8. Chen, X., Tao, Y., Wang, G., Kang, R., Grossman, T., & Coros, S. (2018). Forte: User-driven generative design. In *Proceedings of the 2018 CHI Conference on Human Factors in Computing Systems* (p. 496). New York: ACM.
9. Chung, F. R. (1997). *Spectral graph theory* (Number 92). Providence: American Mathematical Society.
10. Defferrard, M., Bresson, X., & Vandergheynst, P. (2016). Convolutional neural networks on graphs with fast localized spectral filtering. In *Advances in Neural Information Processing Systems* (pp. 3844–3852).
11. Garcia-Gasulla, D., Ayguadé, E., Labarta, J., Béjar, J., Cortés, U., & Suzumura, T. (2017). A visual embedding for the unsupervised extraction of abstract semantics. *Cognitive Systems Research, 42*, 73–81.
12. *Geometric deep learning* (2018). http://geometricdeeplearning.com/
13. Glorot, X., & Bengio, Y. (2010). Understanding the difficulty of training deep feedforward neural networks. In *Proceedings of the Thirteenth International Conference on Artificial Intelligence and Statistics* (pp. 249–256).
14. Goodfellow, I., Pouget-Abadie, J., Mirza, M., Xu, B., Warde-Farley, D., & Ozair, S. (2014). Generative adversarial nets. In *Advances in Neural Information Processing Systems* (pp. 2672–2680).
15. Goyal, P., Chhetri, S. R., & Canedo, A. (2018). dyngraph2vec: Capturing network dynamics using dynamic graph representation learning. arXiv preprint arXiv:1809.02657.
16. Goyal, P., & Ferrara, E. (2018). Graph embedding techniques, applications, and performance: A survey. *Knowledge-Based Systems, 151*, 78–94.
17. Grover, A., & Leskovec, J. (2016). node2vec: Scalable feature learning for networks. In *Proceedings of the 22nd ACM SIGKDD International Conference on Knowledge Discovery and Data Mining* (pp. 855–864). New York: ACM.
18. Hadsell, R., Chopra, S., & LeCun, Y. (2006). Dimensionality reduction by learning an invariant mapping. In *IEEE Computer Society Conference on Computer Vision and Pattern Recognition, 2006* (Vol. 2, pp. 1735–1742). Piscataway: IEEE.
19. Haik, Y., Sivaloganathan, S., & Shahin, T. M. (2018). *Engineering design process*. Toronto: Nelson Education.
20. Hamilton, W., Ying, Z., & Leskovec, J. (2017). Inductive representation learning on large graphs. In *Advances in Neural Information Processing Systems* (pp. 1025–1035).
21. Henaff, M., Bruna, J., & LeCun, Y. (2015). Deep convolutional networks on graph-structured data. arXiv preprint arXiv:1506.05163.
22. Kapllanaj, E. (2018). 3D printed robotic arm with gripper. https://grabcad.com/library/3d-printed-robotic-arm-with-gripper-1. Retrieved November 14, 2018.
23. Kazi, R. H., Grossman, T., Cheong, H., Hashemi, A., & Fitzmaurice, G. W. (2017). Dreamsketch: Early stage 3d design explorations with sketching and generative design. In *UIST* (pp. 401–414).
24. Kingma, D. P., & Ba, J. (2014). Adam: A method for stochastic optimization. arXiv preprint arXiv:1412.6980.
25. Kipf, T. N., & Welling, M. (2016). Semi-supervised classification with graph convolutional networks. arXiv preprint arXiv:1609.02907.
26. Kumar, A. (2018). V6 car engine. https://grabcad.com/library/v6-car-engine-7. Retrieved November 17, 2018.
27. Levie, R., Monti, F., Bresson, X., & Bronstein, M. M. (2017). Cayleynets: Graph convolutional neural networks with complex rational spectral filters. arXiv preprint arXiv:1705.07664.

28. Li, B., & Franzon, P. D. (2016). Machine learning in physical design. In *IEEE 25th Conference on Electrical Performance of Electronic Packaging and Systems (EPEPS), 2016* (pp. 147–150). Piscataway: IEEE.
29. Mikolov, T., Sutskever, I., Chen, K., Corrado, G., & Dean, J. (2013). Distributed representations of words and phrases and their compositionality. In *Proceedings of the 26th International Conference on Neural Information Processing Systems - Volume 2*, NIPS'13, USA (pp. 3111–3119). New York: Curran Associates Inc.
30. Mohammad (2018). Go kart. https://grabcad.com/library/go-kart-123. Retrieved November 14, 2018.
31. Narayanan, A., Chandramohan, M., Chen, L., Liu, Y., & Saminathan, S. (2016). subgraph2vec: Learning distributed representations of rooted sub-graphs from large graphs. arXiv preprint arXiv:1606.08928.
32. Niazi, K. (2018). Pegasus UAV. https://grabcad.com/library/pegasus-uav-1. Retrieved November 14, 2018.
33. Pandey, M. (2018). Machine learning and systems for building the next generation of EDA tools. In *Proceedings of the 23rd Asia and South Pacific Design Automation Conference* (pp. 411–415). Piscataway: IEEE Press.
34. Pu, Y., Gan, Z., Henao, R., Yuan, X., Li, C., & Stevens, A. (2016). Variational autoencoder for deep learning of images, labels and captions. In *Advances in Neural Information Processing Systems* (pp. 2352–2360).
35. Qi, W. (2017). *IC Design Analysis, Optimization and Reuse via Machine Learning*. Raleigh: North Carolina State University.
36. Ribeiro, L. F. R., Saverese, P. H. P., & Figueiredo, D. R. (2017). struc2vec: Learning node representations from structural identity. In *Proceedings of the 23rd ACM SIGKDD International Conference on Knowledge Discovery and Data Mining* (pp. 385–394). New York: ACM.
37. Roweis, S. T., & Saul, L. K. (2000). Nonlinear dimensionality reduction by locally linear embedding. *Science, 290*(5500), 2323–2326.
38. Samoilovskikh, P. (2018). Trailer wheel. https://grabcad.com/library/trailer-wheel-2
39. Schuhmacher, M., & Ponzetto, S. P. (2014). Knowledge-based graph document modeling. In *Proceedings of the 7th ACM International Conference on Web Search and Data Mining* (pp. 543–552). New York: ACM.
40. Shervashidze, N., Schweitzer, P., Leeuwen, E. J. v., Mehlhorn, K., & Borgwardt, K. M. (2011). Weisfeiler-Lehman graph kernels. *Journal of Machine Learning Research, 12*(Sep), 2539–2561.
41. Tenenbaum, J. B., De Silva, V., & Langford, J. C. (2000). A global geometric framework for nonlinear dimensionality reduction. *Science, 290*(5500), 2319–2323.
42. Vishwanathan, S. V. N., Schraudolph, N. N., Kondor, R., & Borgwardt, K. M. (2010). Graph kernels. *Journal of Machine Learning Research, 11*(Apr), 1201–1242.
43. Wan, J., Pollard, B. S., Chhetri, S. R., Goyal, P., Faruque, M. A. A., & Canedo, A. (2018). Future automation engineering using structural graph convolutional neural networks. arXiv preprint arXiv:1808.08213.
44. Yi, L., Su, H., Guo, X., & Guibas, L. J. (2017). SyncSpecCNN: Synchronized spectral CNN for 3D shape segmentation. In *CVPR* (pp. 6584–6592).

Chapter 10
Design Automation for Energy Storage Systems

Swaminathan Narayanaswamy, Sangyoung Park, Sebastian Steinhorst, and Samarjit Chakraborty

10.1 Electrical Energy Storage (EES) Systems

Global warming and associated climate change have already had observable serious effect on the environment. NASA's analysis in [1] says that 2017 was the second hottest year on the record where the average global temperature increased by $0.9\,°C$ above the average value from 1951 to 1980. With similar trends global temperatures could break the internationally agreed upper $1.5\,°C$ limit within the next 5 years as predicted by [2]. Effects of such increased global warming can be observed by extreme weather events and natural calamities causing floods, hurricanes, famines, and water scarcity all over the world. Towards this, both developed and also developing countries like China and India have created joint road maps to address the problem of climate change.

Power and transportation sectors are identified as the major sources of CO_2 and other greenhouse gas emissions resulting in increased global warming. For instance, 71% of CO_2 emissions in India is from the energy sector which is predominantly dominated by coal-based thermal power plants [3]. Similarly, the transportation sectors account for nearly 28.5% of greenhouse gas emissions in the USA in 2016, predominantly from the passenger cars [4]. Towards this multiple steps have been planned to minimize the dependency on the usage of fossil fuels. In the power sector, renewable energy sources generating green electricity from natural resources such as solar, wind, and hydro have been proposed. Likewise, Electric vehicles (EVs) and Hybrid electric vehicles (HEVs) are considered as a pollution-free transportation

S. Narayanaswamy (✉) · S. Steinhorst · S. Chakraborty (✉)
Technical University of Munich, Munich, Germany
e-mail: swaminathan.narayanaswamy@tum.de; sebastian.steinhorst@tum.de; samarjit@tum.de

S. Park
Technical University of Berlin, Einstein Center Digital Future, Berlin, Germany
e-mail: sangyoung.park@tu-berlin.de

© Springer Nature Switzerland AG 2019
M. A. Al Faruque, A. Canedo (eds.), *Design Automation of Cyber-Physical Systems*, https://doi.org/10.1007/978-3-030-13050-3_10

option compared to the gasoline driven cars. This is evident with the increasing deployment trend of renewable energy sources and sales rate of EVs [5].

10.1.1 Challenges with Alternative Green Technologies

The major challenge towards successful integration of different renewable energy sources into the electricity grid is their volatility. Their high dependence on favorable weather conditions to generate electrical energy makes them unpredictable and increases the complexity of grid control leading to instability. For instance, rooftop solar Photovoltaic (PV) panels generate majority of their electricity during the middle of the day when the sunshine is at maximum. However, the electricity demand of a typical household peaks during the early morning and late evening periods creating a mismatch between the generation and usage time frames. This results in significant economic losses for large-scale renewable energy systems. For instance, the Comptroller and Auditor General of India estimated approximately 2.4 billion INR losses due to lack of power evacuation from the wind farms during the period 2007–2014 [6]. In addition, the electricity grid in many of the developing countries such as India and China is not fully mature, resulting in several challenges to integrate the renewable energy sources from different geographical locations. Therefore, Electrical energy storage (EES) systems become necessary to store the electrical energy produced during favorable weather conditions and reuse them whenever required.

On the other hand, EVs and HEVs are also facing similar challenges with their energy storage technologies. The battery pack, which is the main source of power in these new green transportation systems, is highly expensive constituting nearly half of the overall vehicle cost. Long charging times and limited driving ranges are the major concerns for these new technologies to capture the market of gasoline-driving Internal combustion engine (ICE) vehicles. Moreover, the battery packs used in EVs and HEVs have to be replaced once their lifetime reaches 70% of their initial value, since they cannot guarantee the same range specifications as a fresh battery pack. Methods for extending this threshold and techniques to reuse these retired battery packs from automotive applications into other less critical domains are required. Furthermore, techniques to minimize the cost of the battery packs and improve the performance in terms of charging time and driving range are crucial.

10.1.2 Electrochemical Battery Packs

There are different forms of EES systems available depending upon the specific requirements of each application [7]. For very high power storage, in the range of several MW to few GW, pumped hydro storage can be used, where the electrical energy is stored by pumping water from a lower reservoir to an upper reservoir during off peak periods. On the other hand, EES formed by double-layer capacitors also called as *supercapacitors* is mainly used in applications where the stored energy has to be retrieved within short time periods in the range of seconds to minutes. Similarly, the electrical energy generated during the low demand time is used to electrolyze water molecule and the released H_2 is used in a fuel cell EES to support the high demand during peak periods.

Among the many types of EES, batteries are widely used to store the electrical energy due to their high energy and power density. Higher energy and power density correlates to lower installation volume required for storing the electrical energy. Moreover, batteries can be tailored to meet the specific requirements of the application such as fast charging or long life or higher power capability or higher energy density. Therefore, they are widely used as EES in the kW and MW power range. In addition, they dominate other technologies in the field of mobile EES such as EVs and HEVs due to their highly compact nature with high specific energy and power.

Batteries are electrochemical storage devices meaning their chemical reaction is coupled with an electron transfer. In general, the battery chemistry is broadly classified into *primary* (non-rechargeable) and *secondary* (rechargeable). The primary, non-rechargeable batteries are designed to be used once and are discarded when the active chemical materials of the battery generating electricity are fully utilized. By contrast, the secondary rechargeable battery types can be charged and discharged multiple times. They perform a reversible chemical reaction, which allows them to store electrical energy (*charging*) and release the stored electrical energy by performing the opposite reactions (*discharging*). In case of EES systems, the secondary, rechargeable battery chemistry is preferred since it allows to store and extract the electrical energy without the necessity for replacing the battery itself.

The secondary rechargeable batteries are further classified based on the chemical composition of the anode, cathode, and the electrolyte materials used for construction. For example, a Nickel-metal hydride (NiMH) rechargeable battery chemistry consists of nickel cathode, a hydrogen absorbing anode, and a potassium hydroxide electrolyte. Similarly, the lead-acid battery chemistry is made of lead-dioxide cathode, a metallic lead anode, and an electrolyte of sulfuric acid solution. Compared to all the secondary rechargeable battery chemistries, the Lithium-ion (Li-Ion) based batteries provide superior performance in terms of energy and power densities, since the electrochemical potential of lithium is higher compared to other materials. Therefore, the cells can be manufactured with smaller size and weight for the same required energy and power requirements. Moreover, the flexibility in designing the cell for high specific energy (energy cell) or to design with a high specific power (power cell) provides a wide range of applications for this battery chemistry. In addition, long cycle life with low self-discharge and having high coulombic efficiency make these Li-Ion cells the most appropriate option of high power EES applications.

10.1.2.1 High Voltage Battery Packs

The current capacity of a single Li-Ion cell depends upon the geometry of the cell. For instance, a Li-Ion cell of "18650" form factor (18 mm in diameter and 65 mm length) has typically a capacity in the range of 3 A h, whereas the large format pouch cells have a capacity in the range of 60 A h. Moreover, the operating voltage of a single Li-Ion cell is in the range of 2.7–4 V. Nevertheless, the voltage and capacity of a single Li-Ion is insufficient to support high power EV and HEV requirements of 450 V and 200 A h. Therefore, battery packs are formed with a number of series- and parallel-connected individual Li-Ion cells. In order to have a higher current capability, multiple Li-Ion cells are connected in parallel and the required higher operating voltage can be obtained by series connection of individual cells. For example, with a single Li-Ion cell with a capacity of 3 A h and a nominal voltage of 4 V, connecting five cells in parallel will result in a capacity of 15 A and then connecting twelve of these parallel-connected cells in series will yield a voltage of 48 V for the battery pack corresponding to a 720 W h EES system.

Example High Voltage Automotive Battery Packs The Tesla Model-S full EV consists of a battery pack of 85 kW h. It is made up of Panasonic "18650" Li-Ion cells each having a capacity of 3.2 W h. The battery pack is divided into 16 series-connected modules and each module has six series-connected groups of 74 parallel-connected individual Li-Ion cells as shown in Fig. 10.1a. The battery pack configuration is represented as 74P6S16S with a total of 7104 individual Li-Ion

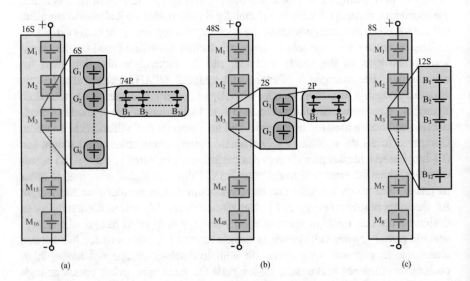

(a)   (b)   (c)

Fig. 10.1 High voltage battery packs. (**a**) Tesla Model-S 85 kW h battery pack, consisting of 96 series-connected modules with 74 parallel-connected individual Li-Ion cells in each module. (**b**) Nissan leaf battery pack with 96 series-connected modules each having 2 sheet shaped 32.5 A h Li-Ion cells in parallel. (**c**) BMW i3 EV battery pack having 96 series-connected 60 A h cells

cell. On the other hand, the battery pack in the Nissan leaf EV is formed by 48 series-connected modules as shown in Fig. 10.1b. Each module consists of 4 cells that are configured in 2P2S fashion, 2 series-connected groups and in each group 2 sheet shaped 32.5 A h Li-Ion cells are connected in parallel. The entire battery pack configuration is represented as 2P2S48S. The BMW i3 EV has 12 series-connected 60 A h prismatic Li-Ion cells forming a module and the battery pack consists of 8 modules connected in series (Fig. 10.1c).

10.1.3 Battery Pack Challenges

The benefits of high energy and power densities offered by Li-Ion cells do not come for free. A comprehensive overview of issues associated with battery packs consisting of Li-Ion cells is provided in [8]. The critical challenges pertaining to high voltage battery packs consisting of multiple series-connected Li-Ion cells are its safety and energy output.

10.1.3.1 Safety

Li-Ion cells have a defined set of safe operating conditions in terms of voltage, current, and temperature. The minimum and maximum operating voltage of most Li-Ion cells are in the range of 2.7 V and 4.2 V, respectively. Charging a Li-Ion cell with a voltage higher than that specified causes excessive current flows inside the cell and increases the internal temperature leading to fire or explosion by thermal runaway. Similarly, discharging a Li-Ion cell below its minimum threshold voltage (over-discharging) results in a gradual breakdown of the internal cell electrodes, reducing their lifetime. Similar constraints also hold for the operating current of a Li-Ion cell. Charging with high currents (*fast charging*), especially in terms of EV applications, is gaining more importance, since a regular charging of an EV battery pack might take hours. However, increasing the charging current significantly increases the temperature of the cell and if adequate control measures are not taken to regulate the battery pack temperature, the lifetime of the battery pack will significantly be reduced. Likewise, discharging the cell with higher currents results in an inherent capacity loss, due to the *rate capacity effect*, which is defined as the reduction in the battery capacity due to the increased discharge current. Finally, temperature of a Li-Ion battery pack is a critical parameter that needs to be maintained within a specific operating range to ensure safety and increase the usable capacity. With very low temperatures the speed of the chemical reactions is very slow and therefore results in a reduced current carrying capacity, both in terms of charging and discharging. Prolonged operation of the battery pack at reduced temperatures, below 0 °C, will result in a premature capacity loss of the battery pack. By contrast, increased temperatures will result in catastrophic effects causing fire or explosion due to thermal runaway. Moreover, with increased temperature

exothermic side reactions take place inside the cell that will severely damage them
and reduce their lifetime. Therefore, the operating parameters of all the cells in a
Li-Ion battery pack must be accurately monitored and closely controlled to ensure
safe operation.

10.1.3.2 Energy Output

As shown in Fig. 10.1, battery packs for high power applications such as EVs,
HEVs are formed by multiple series-connected Li-Ion cells to achieve the required
operating voltage. The discharging or charging process of such a series-connected
battery pack must be stopped when any cell in the pack reaches the lower or upper
operating thresholds, respectively. In an ideal case all cells forming the battery pack
are required to be uniform, thereby reaching the top and bottom threshold limits at
the same time to fully utilize the available capacity of the battery pack. However,
in reality, manufacturing differences and varying temperature distribution along the
battery pack lead to variations in the State-of-charge (SoC) of individual cells in the
pack. As a result, the usable capacity of the battery pack is reduced since a series-
connected pack can only be discharged till any cell in the pack reaches its lower
SoC threshold. Subsequently, the charging process is also affected by the charge
variations since any cell reaching the top threshold will halt the charging process.

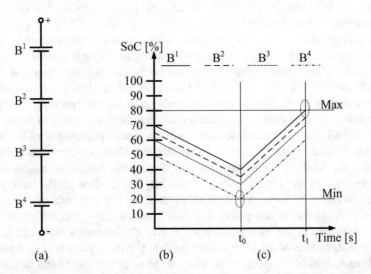

Fig. 10.2 (a) Motivating example consisting of four series-connected Li-Ion cells to show the
impact of cell-to-cell SoC variations. (b) While discharging, cell B^4 with a lower SoC compared
to others will stop the discharging process as soon as it reaches the bottom threshold value. (c)
While charging, cell B^1 with a higher SoC compared to others will stop the charging process once
the top threshold is reached

Figure 10.2 shows the impact of such cell-to-cell variations with an example of 4 cells connected in series. Due to manufacturing variations and temperature distribution, the SoC of cell B^1 is at 70% and that of cell B^4 is at 50%. As soon as a load current is drawn from the battery pack at time 0, the SoC of all cells start to decrease depending upon the load current value. A series-connected battery pack can only be discharged till the SoC of any cell in the pack reaches the bottom threshold value. In this example, cell B^4 is the weakest cell compared to others and will reach the lower threshold value faster than other cells in the pack. Therefore, once cell B^4 is discharged to the lower threshold value, see Fig. 10.2b, the discharging process has to be stopped, even though other cells in the pack have active energy stored in them. Similar loss in usable capacity is observed during the charging process which starts at time t_0. Since all cells are connected in series, the same charging current flows through all of them and their SoC starts to increase depending upon the current value. The charging process continues till any one cell in the pack reaches the top threshold value. In the example here, cell B^1 is the strongest cell and therefore reaches the top SoC threshold earlier than other cells in the pack. Once cell B^1 is fully charged (at time t_1), the charging process has to be stopped (see Fig. 10.2c), even though remaining cells in the pack are not fully charged. This results in a battery pack consisting of cells that are unevenly charged. Repeated charging and discharging of such an imbalanced battery pack will result in a situation where the cell with the low SoC value (B^4) stops the discharging process and the cell with the high SoC (B^1) halts the charging process. This leads to an unusable battery pack and therefore it is required to equalize the SoC of individual cells in the battery pack in order to fully utilize the usable capacity.

10.2 Battery Management System

In order to address the above-mentioned challenges associated with high voltage battery packs, a sophisticated Battery management system (BMS) is required to maintain safe operating conditions and to maximize the usable capacity of the battery pack [9]. The BMS monitors the parameters such as voltage, current, and temperature of individual cells and controls them within their safe operating limits. In addition, the BMS accurately calculates the cell states such as SoC and State-of-health (SoH), which are required to estimate the driving range and lifetime of the battery pack, respectively. Moreover, the BMS controls the cell balancing, which is the process of equalizing the charge levels of the individual cells in the series-connected battery pack and thereby improves its usable capacity.

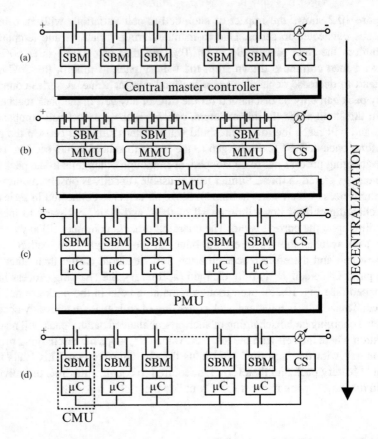

Fig. 10.3 Trends in BMS topologies. (**a**) Centralized, (**b**) hierarchical, (**c**) partially distributed, and (**d**) fully decentralized (smart cells) [10]

10.2.1 BMS Topologies

Topology of a BMS is defined as the electrical and logical arrangement of modules that perform sensing of cell parameters such as voltage and temperature, computation of cell states such as SoC and SoH and control of balancing circuits. Figure 10.3 shows the trend of the BMS topologies in the literature.

10.2.1.1 Conventional BMS Topologies

Conventional approaches are either centralized or hierarchical as shown in Fig. 10.3a, b, respectively. In the centralized system, each cell is associated with a Sensing and balancing module (SBM) as shown in Fig. 10.3a measuring the voltage and temperature of each cell [11]. A single Current sensor (CS) either at the positive or at the negative terminal of the pack can be used to measure the pack current since the current flowing through all the cells will be equal as they are connected in series. The balancing part of the SBM will typically be a simple high power resistor in series with a power transistor realizing a passive balancing approach. The individual SBMs are controlled by a single master controller, which in addition to maintaining safe operation of each cell also implements the pack-level functions such as pack SoC and SoH calculations. Alternatively, the hierarchical approach groups typically four to six series-connected cells as a *module* and multiplexes a single SBM with each cell in the module as shown in Fig. 10.3b [9]. Additionally, an intermediate control layer in the form of Module management unit (MMU) is introduced to individually manage the respective module and uses the master Pack management unit (PMU) controller for only pack-level functions. The BMS of a Tesla Model-S is an example for this type of BMS topology.

Challenges   Even though the centralized BMS topology is cost effective to implement, there are significant challenges faced by this approach due to the growing complexity of battery packs and a huge demand to reduce the time-to-market especially in terms of EV applications.

- **Modularity and Scalability:** *Scalability* of these conventional approaches is significantly limited. The design of the electrical architecture of the SBM and the MMU is highly integrated with the underlying cell and its parameter specifications. Similar dependence is also observed in the management algorithms of the master controller in both centralized and hierarchical approaches, which have to be modified depending upon the application scenario. Moreover, addition of new cells to the pack will not be easily supported and requires a complete redesign since the computational capability and the input/output performance of the master and the MMUs are limited that do not scale with the number of cells.
- **Wiring and Control:** With a central master controller monitoring parameters of each individual cells, there exists a huge amount of wiring between cells and the controller. This significantly increases the wiring harness and weight, which in EV and HEV applications directly impacts the driving range. In addition, the *balancing* capability of these BMS topologies is often limited to energy-inefficient, dissipative passive techniques, since the energy-efficient active equalization approaches consist of a dense switching network requiring a complex control scheme that cannot be satisfied by a single master controller while performing other critical pack-level BMS functions.
- **Reliability:** The master controller in these conventional approaches represents a potential single point of failure in the system. Any fault in the controller will isolate the battery pack from the application unit and improper isolation in certain scenarios might lead to catastrophic accidents, especially in terms of EVs. Moreover, the excessive wiring from the cell sensors to the master controller in the centralized topology also increases the probability of failures in the system. It is highly difficult for the central master controller to distinguish between a fault in sensor cable from a fault in the cell. If a signal wire from the sensors of any

cell breaks, the master controller will shut down the application unit by isolating the battery pack.

10.2.1.2 Emerging Decentralized BMS Topologies

With increasing applications of battery packs and the demand for shorter time to market of the applications using these high power battery packs, the system integration aspects of battery systems are gaining more attention. Here methods for customization-free plug-and-play integration providing high degree of scalability and reliability are of paramount importance. At the same time, the system architecture must enable implementation of energy-efficient active cell balancing approaches and should consist of homogeneous modules at all abstraction-levels supporting mass production. This trend resulted in distributing the control and computational units close to each cell adding more intelligence at the cell-level.

First approaches for decentralization were proposed in [12] as shown in Fig. 10.3c, where each cell is monitored with a dedicated cell-level control unit that is in turn connected to a light-weight master controller. Here, the local controllers perform the cell-level functions of the BMS such as cell voltage, temperature measurements, SoC and SoH calculations, and control of the individual balancing units, while the light-weight master only performs system-level BMS functions. By contrast, Steinhorst et al. [13] proposes a fully decentralized system topology as shown in Fig. 10.3d, where the local cell-level controllers together with the SBM form an autonomous Cell management unit (CMU), thereby managing all the parameters of the cell it is attached to. The cell along with this CMU is termed as the *smart cell*, and the battery pack is formed by interconnection of these individual smart cells that perform all the pack-level functions such as cell balancing or pack SoC calculation in a cooperative fashion adopting techniques from the domain of self-organizing distributed systems.

Benefits of Decentralization Having homogeneous electrical circuit architecture and algorithms for the cell-level controller favors mass production and customization-free integration thereby, significantly increases the scalability of the system. Furthermore, adding more intelligence to each cell enables accurate monitoring and control of cell parameters within their allowable limit and timely reaction to faults in the system. Moreover, the decentralized approaches do not suffer from single point of failures commonly experienced in the conventional architectures. Failure of a single cell-level controller will only render the associated cell unusable and failure of the master controller in case of partially distributed topology will not be catastrophic since the individual cell-level controllers can still function without the supervision of the master. Finally, increasing the computational and controlling capability at the cell-level promotes the realization of complex active cell balancing approaches and reconfigurable interconnection schemes that vastly improve the overall usable capacity of the battery pack. Owing to these benefits, decentralized BMS approaches are gaining more importance and in the remainder of this chapter we will focus on their design challenges and introduce our hardware/software development platform that will help in faster design and verification of decentralized circuit architectures and distributed algorithms.

10.2.1.3   Distributed BMS Challenges

Even though distributed BMS topologies significantly address the challenges associated with the centralized approach, they require a paradigm shift in their design methodology. Design challenges in distributed BMSs that has to be addressed for an efficient implementation are classified into three levels as follows.

Cell-Level Low-power consumption is the key design criteria for the cell-level controllers in a distributed BMS as they are powered from the battery cell itself. Moreover, in a series-connected battery pack, the DC potential of each cell varies with respect to the negative terminal of the battery pack. For instance, the voltage across the terminals of cell 2 is 4 and 8 V and that of cell 100 would be 396 and 400 V, with respect to the negative terminal of the battery pack, respectively. Since the individual cell controllers are powered directly from their respective battery cells, the ground potential of each controller varies while charging and discharging the pack. As a result, the sensing module in each cell-level controller has to be designed in such a way to overcome the measurement inaccuracies and the high level of common-mode noise introduced by the varying DC potential. In addition, conventional approaches have a fixed electrical interconnection between the cells to form a battery pack. Recently, reconfiguration architectures have been proposed where the electrical interconnection scheme of the battery pack can be modified at runtime to improve the performance [14]. Even though the decentralized topologies can support the additional complexity introduced by these reconfiguration approaches, the high power dissipation across the switches used in these reconfiguration architectures is still a challenge that requires sophisticated power electronic design.

Module-Level The module-level comprises of the balancing circuit architectures that are required to equalize the SoC of each cell in the pack. Compared to conventional *passive* techniques [15], where the excess charge in cells with high SoC is dissipated as heat across a resistor, energy-efficient active cell balancing techniques are an emerging alternative. Here, the SoC variation among cells is minimized by performing charge transfers using temporary energy storage elements such as inductors, transformers, and capacitors coupled with a power Metal-oxide-semiconductor field-effect transistor (MOSFET) switching network [16, 17]. Moreover, decentralization of the BMS topology enables implementation of complex energy-efficient active cell balancing approaches that require sophisticated control scheme with multiple high frequency Pulse width modulated (PWM) signals having strict timing requirements. However, this decentralization shifts the focus now towards the electrical architecture design and their verification in order to satisfy the modularity and homogeneity requirements of the decentralized BMS topologies [18]. The active cell balancing modules must have a homogeneous electrical architecture that can be modularized into identical units that can be attached with each cell and a system-level balancing architecture can be easily formed by extending them without requiring custom modifications. Moreover, all high frequency control signals required for performing charge transfers must be generated locally from the cell-level controller without requiring any high frequency synchronization with other controllers. Verifying the switching scheme for such complex active cell balancing architectures becomes a non-trivial task and cannot be performed manually, requiring automated design techniques. Finally, optimal component selection for these active cell balancing architectures is an essential task to ensure high energy efficiency, lower installation volume, and faster equalization.

Pack-Level The computation, control, and communication aspects of the distributed BMS constitute the pack-level. Computation of cell states such as SoC, SoH, and cell aging are the main tasks performed by the BMS. This involves solving complex analytical equations and applying sophisticated filtering techniques such as Kalman filtering or Extended Kalman filtering (EKF) [19]. While the cell-level controller in a distributed BMS topology is typically a microcontroller with limited computational capability, software implementation of these mathematical tasks become challenging. There exists a trade-off in selecting this computational platform since increasing the computation capability of the cell-level controller improves the efficient implementation of complex state estimation techniques, while on the other hand will also increase their power consumption from the cell. Moreover, as the individual cell-level controllers are only responsible for monitoring the status of their associated cells, the system-level properties of the battery pack such as pack SoC and pack voltage are calculated by communicating with other cell-level controllers. Therefore, the communication channel must support high bandwidth and also the protocol must enable the individual controllers with high priority to access the communication channel to quickly broadcast in case of any fault in the cell. The controlling aspect of the BMS involves the control of the underlying active cell balancing architectures. Here complex switching schemes with multiple high frequency PWM signals having strict timing requirements must be generated by the cell-level controller, requiring a sophisticated timing module and sufficient input/output port performance. Moreover, the strategy for balancing the SoC variations in the pack, that is identifying the source and destination cells for the charge transfer, is a complex task. Finding an optimal strategy that will minimize both the energy dissipation and the time to equalize the pack is a non-trivial optimization problem that requires sophisticated design automation techniques. Moreover, developing these energy-efficient strategies for a decentralized BMS is significantly challenging since there is no master controller with a global knowledge is present in such systems.

Table 10.1 Cell-level and pack-level functions of a distributed BMS topology

	Input							Pack			
	Cell										
Output	Voltage	Temperature	SoC	SoH	Resistance	Balancing current	Pack current	SoC	SoH
Cell									
Calculate: SoC* $\in \mathbb{R}$	✓	✓				✓	✓		
Calculate: SoH* $\in \mathbb{R}$	✓	✓					✓		
Calculate: Resistance* $\in \mathbb{R}$	✓				✓		✓		
Detect: Fault $\in \{0, 1\}$	✓	✓	✓	✓					
Detect: Over charge or discharge $\in \{0, 1\}$	✓		✓						
Pack									
Calculate: SoC*{min, max} $\in \{\mathbb{R} \times \mathbb{R}\}$			✓					✓	
Calculate: SoH* $\in \mathbb{R}$				✓					
Perform: Balancing $\in \{0, 1\}$	✓		✓			✓		✓	
Calculate: {maxPower, maxTime} $\in \mathbb{R} \times \mathbb{R}$	✓	✓	✓	✓			✓	✓	✓

Inputs specified in the columns are used to calculate cell and pack-level outputs mentioned in the rows. Output of certain functions that are marked with * is used as inputs for calculations in other functions

10.2.2   Need for Design Automation

In order to address the above-mentioned challenges associated with the decentralized BMS topologies novel design automation techniques are required to be developed [20]. In contrast to the well-established design automation techniques for designing integrated circuits, the tools required for efficient design of BMSs are significantly different. The BMS falls under the domain of cyber-physical systems, where the measurement of parameters such as voltage, temperature, and current of the battery cells forms the physical part and the cyber part is constituted with the computation and control functionalities. For instance, Table 10.1 shows typical functions that are performed in a distributed BMS topology. Inputs to the functions are specified in the columns and the functions along with their outputs are specified in the rows of the table.

Cell-Level Functions Cell-level functions are implemented in their respective local control units since these functions are independent of other cells, meaning the status and parameters of other cells in the battery pack are not required to be communicated to implement these cell-level functions. The inputs required for each function calculation are marked with √ along the respective columns in Table 10.1. For example, the function that calculates the SoC of the individual cell requires the cell voltage, temperature, balancing current, and pack current as inputs. For calculating the individual cell SoC, the local controller does not require data from other cells in the battery pack.

Pack-Level Functions In contrast to cell-level functions that are independently implemented in each cell-level control unit, pack-level functions are realized in a distributed fashion. Here all individual cell controllers collectively exchange their cell data through the communication channel, to compute the pack parameters in a cooperative manner. They can either be a measured parameter such as voltage and temperature or a computed result of a certain cell-level function such as SoC and SoH. Outputs of such cell-level functions that are in turn used as inputs for calculating battery pack-level parameters are marked with a * in Table 10.1. For example, to calculate the minimum and maximum SoC of the battery pack, the individual SoC of each cell has to be communicated to other cells through the communication channel.

Therefore, special set of design automation methods and tools are required for efficient design of these complex cyber-physical systems where the modeling of the physical process and the design of control algorithms that control these processes are all performed in an integrated fashion.

10.3   Design Automation Techniques

In this section, we present an overview of our hardware and software development platform that are available to assist in solving the challenges associated with distributed BMSs. Such a platform can be used to perform Hardware-in-the-loop (HIL) studies for evaluating different decentralized circuit architectures, active cell balancing topologies, and distributed algorithms. All design files of the development platform are made open source by uploading them in an online repository [21] along with detailed instructions for duplicating them. This enables the scientific community to kick start their work on developing distributed algorithms by providing a testbed where they can evaluate their developed algorithms, saving the time and effort to develop custom prototyping platforms.

10.3.1   Hardware Development Platform

A major step towards the development of decentralized BMS algorithms and novel active cell balancing architectures and their control strategy is prototyping. This involves several steps such as the development of hardware implementations for the BMS controller, active cell balancing architecture, their integration, software development, hardware and software verification, leading to a huge overhead in terms of time and cost. Towards this we propose a distributed BMS hardware development platform as shown in Fig. 10.4. With minor adjustments to their connection scheme, the proposed platform could be reconfigured to emulate different types of BMS topologies, thereby enabling rapid prototyping and fast development and validation of BMS algorithms. Here each cell is associated with an individual cell-level controller and an active cell balancing module. The cell-level controllers monitor the parameters of the associated cell and maintain them within safe operating range and the active cell balancing modules perform charge transfers between the cells to minimize the SoC variation among them. As such this development platform directly emulates the fully decentralized *smart cell* BMS topology shown in Fig. 10.3d. Moreover, by making one of the cell-level controllers as a master control unit, the partially distributed BMS topology shown in Fig. 10.3c is obtained.

10.3.1.1   Cell-Level Controller

With the functions performed in a typical distributed BMS as listed in Table 10.1, the necessary modules that are required in the cell-level controller are as follows:

- **Sensing:** The sensing module is used to measure the cell parameters such as cell voltage, temperature, balancing current, and pack current, which are used as inputs for calculating the cell and pack-level functions of the BMS.
- **Computation:** The computation module implements all the cell-level and pack-level functions of the distributed BMS topologies. It takes the measured parameters from the sensing module as inputs for calculating the cell-level functions and the pack-level functions are performed by communicating with other cell-level controllers.

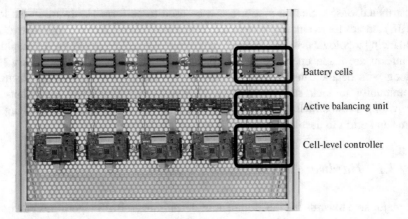

Fig. 10.4 A development platform for distributed BMSs consisting of 5 series-connected Li-Ion cells and their respective control units. Each unit consists of 3 parallel-connected Li-Ion cells that is monitored and controlled by the dedicated cell-level controller and individual active cell balancing unit

- **Communication:** Data exchange between the individual cell-level controllers or to the master controller is facilitated by the communication module.
- **Cell balancing:** Equalization of individual cell capacities is performed by the cell balancing module to improve the usable capacity of the battery pack.
- **Power supply:** Power supply module provides a constant regulated supply voltage, which is required for efficient functioning of other modules in the cell-level controller unit.

Figures 10.5 and 10.6 show the top and bottom side of the cell-level controller board. Implementation of each individual module was performed using commercial off-the-shelf components that are combined in a custom designed Printed circuit board (PCB) that can be directly powered from a battery cell. The design of the PCB was focused towards facilitating extensive debugging and obtaining high accuracy measurements. Inputs to each module can be actuated with test signals and their corresponding outputs can be measured separately without involving other modules. This facilitates functional verification of each module and also enables to characterize their performance individually in terms of energy consumption. Communication between the individual cell controllers is established using an isolated Controller area network (CAN) bus topology through which the functions of a distributed BMS are performed by negotiations between the individual cell controllers. Since the cell-level controllers are powered by their respective battery cells, a galvanically isolated communication channel is used to avoid potential short circuits between cells. In addition to the bus-based communication architecture, a galvanically isolated daisy chain communication topology between the individual cell-level controllers is also provided. This enables performance comparison of different communication architectures with respect to evaluating the pack-level functions.

The modular design of the development platform enables easy interfaces with external test equipment and Data acquisition (DAQ) systems. This facilitates functional verification of different distributed BMS functions and also to obtain high accuracy measurements for model validation purposes. Moreover, the isolated CAN communication bus can also be tapped and connected to a PC using suitable adapter. This enables the individual cell-level controllers to be operated using a system-level algorithm that is running on the PC, thereby facilitating HIL simulations. All hardware design files and the firmware of the cell-level controller are uploaded in an online repository [21] and made publicly accessible. Details regarding the implementation of each module in the cell-level controller and techniques to fabricate multiple copies of the controller can be found in [21]. Such an open source implementation enables easy reproduction of the development platform with minimal integration efforts and facilitate the scientific community in rapid development of distributed BMS functions and algorithms.

10.3.1.2 Active Balancing Unit

In addition to the cell-level controller, development boards for evaluating different active cell balancing architectures are also available. Active cell balancing involves

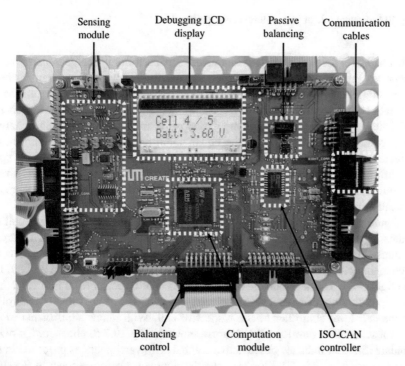

Fig. 10.5 Top side of the cell-level controller board

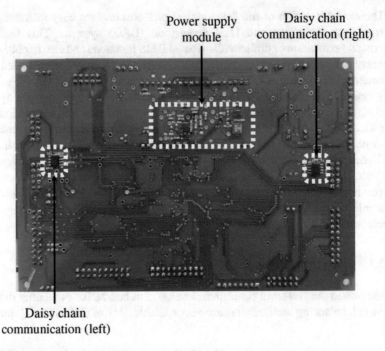

Fig. 10.6 Bottom side of the cell-level controller board

equalization of the SoC variations among cells by performing charge transfers using temporary energy storage elements such as capacitors, inductors, and transformers. Even though capacitor-based approaches [22–24] have a simpler control scheme and reduced installation area, they can only achieve a maximum of 50% energy efficiency, due to the inherent energy dissipation of the capacitor being charged by a voltage source. Moreover, the equalization speed of these capacitor-based approaches is significantly smaller, since the balancing current value depends upon the difference between the cell voltages, which in case of Li-Ion cells is very small. Therefore, active cell balancing techniques that use inductors and transformers are typically preferred due to their high energy efficiency and balancing speed.

Figure 10.7 shows a development platform that will be attached to each cell for evaluating the performance of different inductor-based active cell balancing architectures. The battery cell is connected at the top of the board and connections for exchanging charge between the adjacent cells in the pack are provided on both left and right sides. The development platform consists of 12 power MOSFETs and their corresponding high frequency gate drivers. The gate drivers are directly controlled by the cell-level controller of the respective cell. With minor adjustments to the control scheme the development platform shown in Fig. 10.7 can be reconfigured to emulate different inductor-based active cell balancing architectures proposed in the literature such as [25–28]. Similarly, the development platform shown in Fig. 10.8 can be used to emulate different types of transformer-based active cell balancing architectures proposed in [18, 29]. These reconfigurations are facilitated by the high speed gate driver arrangement, which forms a crucial part of the active cell balancing unit as explained in the following.

High Frequency Gate Drive Power MOSFETs in the active cell balancing architectures are connected with the power line of the battery pack and therefore they cannot be actuated directly from the computation module. Moreover, the voltage of the control signals from the computation module is in the order of 3–5 V which is less compared to the higher gate drive voltage required for actuating the power MOSFETs. Therefore, external gate drive units are required to interface the MOSFETs to the computation module. Gate drive for MOSFETs that are actuated with either *ON* or *OFF* DC signals can be accomplished using a photovoltaic gate drive units. However, the turn-*ON* and turn-*OFF* times of the photovoltaic gate drive components are relatively slow compared to the requirements of the high frequency actuation signals used in active cell balancing purposes that are in the range of 10–100 kHz. Therefore, they cannot be employed as gate drive units for actuating MOSFETs that are actuated with high frequency control signals. As a result, special type of gate drive arrangement is required for actuating MOSFETs that are operated with high frequency control signals.

An optocoupler-based MOSFET gate drive unit is used to boost the low voltage control signals that are generated from the computation module of the cell-level

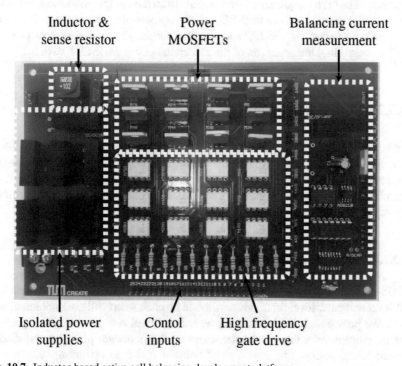

Fig. 10.7 Inductor-based active cell balancing development platform

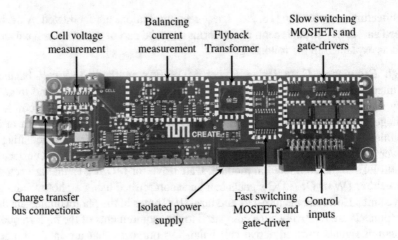

Fig. 10.8 Transformer-based active cell balancing development platform

controller. Moreover, the optocoupler gate drive unit isolates the high voltage battery and active cell balancing circuitries from the low voltage computation units. In addition, an isolated power supply unit powered from the battery cell is used to obtain a higher supply voltage required for the optocoupler-based MOSFET gate drive unit. The high frequency control signals from the computation module actuate the input Light emitting diode (LED) of the optocoupler gate drive unit and the illumination created by the LED makes the output phototransistors to conduct, thereby controlling the actuation of the associated power MOSFET switch.

10.3.2 Software Development Platform

In this section, we provide an overview of the software tools that are available to interact with the hardware development platform explained above. This includes the firmware that is implemented on the individual cell-level controllers and the CAN controller framework that is used to visualize, control, and record the status of the individual cells in the development platform.

10.3.2.1 Real-Time Operating System

μC/OS-III from Micrium [30] is used as a real-time operating system on which both cell-level and pack-level functions performed by each smart cell are implemented as tasks. We provide a brief explanation regarding the μC/OS-III real-time operating system, current tasks that are implemented on the hardware platform and discuss its scheduling policy. The objective of using a real-time operating system is to split each function into different *tasks*, which are then scheduled to run on the

Table 10.2 List of tasks implemented in the µC/OS-III real-time operating system

Task ID	Task name	Priority	Trigger type
T0	*ProcessCANMsg*	1	Event
T1	*PackMonitoring*	2	Periodic
T2	*ChargeRequesting*	2	Periodic
T3	*ChargeAcknowledge*	2	Periodic
T4	*CurrentSampling*	3	Periodic
T5	*VoltageAverage*	3	Periodic
T6	*StatusMessage*	3	Periodic

The priority and the type of trigger for each task are also specified

computational unit depending upon their priority-levels. The µC/OS-III real-time operating system supports *multi-tasking*, a process of scheduling and switching the computational module between several tasks. This facilitates an application programmer to implement a complex function into multiple modular tasks that are then periodically executed by the operating system. Moreover, by assigning different priorities to the individual tasks, we ensure that the critical functions of the BMS are executed with high priority, in a timely manner meeting their deadlines, which would have been difficult to implement in other software programming architectures such as super-loop.

Tasks Table 10.2 provides the list of tasks that are currently implemented in the µC/OS-III operating system, along with their priority-level and the type of trigger. All tasks can be broadly classified into periodic and event-driven tasks. The periodic tasks are executed in a time-triggered manner while the event-driven tasks are performed on occurrence of a certain event. The functions performed by each task are as follows:

- *ProcessCANMsg:* This task is the highest priority compared to all the other implemented tasks and it is triggered with an incoming CAN message through the communication channel. It processes the received CAN message and executes certain functions or calls other tasks depending upon the type of message.
- *PackMonitoring:* In this task, the individual smart cell monitors its own SoC with the SoC of other cells and triggers a balancing request if the deviation in the SoC is above a certain threshold.
- *ChargeRequesting:* This task pertains to active cell balancing. The smart cell checks its own SoC with other cells and requests charge if its SoC is lower compared to other cells.
- *ChargeAcknowledge:* In contrast to the *ChargeRequesting* task, in this task, the smart cell decides to give charge to other cells if its SoC is higher than other cells.
- *CurrentSampling:* Measured balancing and pack currents from the sensing module are obtained periodically in this task.
- *VoltageAverage:* Multiple readings of the cell voltage are taken from the sensing module and are filtered using a digital Infinite impulse response (IIR) filter to obtain the cell voltage accurately.

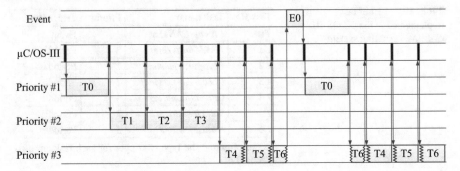

Fig. 10.9 Preemptive, round-robin scheduling algorithm in μC/OS-III implementing the individual tasks. High-priority tasks are executed until they are done and the low priority tasks can be interrupted by the high-priority tasks when the specific events occur [30]

- *StatusMessage:* The smart cell periodically broadcasts its individual cell voltage and SoC through the communication channel to other smart cells in the battery pack. Also the received information from other smart cells is used to update the battery pack parameters.

Scheduling Policy μC/OS-III follows a *preemptive, priority-based* scheduling, meaning the high-priority tasks if it's ready-to-run will preempt the execution of the low-priority tasks. Moreover, for tasks with equal priority, a *round-robin* scheduling algorithm is employed, where each ready-to-run task of the same priority is executed for a defined time period. After this time period, the operating system executes the next ready-to-run task at the same priority. Figure 10.9 shows the timing diagram of the scheduling process of the μC/OS-III operating system. The entire scheduling process can be visualized as a state diagram consisting of states representing the status of each task (for the complete state diagram, please refer to page 95 of [30]).

The tasks that are ready-to-run are placed in the *ready* state from where they are moved to *running* state in order to be executed. If it is an event-driven task and if the corresponding event has not occurred, it will move from the *ready* state to the *pending* state, where it waits for the event to take place. Tasks while waiting for an event in the *pending* state do not consume any controller operation time. As soon as an event occurs, the corresponding task in the *pending* state moves to the ready state, notifying the operating system. The μC/OS-III operating system checks the priority-level of the newly readied task and preempts the currently running task, if the ready-to-run task is of higher priority.

This is shown in Fig. 10.9, where the higher priority task $T0$ (*ProcessCANMsg*) gets executed preempting the low priority task $T6$ (*StatusMessage*) as soon as the event $E0$ occurs, which in this case is an incoming CAN message. The preempted low priority task moves to the *ready* state and begins execution from the same point where it was preempted, only after the processing of high-priority task is finished. Moreover, Fig. 10.9 also shows the *round-robin* scheduling example, where equal priority tasks $T4$ (*CurrentSampling*), $T5$ (*VoltageAverage*), and $T6$ (*StatusMessage*) do not run for completion once executed. Instead, each task runs for the same amount of time period and then the next ready-to-run equal priority task is executed. This maintains equal resource utilization for all tasks that are at the same priority-levels.

10.3.2.2   CAN Visualization Framework

Interactions with the hardware development platform shown in Fig. 10.4 are facil-itated with a software framework. The isolated CAN communication bus in the hardware development platform is connected to an external PC through a CAN adapter. A custom-designed CAN visualization framework is developed in Python environment. Figure 10.10 shows the screenshot of the visualization framework that is used to control, record, and analyze the communication messages and in turn the behavior of the distributed BMS hardware development platform. A live view of the transmitted cell parameters is shown on the right side, while update rate and a list of the sent messages are displayed in the mid and left sections, respectively. The framework can be used to visualize and record CAN messages that are sent between the controllers in the hardware development platform. Moreover, the recorded messages can be exported to a database for structured storage and analysis at a later point in time. In addition to the CAN messages, the framework can also be used to measure and store the cell status with high precision that can be used for model validation purposes.

The lower part of the framework is used for controlling the parameters for broadcasting and monitoring, active balancing status, sending debug-messages, and recording messaging sessions. The control part of the framework shown on the bottom side involves setting of timing parameters for the balancing control signals, initiating a charge transfer between any two cells in the platform. Active balancing can be enabled and disabled globally and single forced transactions to test the transfer of charge between specified cells can also be triggered. In addition, the framework also comes with predefined active cell balancing strategies (bottom right) proposed in [31]. The hardware platform can be controlled using the CAN visualizing framework to perform any of the predefined strategies thereby enabling to compare the performance of the individual algorithms. In combination with the hardware platform, the CAN visualization framework facilitates in performing HIL simulation studies to evaluate the performance of different balancing architectures, equalization strategies, and several distributed battery management algorithms.

10.4   Summary

There is a huge overhead involved in terms of time and cost to develop circuit architectures and software algorithms for distributed BMSs, as it requires custom development platforms to verify each of the interested functionality. In this chapter,

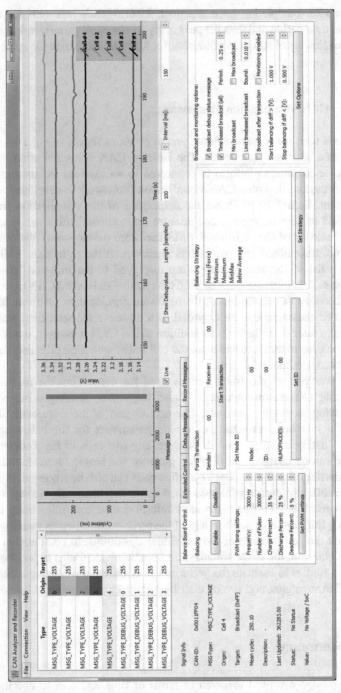

Fig. 10.10 Screenshot of the CAN visualization framework that is used to control the distributed BMS hardware development platform

we presented a hardware/software development platform for accelerating the design and verification of distributed BMSs. The hardware development platform can be flexibly reconfigured to emulate different types of distributed BMS topologies. Moreover, development boards for evaluating active cell balancing functionality are also introduced. Interactions with the hardware development platform are supported by the proposed software framework, which facilitates to visualize, control, and record the status of the individual cells in the development platform. Together these hardware and software development platforms enable to perform HIL simulation studies for functional verification of active cell balancing architectures, model validation and evaluation of distributed active cell balancing algorithms. All design files regarding the hardware development platform and the software framework are uploaded in an online repository [21] and made publicly available. This enables easy replication of the development platforms with minimal integration efforts and facilitates the scientific community in rapid development of distributed BMS circuit architectures, active cell balancing topologies, and their equalization algorithms.

References

1. Wendel, J. (2018). Global average temperatures in 2017 continued upward trend. https://eos.org/articles/global-average-temperatures-in-2017-continued-upward-trend, January 2018. Last accessed 16 July 2018.
2. Guardian. (2018). Met office warns of global temperature rise exceeding 1.5c limit. https://www.theguardian.com/science/2018/jan/31/met-office-warns-of-global-temperature-rise-exceeding-15c-limit, January 2018. Last accessed 16 July 2018.
3. Shearer, C., Fofrich, R., & Davis, S. J. (2017). Future CO2 emissions and electricity generation from proposed coal-fired power plants in India. *Earth's Future, 5*(4), 408–416.
4. EPA. (2018). *Inventory of US greenhouse gas emissions and sinks: 1990–2016* (pp. 1–655). Washington, DC: EPA.
5. Cazzola, P., Gorner, M., Munuera, L., Schuitmaker, R., Maroney, E., & Gorner, M. (2017). *Global EV outlook 2017* (pp. 1–71). Paris: International Energy Agency.
6. Business Standard. (2015, December). Revenue loss of Rs 2,040 cr in Tamil Nadu due to backing down of wind power, says CAG. https://www.business-standard.com/article/economy-policy/revenue-loss-of-rs-2-040-cr-in-tamil-nadu-due-to-backing-down-of-wind-power-says-cag-115120800923_1.html. Last accessed 16 July 2018.
7. International Electrotechnical Commission. (2010). Electrical energy storage. *IEC White papers and Technology reports* (pp. 1–78).
8. Lu, L., Han, X., Li, J., Hua, J., & Ouyang, M. (2013). A review on the key issues for lithium-ion battery management in electric vehicles. *Journal of Power Sources, 226*, 272–288.
9. Brandl, M., Gall, H., Wenger, M., Lorentz, V., Giegerich, M., Baronti, F., et al. (2012, March). Batteries and battery management systems for electric vehicles. In *Proceedings of Design, Automation Test in Europe Conference Exhibition (DATE)* (pp. 971–976).
10. Narayanaswamy, S., Park, S., Steinhorst, S., & Chakraborty, S. (2018). Design automation for battery systems. In *Proceedings of the International Conference on Computer-Aided Design (ICCAD'18)* (pp. 27:1–27:7).
11. Andrea, D. (2010). *Battery management systems for large lithium ion battery packs*. Boston: Artech House.
12. Baronti, F., Fantechi, G., Roncella, R., & Saletti, R. (2012). Intelligent cell gauge for a hierarchical battery management system. In *Proceedings of Transportation Electrification Conference and Expo (ITEC)* (pp. 1–5). Piscataway: IEEE.

13. Steinhorst, S., Lukasiewycz, M., Narayanaswamy, S., Kauer, M., & Chakraborty, S. (2014, August). Smart cells for embedded battery management. In *Proceedings of Cyber-Physical Systems, Networks, and Applications (CPSNA)* (pp. 59–64).
14. Ci, S., Lin, N., & Wu, D. (2016). Reconfigurable battery techniques and systems: A survey. *IEEE Access, 4*, 1175–1189.
15. Cao, J., Schofield, N., & Emadi, A. (2008, September). Battery balancing methods: A comprehensive review. In *Proceedings of IEEE Vehicle Power and Propulsion Conference (VPPC)* (pp. 1–6).
16. Moore, S. W., & Schneider, P. J. (2001). A review of cell equalization methods for lithium ion and lithium polymer battery systems. SAE Publication, 2001-01-0959.
17. Daowd, M., Omar, N., Bossche, P. V. D., & Mierlo, J. V. (2011, September). Passive and active battery balancing comparison based on MATLAB simulation. In *Proceedings of IEEE Vehicle Power and Propulsion Conference (VPPC)* (pp. 1–7).
18. Narayanaswamy, S., Kauer, M., Steinhorst, S., Lukasiewycz, M., & Chakraborty, S. (2017, March). Modular active charge balancing for scalable battery packs. *IEEE Transactions on Very Large Scale Integration (VLSI) Systems, 25*, 974–987.
19. Plett, G. L. (2004). Extended Kalman filtering for battery management systems of LiPB-based HEV battery packs: Part 3. State and parameter estimation. *Journal of Power Sources, 134*(2), 277–292.
20. Chang, N., Faruque, M. A. A., Shao, Z., Xue, C. J., Chen, Y., & Baek, D. (2018). Survey of low-power electric vehicles: A design automation perspective. *IEEE Design Test, 35*(6), 44–70.
21. Hardware/software design files of the smart cell development platform. (2015). https://github.com/Swaminara/Smart-Cell-Development-Platform.git.
22. Pascual, C., & Krein, P. (1997, February). Switched capacitor system for automatic series battery equalization. In *Proceedings of Applied Power Electronics Conference (APEC)* (Vol. 2, pp. 848–854).
23. Fukui, R., & Koizumi, H. (2013, November). Double-tiered switched capacitor battery charge equalizer with chain structure. In *Proceedings of Annual Conference of IEEE Industrial Electronics Society (IECON)* (pp. 6715–6720).
24. Baughman, A., & Ferdowsi, M. (2008, June). Double-tiered switched-capacitor battery charge equalization technique. *IEEE Transactions on Industrial Electronics, 55*, 2277–2285.
25. Kauer, M., Naranayaswami, S., Steinhorst, S., Lukasiewycz, M., Chakraborty, S., & Hedrich, L. (2013, May). Modular system-level architecture for concurrent cell balancing. In *Proceedings of Design Automation Conference (DAC)* (pp. 1–10).
26. Lukasiewycz, M., Steinhorst, S., & Narayanaswamy, S. (2014). Verification of balancing architectures for modular batteries. In *Proceedings of the 12th International Conference on Hardware/Software Codesign and System Synthesis (CODES+ISSS 2014)* (pp. 1–10).
27. Kauer, M., Narayanaswamy, S., Steinhorst, S., Lukasiewycz, M., & Chakraborty, S. (2015). Many-to-many active cell balancing strategy design. In *Proceedings of the 20th Asia and South Pacific Design Automation Conference (ASP-DAC 2015)* (pp. 267–272).
28. Kutkut, N. H. (1998). A modular nondissipative current diverter for EV battery charge equalization. In *Proceedings of Applied Power Electronics Conference (APEC)* (Vol. 2, pp. 686–690).
29. L. Technology. (2013, April). LTC3300-1 high efficiency bidirectional multicell battery balancer. http://www.linear.com/product/LTC3300-1. Last accessed 16 July 2018.
30. Labrosse, J. J. (2011). *uC/OS-III: The real-time kernel for the STM32 ARM Cortex-M3*. Micrium.
31. Steinhorst, S., Kauer, M., Meeuw, A., Narayanaswamy, S., Lukasiewycz, M., & Chakraborty, S. (2016). Cyber-physical co-simulation framework for smart cells in scalable battery packs. *ACM Transactions on Design Automation of Electronic Systems (TODAES), 21*, 1–26.

Index

© Springer Nature Switzerland AG 2019
M. A. Al Faruque, A. Canedo (eds.), *Design Automation of Cyber-Physical Systems*, https://doi.org/10.1007/978-3-030-13050-3

Printed in the United States
By Bookmasters